THE CELL METHOD

THE CELL METHOD

A PURELY ALGEBRAIC
COMPUTATIONAL METHOD
IN PHYSICS AND ENGINEERING

ELENA FERRETTI

MOMENTUM PRESS

MOMENTUM PRESS, LLC, NEW YORK

First published by Momentum Press®, LLC
222 East 46th Street, New York, NY 10017
www.momentumpress.net

ISBN-13: 978-1-60650-604-2 (hardcover)
ISBN-10: 1-60650-604-8 (hardcover)
ISBN-13: 978-1-60650-606-6 (e-book)
ISBN-10: 1-60650-606-4 (e-book)

DOI: 10.5643/9781606506066

Cover design by Jonathan Pennell
Interior design by Exeter Premedia Services Private Ltd.
Chennai, India

10 9 8 7 6 5 4 3 2 1

Printed in the United States of America

CONTENTS

ACKNOWLEDGMENTS

The author wishes to acknowledge those who, through the years, contributed to the research activity that has allowed the realization of this book.

Among the Professors of the School of Engineering and Architecture – Alma Mater Studiorum University of Bologna (Italy), grateful thanks go to Prof. Angelo Di Tommaso, for having competently oriented my academic education, Prof. Antonio Di Leo, for his continuous valuable suggestions and encouragement since I was a Ph.D. student, and Prof. Erasmo Viola, for sharing his vast knowledge and experience with me each day.

Prof. Enzo Tonti, with his passionate seminarial activity on the role of classification in physics, sparked my interest for the fascinating topic of algebraic formulation. This led me to seek and deepen the mathematical foundations of the Cell Method, the major factor motivating this book. A precious help in this process was the possibility to preview some drafts of the papers and books that Prof. Tonti wanted to share with me, in confidence.

I am also grateful to Prof. Satya N. Atluri, founder and Editor-in-Chief of many journals, for having appreciated my contributions to the development of the Cell Method for computational fracture mechanics, in many circumstances. His estimate has been a source of renewed enthusiasm in approaching this innovative subject.

A special thanks goes to Joel Stein for having had trust in this project and to Millicent Treloar for her swift feedback and guidance to improve this book.

Elena Ferretti
Bologna, October 2013

PREFACE

The computational methods currently used in physics are based on the discretization of the differential formulation, by using one of the many methods of discretization, such as the finite element method (FEM), the boundary element method (BEM), the finite volume method (FVM), the finite difference method (FDM), and so forth. Infinitesimal analysis has without doubt played a major role in the mathematical treatment of physics in the past, and will continue to do so in the future, but, as discussed in Chapter 1, we must also be aware that several important aspects of the phenomenon being described, such as its geometrical and topological features, remain hidden, in using the differential formulation. This is a consequence not of performing the limit, in itself, but rather of the numerical technique used for finding the limit. In Chapter 1, we analyze and compare the two most known techniques, the iterative technique and the application of the Cancelation Rule for limits. It is shown how the first technique, leading to the approximate solution of the algebraic formulation, preserves information on the trend of the function in the neighborhood of the estimation point, while the second technique, leading to the exact solution of the differential formulation, does not. Under the topological point of view, this means that the algebraic formulation preserves information on the length scales associated with the solution, while the differential formulation does not. On the basis of this observation, it is also proposed to consider that the limit provided by the Cancelation Rule for limits is exact only in the broad sense (i.e., the numerical sense), and not in the narrow sense (involving also topological information). Moreover, applying the limit process introduces some limitations as regularity conditions must be imposed on the field variables. These regularity conditions, in particular those concerning differentiability, are the price we pay for using a formalism that is both very advanced and easy to manipulate.

The Cancelation Rule for limits leads to point-wise field variables, while the iterative procedure leads to global variables (Section 1.2), which, being associated with elements provided with an extent, are set functions (Section 1.3). The use of global variables instead of field variables allows us to obtain a purely algebraic approach to physical laws (Chapter 4, Chapter 5), called the direct algebraic formulation. The term "direct" emphasizes that this formulation is not induced by the differential formulation, as is the case for the so-called discrete formulations that are often compared to it (Section 1.4). By performing densities and rates of the global variables, it is then always possible to obtain the differential formulation from the direct algebraic formulation.

Since the algebraic formulation is developed before the differential formulation, and not vice-versa, the direct algebraic formulation cannot use the tools of the differential formulation for describing physical variables and equations. Therefore, the need for new suitable tools arises, which allows us to translate physical notions into mathematical notions through the intermediation of topology and geometry. The most convenient mathematical setting where to formulate a geometrical approach of physics is algebraic topology, the branch of the mathematics that develops notions corresponding to those of the differential formulations, but based on global variables instead of field variables. This approach leads us to use algebra instead of differential calculus. In order to provide a better understanding of what using algebra instead of differential calculus means, Chapter 2 deals with exterior algebra (Section 2.1) and geometric algebra (Section 2.2), the two fundamental settings for the geometric study of spaces not just of geometric vectors, but of other vector-like objects such as vector fields or functions. Algebraic topology and its features are then treated in Chapter 3.

The Cell Method (CM) is the computational method based on the direct algebraic formulation developed by Enzo Tonti.[1] Tonti's first papers on the direct algebraic formulation date back to 1974 (see Reference Section). The main motivation of these early works is that physical integral variables are naturally associated with geometrical elements in space (points, lines, surfaces, and volumes) and time elements (time instants and time intervals), an observation that also allows us to answer the question on why analogies exist between different physical theories:

> *"Since in every physical theory there are integral variables associated with space and time elements it follows that there is a correspondence between the quantities and the equations of two physical theories in which the homologous quantities are those associated with the same space-time elements."*

The CM was implemented starting from the late '90s. The first theory described by means of the direct algebraic formulation was electromagnetism in 1995, followed by solid mechanics and fluids.

The strength of the CM is that of associating any physical variable with the geometrical and topological features (Chapter 1, Chapter 4), usually neglected by the differential formulation. This goal is achieved by abandoning the habit to discretize the differential equations. The governing equations are derived in algebraic manner directly, by means of the global variables, leading to a numerical method that is not simply a new numerical method among many others. The CM offers an interdisciplinary approach, which can be applied to the various branches of classical and relativistic physics. Moreover, giving an algebraic system of physical laws is not only a mathematical expedient, needed in computational physics because computers can only use a finite number of algebraic operators. The truly algebraic formulation also provides us with a numerical analysis that is more adherent to the physical nature of the phenomenon under consideration (Chapter 1). Finally, differently from their variations, the global variables are always continuous through the interface of two different media and in presence of discontinuities of the domain or the sources of the problem (Section 1.2). Therefore, the CM can be usefully employed in problems with domains made of several materials, geometrical discontinuities (corners), and concentrated sources. It also allows an easy computation in contact problems.

Even if having shown the existence of a common mathematical structure underlying the various branches of physics is one of the most relevant key-points of the direct algebraic formulation, the purpose of this book is not that of explaining the origin of this common structure,

---

[1] **Enzo Tonti** (born October 30, 1935) is an Italian mathematical physicist, now emeritus professor at the University of Trieste (Italy). He began his own scientific activity in 1962, working in the field of Mathematical Physics, the development of mathematical methods for application to problems in physics.

as already extensively done by Tonti, in his publications. Our focus will be above all on giving the mathematical foundations of the CM, and highlighting some theoretical features of the CM, not yet taken into account or adequately discussed previously. To this aim, the basics of the CM will be exposed in this book only to the extent necessary to the understanding of the reader.

One of the contributions given in this book to the understanding of the CM theoretical foundations is having emphasized that the Cancelation Rule for limits acts on the actual solution of a physical problem as a projection operator, as we have already pointed out. In Section 1.1.3, this new interpretation of the Cancelation Rule for limits is discussed in the light of the findings of non-standard calculus, the modern application of infinitesimals, in the sense of non-standard analysis, to differential and integral calculus. It is concluded that the direct algebraic approach can be viewed as the algebraic version of non-standard calculus. In fact, the extension of the real numbers with the hyperreal numbers, which is on the basis of non-standard analysis, is equivalent to providing the space of reals with a supplementary structure of infinitesimal lengths. In other words, it is an attempt to recover the loss of length scales due to the use of the Cancelation Rule for limits, in differential formulation. For the same reasons, the CM can be viewed as the numerical algebraic version of those numerical methods that incorporate some length scales in their formulations. This incorporation is usually done, explicitly or implicitly, in order to avoid numerical instabilities. Since the CM does not need to recover the length scales, because the metric notions are preserved at each level of the direct algebraic formulation, the CM is a powerful numerical instrument that can be used to avoid some typical spurious solutions of the differential formulation. The problem of the numerical instabilities is treated in Chapter 6, with special reference to electromagnetics, electrodynamics, and continuum mechanics. Particular emphasis is devoted to the associated topic of non-locality in continuum mechanics, where the classical local continuum concept is not adequate for modeling heterogeneous materials in the context of the classical differential formulation, causing the ill-posedness of boundary value problems with strain-softening constitutive models. Further possible uses of the CM for the numerical stability in other physical theories are under study, at the moment.

Some other differences and improvements, with respect to the papers and books on the CM by other Authors, include:

- The CM is viewed as a geometric algebra, which is an enrichment (or more precisely, a quantization) of the exterior algebra (Section 2.2.1). Since the geometric algebra provides compact and intuitive descriptions in many areas, including quantum mechanics, it is argued (Section 4.1) that the CM can be used even for applications to problems of quantum mechanics, a field not yet explored, at the moment.
- The p-space elements and their inner and outer orientations are derived inductively, and not deductively. They are obtained from the outer product of the geometric algebra and the features of p-vectors (Section 2.2.2). It is shown that it is possible to establish an isomorphism between the orthogonal complement and the dual vector space of any subset of vectors, which extends to the orientations. Some similarities with the general Banach spaces are also highlighted. It is concluded that the notions of inner and outer orientations are implicit in geometric algebra.
- Each cell of a plane cell complex is viewed as a two-dimensional space, where the points of the cell, with their labeling and inner orientation, play the role of a basis scalar, the edges of the cell, with their labeling and inner orientation, play the role of basis vectors, and the cell itself, with its inner orientation, plays the role of basis bivector (Section 3.5).

- Space and time global variables are treated in a unified four-dimensional space/time cell complex, whose elementary cell is the tesseract (Sections 3.8, 5.1.2-5.1.4). The resulting approach shows several similarities with the four-dimensional Minkowski spacetime. Moreover, the association between the geometrical elements of the tesseract and the "space" and "time" global variables allows us to provide an explanation (Section 4.4) of why the possible combinations between oriented space and oriented time elements are in number of 32, as observed by Tonti and summarized in Section 4.1. It is also shown how the coboundary process on the discrete p-forms, which is the tool for building the topological equations in the CM, generalizes the spacetime gradient in spacetime algebra (Section 5.1.2).
- The configuration variables with their topological equations, on the one hand, and the source variables with their topological equations, on the other hand, are viewed as a bialgebra and its dual algebra (Section 4.1). This new point of view allows us to give an explanation of why the configuration variables are associated with space elements endowed with a kind of orientation and the source variables are associated with space elements endowed with the other kind orientation.
- The properties of the boundary and coboundary operators are used in order to find the algebraic form of the virtual work theorem (Section 4.2).
- It is made a distinction between the three coboundary operators, δ^D, δ^C, and δ^G, which, being tensors, are independent of the labeling, the three incidence matrices, $\mathbf{D}$, $\mathbf{C}$, and $\mathbf{G}$, whose incidence numbers depend on the particular choice of labeling, and the three matrices, $\mathbf{T}^D$, $\mathbf{T}^C$, and $\mathbf{T}^G$, which represent the coboundary operators for the given labeling of the cell complex (Section 5.1). In the special case where all the 1-cells of the three-dimensional cell complex are of unit length, all the 2-cells are of unit area, and all the 3-cells are of unit volume, $\mathbf{T}^D$, $\mathbf{T}^C$, and $\mathbf{T}^G$ equal $\mathbf{D}$, $\mathbf{C}$, and $\mathbf{G}$, respectively. If this is not the case, $\mathbf{T}^D$, $\mathbf{T}^C$, and $\mathbf{T}^G$ are obtained with a procedure of expansion and assembling of local matrices, which is derived from the procedure of expansion and assembling of the stiffness matrix. The rows of $\mathbf{D}$, $\mathbf{C}$, and $\mathbf{G}$ give the right operators in the expansion step.
- Possible developments of the CM are investigated for the representation of reality through a purely algebraic unifying gravitational theory, theorized by Einstein during the last decades of his life (Section 6.4).

Elena
Bologna, October 2013

KEYWORDS

Cell method, heterogeneous materials, non-local models, non-standard analysis, bialgebra, Clifford algebra, discrete formulations, fracture mechanics, electromagnetics, electrodynamics, solid mechanics, fluid mechanics, space-time continuum, numerical instabilities, topological features of variables, graph theory, coboundary process, finite element method, boundary element method, finite volume method, finite difference method.

CHAPTER 1

A COMPARISON BETWEEN ALGEBRAIC AND DIFFERENTIAL FORMULATIONS UNDER THE GEOMETRICAL AND TOPOLOGICAL VIEWPOINTS

In this chapter, we analyze the difference between the algebraic and the differential formulation from the mathematical point of view.

The basis of the differential formulation is discussed in Section 1.1. Particular attention is devoted to the computation of limits—by highlighting how the numerical techniques used for performing limits may imply a loss of information. The main motivation for the most commonly used numerical technique in differential formulation, the Cancelation Rule for limits, is to avoid the iterative computation of limits, which is implicit in the definition itself of a limit (the $\varepsilon - \delta$ definition of a limit). The reason for this is that iterations necessarily involve some degree of approximation, while the purpose of the Cancelation Rule for limits is to provide a direct exact solution. Nevertheless, this exact solution is only illusory, since we pay the direct computation of the Cancelation Rule for limits by losing information on the trend of the function in the neighborhood of the estimation point. Conversely, by computing the limit iteratively, with the dimension of the neighborhood that decreases at each iteration, leading also the error on the solution to decrease, we conserve information on the trend of the function in the neighborhood of the estimation point. This second way to operate, where the dimension of the neighborhood approaches zero but is never equal to zero, follows from the $\varepsilon - \delta$ definition of a limit directly and leads to the algebraic formulation. When the Cancelation Rule for limits is used for finding densities and rates, we also lose information on the space and time extent of the geometrical and temporal objects associated with the variables we are computing, obtaining point- and instant-wise variables. By using the algebraic formulation, on the contrary, we preserve both the length and the time scales. Consequently, the physical variables of the algebraic formulation maintain an association with the space and time multi-dimensional elements. In Section 1.1.3, we discuss how the Cancelation Rule for limits acts on the actual solution of a physical problem as a projection operator. The consequence is that the algebraic formulations is to the differential formulation as the actual solution of a physical problem is to the projection of the actual solution on the tangent space of degree 0, where each physical phenomenon is described in terms of space elements of degree 0, the points, and time elements of degree 0, the time instants. In other words, the differential solution is the shadow of the algebraic solution in the tangent space of degree 0. In Section 1.1.3, we also discuss how using the algebraic formulation, instead of the differential formulation, is similar to performing non-standard calculus, the modern application of infinitesimals to differential and integral calculus, instead of standard calculus. In this sense, the derivative of a function can be viewed as the standard part, or the shadow, of the difference quotient. The extension of real numbers, which leads to non-standard calculus, is indeed an attempt to recover the loss of length scales. As for other techniques that will be discussed in Chapter 6, the enrichment with a length scale has a regularization effect on the solution.

In Section 1.2, the features of the algebraic variables (global variables) are compared with those of the differential variables (field variables). Then, with reference to the spatial description, we introduce the association between the global physical variables and the four space elements (point P, line L, surface S, and volume V) and/or the two time elements (time instant I and time interval T). It is also discussed how the association between global variables and space elements in dimensions 0, 1, 2, and 3 requires a generalization of the coordinate systems and time axes, in order for the global variables to be used in numerical modeling. The suitable reference structures are cell complexes, whose elements, properly labeled, are endowed with spatial or time extents. The algebraic formulation then uses notations of algebraic topology, which develops notions corresponding to those of the differential formulations, but based on global variables instead of field variables. This allows us to use algebra instead of differential calculus, for modeling physics.

In Section 1.3, we give the definition of set functions and recognize in the global variables a special case of set functions, due to the association between global variables and elements provided with an extent.

Finally, in Section 1.4, we compare the cell method (CM) with other so-called discrete methods. The comparison shows how the CM is actually the only numerical method being truly algebraic, at the moment.

1.1 Relationship Between How to Compute Limits and Numerical Formulations in Computational Physics

1.1.1 Some Basics of Calculus

In order to explain why the algebraic approach of the cell method (CM) is a winning strategy, if compared to that of the differential formulation, let's start with a brief excursus on the foundation of the differential formulation, calculus.

As is well known, calculus is the mathematical study of how things change and how quickly they change. Calculus uses the concept of **limit** to consider end behavior in the infinitely large and to provide the behavior of the output of a function as the input of that function gets closer and closer to a certain value. The second type of behavior analysis is similar to looking at the function through a microscope and increasing the power of the magnification so as to zoom in on a very small portion of that function. This principle is known as **local linearity** and guarantees that the graph of any continuous[2] smooth function looks like a line, if you are close enough to any point **P** of the curve. We will call this line the **tangent line** (from the Latin word *tangere*, "to touch"). **P** is the **point of tangency**.

Calculus has two major branches, differential calculus and integral calculus, related to each other by the fundamental theorem of calculus.[3] Differential calculus concerns rates of change and slopes of curves, while integral calculus concerns accumulation of quantities and the areas under curves.

As far as differential calculus is concerned, there are two kinds of rate of change, **average rates of change** and **instantaneous rates of change**.

If a quantity changes from a value of m to a value of n over a certain interval from a to b, then the average rate of change is the change, $n - m$, divided by the length of the interval:

$$\text{Average rate of change} = \frac{n - m}{b - a}. \tag{1.1.1}$$

The instantaneous rate of change at a point on a curve describing the change is the slope of the curve at that point, where the slope of a graph at a point is the slope of the tangent line at that point (provided that the slope exists). Unless a function describing the change is continuous and smooth at a point, the instantaneous rate of change does not exist at that point.

By summarizing the differences between the two rates of change, average rates of change

- measure how rapidly (on average) a quantity changes over an interval,
- are a difference of output values,

---

[2] A function f is said to be **continuous at a number** c under the following threefold condition of continuity at c:

- $f(c)$ is defined,
- $\lim_{x \to c} f(x)$ exists,
- $\lim_{x \to c} f(x) - f(c)$.

[3] The first part of the fundamental theorem of calculus, sometimes called the **first fundamental theorem of calculus**, is that an indefinite integration can be reversed by a differentiation. This part of the theorem is also important because it guarantees the existence of **antiderivatives**, or **primitive integrals**, or **indefinite integrals**, for continuous functions, where the antiderivative of a function f is a differentiable function F whose derivative is equal to f.

$$F' = f.$$

The second part, sometimes called the **second fundamental theorem of calculus**, is that the definite integral of a function can be computed by using any one of its infinitely many antiderivatives. This part of the theorem has invaluable practical applications, because it markedly simplifies the computation of definite integrals. The process of solving for antiderivatives is called **antidifferentiation** (or **indefinite integration**). Its opposite operation is called **differentiation**, which is the process of finding a derivative.

- can be obtained by calculating the slope of the secant line (from the Latin word *secare*, "to cut") between two points, the line that passes through the two points on the graph,
- require data points or a continuous curve[4] to calculate.

Instantaneous rates of change (or **rates of change** or **slopes of the curve** or **slope of the tangent line** or **derivatives**)

- measure how rapidly a quantity is changing at a point,
- describe how quickly the output is increasing or decreasing at that point,
- can be obtained by calculating the slope of the tangent line at a single point,
- require a continuous, smooth curve to calculate.

The line tangent to a graph at a point **P** can also be thought of as the limiting position of nearby secant lines – that is, secant lines through **P** and nearby points on the graph. The slope function is continuous as long as the original function is continuous and smooth.

Equations involving derivatives are called **differential equations** and a numerical formulation using differential equations is called **differential formulation**.

Limits give us the power to evaluate the behavior of a continuous function at a point. In particular, limits may be used in order to evaluate the behavior of the function giving the slope of another function, when this slope is a continuous function.

1.1.2 The $\varepsilon - \delta$ Definition of a Limit

The $\varepsilon - \delta$ definition of a limit is the formal mathematical definition of a limit.

Let f be a real-valued function defined everywhere on an open interval containing the real number c (except possibly at c) and let L be a real number. The statement:[5]

$$\lim_{x \to c} f(x) = L, \tag{1.1.2}$$

---

[4] A function is **continuous on an (open) interval** if the output of the function is defined at every point on the interval and there are no breaks, jumps, or holes in the function output. By extending the assiomatic definition of continuity at a number c, given in Footnote (2), to continuity on an interval, a function f is said to be continuous:

- on an open interval (a, b), if it is continuous at every number in the interval,
- on a closed interval $[a, b]$, if it is continuous on (a, b) and, in addition,

$$\lim_{x \to a^+} f(x) = f(a) \text{ and } \lim_{x \to b^-} f(x) = f(b).$$

If the right-hand limit condition $\lim_{x \to a^+} f(x) = f(a)$ is satisfied, we say that f is continuous from the right at a. Analogously, if $\lim_{x \to b^-} f(x) = f(b)$, then f is continuous from the left at b.

[5] The modern notation of placing the arrow below the limit symbol is due to Godfrey Harold "G. H." Hardy (7 February 1877–1 December 1947), in his book *A Course of Pure Mathematics* (1908).

means that, for every real $\varepsilon > 0$,[6] there exists a real $\delta > 0$[7] such that, for all real x, if $0 < |x - c| < \delta$, then $|f(x) - L| < \varepsilon$. Symbolically:[8]

$$\forall \varepsilon > 0 \; \exists \delta > 0 : \; \forall x \left(0 < |x - c| < \delta \Rightarrow |f(x) - L| < \varepsilon \right). \tag{1.1.3}$$

Eq. (1.1.2) can be read as "the limit of $f(x)$, as x approaches c, is L".
The above definition of a limit is true even if

$$f(c) \neq L. \tag{1.1.4}$$

Indeed, the function f does not need to be defined at c. The value of the limit does not depend on the value of $f(c)$, nor even that c be in the domain of f.

The absolute value $|x - c|$ in Eq. (1.1.3) means that x is taken sufficiently close to c from either side (but different from c). The limit value of $f(x)$ as x approaches c from the left, $x \to c^-$, is denoted as **left-hand limit**, and the limit value of $f(x)$ as x approaches c from the right, $x \to c^+$, is denoted as **right-hand limit**. Left-handed and right-handed limits are called **one-sided limits**. A limit exists only if the limit from the left and the limit from the right are equal. Consequently, the limit notion requires a smooth function.

The derivative $f'(x)$ of a continuous function $f(x)$ is defined as either of the two limits (if they exist):[9]

$$f'(x) \triangleq \lim_{s \to x} \frac{f(s) - f(x)}{s - x}, \tag{1.1.5}$$

and

$$f'(x) \triangleq \lim_{h \to 0} \frac{f(x+h) - f(x)}{h}, \; h > 0, \tag{1.1.6}$$

---

[6] ε is a small positive number. It represents the **error** in the measurement of the value at the limit. The $\varepsilon - \delta$ definition of a limit assures us that the error can be made as small as desired by reducing the distance, δ, to the limit point, c.

[7] δ is a positive number. It represents the **distance** to the limit point, c. Since how close to c one must make the independent variable x depends on how close to L we desire to make $f(x)$, δ depends on ε.

[8] The real inequalities exploited in the $\varepsilon - \delta$ definition of a limit were pioneered by Bernhard Placidus Johann Nepomuk Bolzano (October 5, 1781–December 18, 1848), in 1817, and Augustin-Louis Cauchy (21 August 1789–23 May 1857), in 1821. However, the work of Bolzano was not known during his lifetime and the contribution given by Cauchy is not often recognized, because he only gave a verbal definition in his *Cours d'Analyse*. The $\varepsilon - \delta$ definition of a limit in the form it is usually written today was subsequently formalized by Karl Theodor Wilhelm Weierstrass (31 October 1815–19 February 1897).

[9] The symbol "$\triangleq$", equivalent to "$\doteq$", "$\overset{def}{=}$", "$\equiv$", "$:=$", "$=:$", and "$:\Leftrightarrow$", means "is equal by definition to". In particular, $x \equiv y$, $x := y$, or $x =: y$, means "x is defined to be another name for y, under certain assumptions taken in context". $P :\Leftrightarrow Q$ means "P is defined to be logically equivalent to Q".

where the argument of the limit in Eq. (1.1.6) is called a difference quotient.[10] Both in Eq. (1.1.5) and in Eq. (1.1.6), the slope of the tangent line is obtained by finding a formula for the slope of a secant line in terms of the length of the interval, $s - x$ and h, respectively, and then determining the limiting value of the formula as the length of the interval approaches zero.

---

[10] The difference between two points is known as their **Delta**, ΔP, as is the difference, Δf, between the function results in the two points. The general preference for the direction of differences formation is the forward orientation, leading to Deltas greater than 0.

The function **forward difference** of the function f at the point P,

$$\Delta f(P) \triangleq f(P + \Delta P) - f(P),$$

divided by the point difference ΔP, with $\Delta P > 0$, is known as the **difference quotient**, or Newton's quotient, named after Isaac Newton (25 December 1642–20 March 1727):

$$\bar{f}(P) \triangleq \frac{\Delta f(P)}{\Delta P} = \frac{f(P + \Delta P) - f(P)}{\Delta P}.$$

If ΔP is infinitesimal, then the difference quotient is a **derivative**, otherwise it is a **divided difference**, the first divided difference of the function f with respect to the points P and $P + \Delta P$.

Assuming all the points in the domain of the function f are equidistant, $\Delta P = \text{const}$, and given the $(k-1)$-th divided differences $f^{k-1}(P, P + \Delta P, ..., P + (k-2)\Delta P)$ and $f^{k-1}(P + \Delta P, P + 2\Delta P, ..., P + (k-1)\Delta P)$ are known, with $k > 1$, one can calculate the k-th divided difference as

$$\bar{f}^k(P, P + \Delta P, ..., P + (k-1)\Delta P)$$

$$\triangleq \frac{\bar{f}^{k-1}(P + \Delta P, P + 2\Delta P, ..., P + (k-1)\Delta P) - \bar{f}^{k-1}(P, P + \Delta P, ..., P + (k-2)\Delta P)}{\Delta P}$$

$$= \frac{\dfrac{\Delta^{k-1} f(P + \Delta P, P + 2\Delta P, ..., P + (k-1)\Delta P)}{\Delta P^{k-1}} - \dfrac{\Delta^{k-1} f(P, P + \Delta P, ..., P + (k-2)\Delta P)}{\Delta P^{k-1}}}{\Delta P}$$

$$= \frac{\Delta^{k-1} f(P + \Delta P, P + 2\Delta P, ..., P + (k-1)\Delta P) - \Delta^{k-1} f(P, P + \Delta P, ..., P + (k-2)\Delta P)}{\Delta P^k}$$

$$= \frac{\Delta^k f(P, P + \Delta P, ..., P + (k-1)\Delta P)}{\Delta P^k}.$$

In case the $k+1$ domain points x_i, x_{i+1}, ..., x_{i+k} are not equidistant, the generalization of the k-th divided difference formula provides

$$\bar{f}^k(x_i, x_{i+1}, ..., x_{i+k}) \triangleq \frac{\bar{f}^{k-1}(x_{i+1}, x_{i+2}, ..., x_{i+k}) - \bar{f}^{k-1}(x_i, x_{i+1}, ..., x_{i+k-1})}{x_{i+1} - x_i}$$

$$= \frac{\Delta^k f(x_i, x_{i+1}, ..., x_{i+k})}{(x_{i+1} - x_i)(x_{i+2} - x_{i+1})...(x_{i+k} - x_{i+k-1})},$$

where the k-th forward difference $\Delta^k f(x_i, x_{i+1}, ..., x_{i+k})$, with $k > 1$, is equal to

$$\Delta^k f(x_i, x_{i+1}, ..., x_{i+k}) = \frac{(x_{i+1} - x_i)\Delta^{k-1} f(x_{i+1}, x_{i+2}, ..., x_{i+k}) - (x_{i+k} - x_{i+k-1})\Delta^{k-1} f(x_i, x_{i+1}, ..., x_{i+k-1})}{x_{i+1} - x_i}.$$

(continues next page)

Because the derivative of a function is the limit of quotients whose numerator comes from values of the function and whose denominator comes from values of the independent variable, the units in which the derivative is measured are the units in which values of the function are measured divided by the units in which the independent variable is measured.

Due to the absolute value $|x - c|$ in the $\varepsilon - \delta$ definition of a limit, the limit can also be evaluated on the backward difference[11] of the function $f(x)$:

$$f'(x) \triangleq \lim_{h \to 0} \frac{f(x+h) - f(x)}{h} = \lim_{h \to 0} \frac{f(x) - f(x-h)}{h}, \, h > 0. \tag{1.1.7}$$

A further way for finding the derivative $f'(x)$ is making use of the central difference[12] of the function $f(x)$:

$$f'(x) = \lim_{h \to 0} \frac{f\left(x + \frac{1}{2}h\right) - f\left(x - \frac{1}{2}h\right)}{h}, \, h > 0. \tag{1.1.8}$$

The ratio in Eq. (1.1.6) is not a continuous function at $h = 0$, because it is not defined there. In fact, the limit (1.1.6) has the indeterminate form $(\to 0)/(\to 0)$ as $h \to 0$, since both the numerator and the denominator approach 0 as $h \to 0$.

(*continues from previous page*)

One finally obtains

$$\bar{f}^k\left(x_i, x_{i+1}, \dots, x_{i+k}\right) = \frac{\left(x_{i+1} - x_i\right)\Delta^{k-1}f\left(x_{i+1}, x_{i+2}, \dots, x_{i+k}\right) - \left(x_{i+k} - x_{i+k-1}\right)\Delta^{k-1}f\left(x_i, x_{i+1}, \dots, x_{i+k-1}\right)}{\left(x_{i+1} - x_i\right)\prod_{n=1}^{k}\left(x_{i+n} - x_{i+n-1}\right)}, \, k > 1.$$

This expression gives the k-th divided difference also for $k = 1$, provided that the 0-th forward difference of the function f at the point x_i is the function evaluation at x_i:

$$\Delta^0 f\left(x_i\right) \triangleq f\left(x_i\right).$$

[11] The function **backward difference** of the function f at the point P is defined as

$$\nabla f(P) \triangleq f(P) - f(P - \Delta P),$$

where $\Delta P > 0$.

[12] The function **central difference** of the function f at the point P is defined as

$$\delta f(P) \triangleq f\left(P + \frac{1}{2}\Delta P\right) - f\left(P - \frac{1}{2}\Delta P\right),$$

where $\Delta P > 0$.

We can compute the limiting value (1.1.6) both in an approximated way, by reducing the error[13] with subsequent iterations, as per the $\varepsilon - \delta$ definition of a limit, or in an exact way, by making use of the Cancellation Rule for limits.[14]

In the perspective of a computational analysis using the differential formulation, it is obvious that the choice falls on the exact, rather than the approximated, computation of limits. In effect, in doing so, one can obtain an exact solution of the physical phenomenon under consideration only in few elementary cases, with simple geometric shapes of the domain and under particular boundary conditions. Anyway, the most important aspect is not that the exact numerical solution is hardly ever attained in real cases, but rather, that the choice itself of the term "exact" for the limit promised by the Cancelation Rule is not entirely appropriate. Actually, in order to provide the solution of the limit directly, the Cancellation Rule for limits reduces the order of zero both in the numerator and the denominator by one. Under the numerical point of view, this reduction is made by canceling a quantity with the order of a length, both in the numerator and in the denominator. Under the topological point of view, we could say that the reduction degrades the solution, in the sense that, being deprived of one length scale, the solution given by the Cancelation Rule provides us with a lower degree of detail in describing the physical phenomenon under consideration. In other words, we pay the direct solution of the Cancelation Rule by losing some kind of information on the solution itself. This is why we can say that the solution provided by the Cancelation Rule is not exact in a narrow sense, but only in a broad sense.

1.1.3   A Discussion on the Cancelation Rule for Limits

Let us see, in more detail, which is the information we lose by reducing the order of zero of numerator and denominator. Due to the Cancelation Rule for limits, we can factor h out of the numerator in Eq. (1.1.6):

$$\left[f\left(x+h\right) - f\left(x\right) \right]\Big|_{x=\bar{x}} = h \cdot g\left(h\right), \tag{1.1.9}$$

and cancel this common factor in the numerator and denominator. Then, we can find the limit by evaluating the new expression at $h = 0$, that is, by plugging in 0 for h, because the new expression is continuous at $h = 0$:

$$\lim_{h \to 0} \frac{h \cdot g\left(h\right)}{h} = g\left(h\right)\Big|_{h=0}, \tag{1.1.10}$$

where the result is a real number.

---

[13] Since f is assumed to be a continuous smooth function, the error on the estimation of the (unknown) value of the limit reduces when the difference between two subsequent approximated computations of the limit, for decreasing δ, decreases.

[14] The **Cancellation Rule for limits** states that, if the numerator and the denominator of a rational function share a common factor, then the new function obtained by algebraically canceling the common factor has all limits identical to those of the original function.

The equality in Eq. (1.1.10), established by the Cancelation Rule, is undoubtedly numerically correct, in the sense that the results of the left- and right-hand-side expressions are actually numerically equal, but the way in which these results are achieved is radically different in the two cases. As a matter of fact, the limit on the left side is defined on the open interval of length h, while the function $g(h)$ is evaluated for a given value of the variable, $h = 0$. This difference, negligible from the purely numerical viewpoint, is instead essential from the topological viewpoint. In effect, it is so much essential that the opportunity of using an algebraic rather than a differential formulation could be discussed just on the basis of the equality between the left- and right-hand-side terms in Eq. (1.1.10).

In order to understand this last statement, we must recall that the $\varepsilon - \delta$ definition of a limit implies choosing an (open) interval, containing the point in which we want to estimate a function, with the aim of making the distance between the points in which we compute the function and the point in which we want to estimate the function as small as we want. In other words, the limit on the left side in Eq. (1.1.10) is strictly bonded to the idea of interval of a point and cannot be separated from it. The result of the limit is the value to which the function output appears to approach as the computation point approaches the estimation point. For evaluating this result, we must enough carefully choose the computation points, in order to derive the trend of the output to a specific degree of approximation. That is, the result we obtain by choosing increasingly close points is only an estimation of the actual result and the approximation of the estimation is as much better (the degree of approximation is as much low) as the computation point is close to the estimation point. In conclusion, the $\varepsilon - \delta$ definition of a limit also bounds the limit to the notions of approximation and degree of approximation, or accuracy.

Completely different is the discussion on the right-hand-side function of Eq. (1.1.10). Actually, the new function $g(h)$ is computed at a point, the point $h = 0$, without any need of evaluating its trend on an interval. The consequence is that the result we obtain is exact (in a broad sense), and we do not need to prefix any desired accuracy for the result itself. This is very useful from the numerical point of view, but, from the topological point of view, we lose information on what happens approaching the evaluation point. It is the same type of information we lose in passing from the description of a phenomenon in a space to the description of the same phenomenon in the tangent space at the evaluation point.

Now, the question is if this is an acceptable loss. In the spirit of the principle of the local linearity, factorization and cancelation of the common factors are usually carried out under the implicit assumption that it is. In particular, as far as numerical modeling is concerned, any possible implication of the Cancellation Rule on the numerical result was so far neglected, leading the researchers to use the differential formulation as if it were the natural formulation for computational physics. Better still, we could say that three centuries of differential formulation accustomed us to think that only a differential equation can provide the exact solution of a problem, in general, and a physical problem, in our peculiar case. This unconditional hope in the exact solution of the differential formulation did not leave space to any other consideration, in particular on what the cancelation of a length scale may involve from the physical and topological viewpoints.

The idea underlying this book is that the Cancellation Rule can actually be employed only in those cases where the specific phenomenon uniquely depends on what happens at the point under consideration. In effect, this happens in few physical problems, while, in most cases, the physical phenomenon under consideration also depends on what happens in a neighborhood centered at the point. By extension of Eq. (1.1.10) to functions of more than one variable, studying the physical phenomenon as if it were a point-wise function means that we are using the right-hand side of Eq. (1.1.10), while studying the physical phenomenon as a function of all the points contained in a neighborhood means that we are using the left-hand side, with h approaching zero but never equal to zero. In the first case, we are facing a differential formulation, while, in the second case, we are facing an algebraic formulation.

Operatively, we are using an algebraic formulation whenever we choose increasingly close points (to both the right and the left) of the estimation point, until the outputs remain constant to one decimal place beyond the desired accuracy for two or three calculations. How much the computation points must be close to the estimation point depends on how fast the result of the limit is approached as we approach the point in which the limit is estimated. Therefore, the dimension of the neighborhood is fixed by the trend of the phenomenon around the point under consideration, or, in other words, the distance δ^7 for the evaluation of $f'(c)$ depends both on the error ε^6 and on $f''(c)$. The information we lose by using the Cancellation Rule lies just in the trend of the phenomenon, that is, in the curvature, since the curvature cannot be accounted for in passing from a space to its tangent space at the evaluation point.

In the differential formulation, the notion of limit is used not only for defining derivatives, but also densities. In this second case, the denominator that tends to zero has the dimensions of a length raised to the power of 1, 2, or 3. The Cancelation Rule for limits can be employed also in this second case, by factorizing and canceling length scales in dimension 1, 2, or 3, respectively. This leads to point-wise variables in any cases, the line, surface, and volume densities.

Finally, the Cancelation Rule for limits is used also for finding rates,[15] by factorizing and canceling time scales in dimension 1. This last time, the limit, which is a time derivative, provides an instant-wise variable.

---

[15] The term "rate" denotes the limit of the ratio of a physical quantity to a time interval, that is, the duration of the interval. Thus, the rate is the limit of the ratio between a quantity and an increment in time, while the derivative is the limit of the ratio between two increments, where the denominator can also be a time increment. In this last case, the derivative is a time derivative. Thus, whenever the denominator of a limit is a time increment, we have a rate rather than a time derivative depending on the nature of the numerator, a quantity rather than an increment, respectively.

By way of example, consider the two ways in which we can define the vector velocity at the time instant t, $\mathbf{v}(t)$, in particle mechanics:

$$\mathbf{v}(t) = \lim_{\Delta t \to 0} \frac{\mathbf{u}(\Delta t)}{\Delta t},$$

where $\mathbf{u}(\Delta t)$ is the displacement vector, function of the time interval Δt, and

$$\mathbf{v}(t) = \lim_{\Delta t \to 0} \frac{\mathbf{r}(t + \Delta t) - \mathbf{r}(t)}{\Delta t} = \lim_{\Delta t \to 0} \frac{\Delta \mathbf{r}(t)}{\Delta t} = \frac{d\mathbf{r}(t)}{dt},$$

where $\mathbf{r}(t)$ is the radius vector, function of the time instant t.

As per what said, the vector velocity is both the rate of the displacement vector and the time derivative of the radius vector.

Generally speaking, it is not possible to perform the time derivative of a variable that depends on time intervals. We need variables that depend on the time instant.

(continues next page)

In conclusion, with reference to the space of the physical phenomena, the differential formulation provides the numerical solution in the tangent space of degree 0, where we can describe each physical phenomenon in terms of the space elements of degree 0, the points, and the time elements of degree 0, the time instants. Conversely, the algebraic formulation of the CM allows us to take account of, we could say, the curvatures in space and time at a point, where a point of the space of the physical phenomena is a given physical phenomenon, in a given configuration, at a given time instant.

In other words, the solution given by the Cancelation Rule for limits is the projection of the actual solution from the multi-dimensional space to the tangent space of degree 0. The cancelation of the common factors in numerator and denominator acts as a projection operator and the equality in Eq. (1.1.6) should more properly be substituted by a symbol of projection. Consequently, the solution of the differential formulation is the shadow of the actual solution in the tangent space of degree 0. On the contrary, the CM, avoiding the projection process, provides us with a higher degree solution, approximated in any case, which is more adherent to the physical nature of the phenomenon under consideration.

It is worth noting that concerns about the soundness of arguments involving infinitesimals date back to ancient Greek mathematics, with Archimedes replacing such proofs with ones using other techniques, such as the method of exhaustion.[16] This discussion also permeated the history of calculus, which is fraught with philosophical debates about the meaning and logical validity of fluxions,[17] Newton's term for differential calculus (fluents was his term for integral calculus).

(*continues from previous page*)

According to the nomenclature[10] introduced by Isaac Newton (25 December 1642–20 March 1727), a dot over a letter denotes a time derivative, not a rate. Thus, we can also write the vector velocity as

$$\mathbf{v}(t) = \dot{\mathbf{r}}(t),$$

while the notation

$$\mathbf{v}(t) = \dot{\mathbf{u}}(\Delta t),$$

is not correct.

[16] The **method of exhaustion** (*methodus exhaustionibus*, or *méthode des anciens*) is a method of finding the area of a shape by inscribing inside it a sequence of polygons, whose areas converge to the area of the containing shape. If the sequence is correctly constructed, the difference in area between the n-th polygon and the containing shape will become arbitrarily small as n becomes large. As this difference becomes arbitrarily small, the possible values for the area of the shape are systematically "exhausted" by the lower bound areas successively established by the sequence members. Democritus (c. 460–c. 370 BC) is the first person recorded to consider seriously the division of objects into an infinite number of cross-sections, but his inability to rationalize discrete cross-sections with a cone's smooth slope prevented him from accepting the idea. Antiphon the Sophist (who lived in Athens probably in the last two decades of the 5th century BC) and later Eudoxus of Cnidus (408 BC–355 BC) are generally credited with implementing the method of exhaustion, which is seen as a precursor to the methods of calculus. An important early intermediate step was Cavalieri's principle, named after the Italian mathematician Bonaventura Francesco Cavalieri (in Latin, Cavalerius) (1598–November 30, 1647), also termed the "method of indivisibles", which was a bridge between the method of exhaustion and full-fledged integral calculus.

[17] The standard way to resolve these debates is to define the operations of calculus using limits, rather than infinitesimals. **Non-standard analysis**, instead, reformulates the calculus using a logically rigorous notion of infinitesimal number. Non-standard analysis was introduced in the early 1960s by the mathematician Abraham Robinson (born Robinsohn; October 6, 1918–April 11, 1974).

By using the language of non-standard analysis, which is a rigorous formalization of calculations with infinitesimals,[18] the infinite and infinitesimal quantities can be treated by the system of **hyperreal numbers,**[19] or **hyperreals**, or **nonstandard reals**. Denoted by $*\mathbb{R}$, the hyperreal numbers are an extension of the real numbers, $\mathbb{R}$, that contains numbers greater than anything of the form:

$$1+1+...+1. \tag{1.1.11}$$

Such a number is infinite, and its reciprocal is infinitesimal.

The hyperreal numbers satisfy the transfer principle, a rigorous version of Leibniz's heuristic Law of Continuity.[20] The transfer principle states that true first order statements about $\mathbb{R}$ are also valid in $*\mathbb{R}$. Therefore, the hyperreals were logically consistent if and only if the reals were. This put to rest the fear that any proof involving infinitesimals might be unsound,[16–18] provided that they were manipulated according to the logical rules which Robinson delineated.

---

[18] The notion of infinitely small quantities was discussed by the Eleatic School. The Greek mathematician Archimedes (c.287 BC–c.212 BC), in *The Method of Mechanical Theorems*, was the first to propose a logically rigorous definition of **infinitesimals**. When Sir Isaac Newton (25 December 1642–20 March 1727) and Gottfried Wilhelm von Leibniz (July 1, 1646–November 14, 1716) invented the calculus, they made use of infinitesimals. The use of infinitesimals was attacked as incorrect by George Berkeley (12 March 1685–14 January 1753), also known as Bishop Berkeley (Bishop of Cloyne), in his work *The Analyst*. Mathematicians, scientists, and engineers continued to use infinitesimals to produce correct results. In the second half of the 19th century, the calculus was reformulated by Baron Augustin-Louis Cauchy (21 August 1789–23 May 1857), Bernhard Placidus Johann Nepomuk Bolzano, or Bernard Bolzano in English, (October 5, 1781–December 18, 1848), Karl Theodor Wilhelm Weierstrass (German: *Weierstraß*; 31 October 1815–19 February 1897), Georg Ferdinand Ludwig Philipp Cantor (March 3 [O.S. February 19] 1845–January 6, 1918), Julius Wilhelm Richard Dedekind (October 6, 1831–February 12, 1916), and others using the $\varepsilon-\delta$ definition of limit and set theory. While infinitesimals eventually disappeared from the calculus, their mathematical study continued through the work of Tullio Levi-Civita, (29 March 1873–29 December 1941) and others, throughout the late 19th and the 20th centuries. In the 20th century, it was found that infinitesimals could serve as a basis for calculus and analysis.
[19] The term "hyper-real" was introduced by Edwin Hewitt (January 20, 1920, Everett, Washington–June 21, 1999) in 1948.
[20] The **Law of Continuity** is a heuristic principle introduced by Gottfried Wilhelm von Leibniz (July 1, 1646–November 14, 1716), based on earlier work by Nicholas of Kues (1401–August 11, 1464), also referred to as Nicolaus Cusanus and Nicholas of Cusa, and Johannes Kepler (December 27, 1571–November 15, 1630). It is the principle that "whatever succeeds for the finite, also succeeds for the infinite." Kepler used it to calculate the area of the circle by representing the latter as an infinite-sided polygon with infinitesimal sides, and adding the areas of infinitely many triangles with infinitesimal bases. Leibniz used the principle to extend concepts such as arithmetic operations, from ordinary numbers to infinitesimals, laying the groundwork for infinitesimal calculus. Abraham Robinson (October 6, 1918–April 11, 1974) argued that the law of continuity of Leibniz is a precursor of the transfer principle. He wrote:

> *"[...] As for the objection [...] that the distance between two distinct real numbers cannot be infinitely small, Gottfried Wilhelm Leibniz argued that the theory of infinitesimals implies the introduction of ideal numbers which might be infinitely small or infinitely large compared with the real numbers but which were to possess the same properties as the latter. [...] However, neither he nor his disciples and successors were able to give a rational development leading up to a system of this sort. As a result, the theory of infinitesimals gradually fell into disrepute and was replaced eventually by the classical theory of limits. [...] It is shown in this book that Leibniz's ideas can be fully vindicated and that they lead to a novel and fruitful approach to classical Analysis and to many other branches of mathematics. The key to our method is provided by the detailed analysis of the relation between mathematical languages and mathematical structures which lies at the bottom of contemporary model theory."*

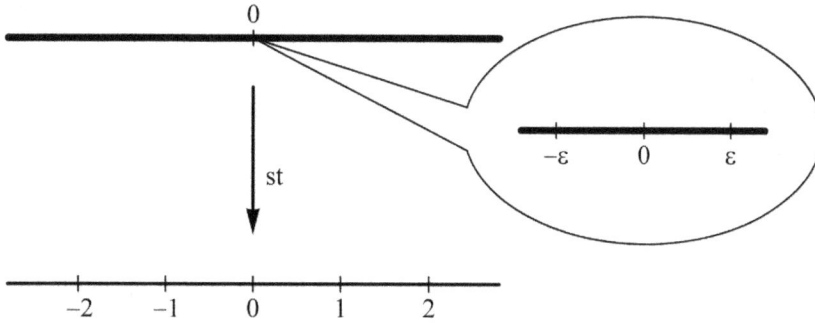

Figure 1.1. The bottom line represents the "thin" real continuum. The line at top represents the "thick" hyperreal continuum. The "infinitesimal microscope" is used to view an infinitesimal neighborhood of 0.

Non-standard analysis deals primarily with the **hyperreal line**, which is an extension of the real line, containing infinitesimals, in addition to the reals (Fig. 1.1). In the hyperreal line every real number has a collection of numbers (called a **monad**, or **halo**) of hyperreals infinitely close to it.

The **standard part function** is a function from the limited (finite) hyperreal to the reals. It associates with a finite hyperreal x, the unique standard real number x_0 which is infinitely close to it (Fig. 1.1):

$$\text{st}(x) = x_0. \tag{1.1.12}$$

As such, the standard part function is a mathematical implementation of the historical concept of adequality introduced by Pierre de Fermat.[21] It can also be thought of as a mathematical

---

[21] Pierre de Fermat (17 August 1601 or 1607–12 January 1665) is recognized for his discovery of an original method of finding the greatest and the smallest ordinates of curved lines, which is analogous to that of the differential calculus, then unknown, and his research into number theory. He used **adequality** (translation of the Latin word *adaequalitas*, as it was used by Fermat) first to find maxima of functions, and then adapted it to find tangent lines to curves. To find the maximum of a term $p(x)$, Fermat did equate (or more precisely adequate) $p(x)$ and $p(x+e)$ and, after doing algebra, he could divide by e, and then discard any remaining terms involving e. To illustrate the method by Fermat's own example, consider the problem of finding the maximum of:

$$p(x) = bx - x^2.$$

Fermat adequated $bx - x^2$ with $b(x+e) - (x+e)^2$, that is, by using the notation "$\sim$", introduced by Paul Tannery (1843–1904), to denote adequality:

$$bx - x^2 \sim bx - x^2 + be - 2ex - e^2.$$

Canceling terms and dividing by e, Fermat arrived at

$$b \sim 2x + e.$$

Removing the terms that contained e, Fermat arrived at the desired result that the maximum occurred when

$$x = \frac{b}{2}.$$

implementation of Leibniz's Transcendental Law of Homogeneity.[22] The standard part function was first defined by Abraham Robinson as a key ingredient in defining the concepts of the calculus, such as the derivative and the integral, in non-standard analysis.

The standard part of any infinitesimal is 0. Thus, if N is an infinite hypernatural, then $1/N$ is infinitesimal, and

$$\mathrm{st}\left(\frac{1}{N}\right) = 0. \tag{1.1.13}$$

The standard part function allows the definition of the basic concepts of analysis, such as derivative and integral, in a direct fashion. The derivative of f at a standard real number x becomes

$$f'(x) = \mathrm{st}\left(\frac{{}^*f(x+\Delta x) - {}^*f(x)}{\Delta x}\right), \tag{1.1.14}$$

where Δx is an infinitesimal, smaller than any standard positive real, yet greater than zero, and ${}^*f$ is the natural extension of f to the hyperreals (* is the transfer operator applied to f). Similarly, the integral is defined as the standard part of a suitable infinite sum.[23]

In this approach, $f'(x)$ is the real number infinitely close to the hyperreal argument of st. For example, the non-standard computation of the derivative of the function $f(x) = x^2$ provides

$$f'(x) = \mathrm{st}\left(\frac{(x+\Delta x)^2 - x^2}{\Delta x}\right) = \mathrm{st}(2x + \Delta x) = 2x, \tag{1.1.15}$$

since

$$2x + \Delta x \approx 2x, \tag{1.1.16}$$

where the symbol "$\approx$" is used for indicating the relation "is infinitely close to." In order to make $f'(x)$ a real-valued function, we must dispense with the final term, Δx, which is the error term. In the standard approach using only real numbers, that is done by taking the limit as Δx tends to zero. In the non-standard approach using hyperreal numbers, the quantity Δx is taken to be an

---

[22] The **Transcendental Law of Homogeneity** (TLH) is a heuristic principle enunciated by Gottfried Wilhelm Leibniz (July 1, 1646–November 14, 1716) most clearly in a 1710 text entitled *Symbolismus memorabilis calculi algebraici et infinitesimalis in comparatione potentiarum et differentiarum, et de lege homogeneorum transcendentali*. It is the principle to the effect that in a sum involving infinitesimals of different orders, only the lowest-order term must be retained, and the remainder discarded.

[23] One way of defining a definite integral in the hyperreal system is as the standard part of an infinite sum on a hyperfinite lattice defined as

$$a,\ a+dx,\ a+2dx,\ \dots,\ a+ndx,$$

where dx is infinitesimal, n is an infinite hypernatural, and the lower and upper bounds of integration are a and $b = a + ndx$.

infinitesimal, a nonzero number that is closer to 0 than to any nonzero real, which is discarded by the standard part function.

The notion of limit can easily be recaptured in terms of the standard part function, st, namely:

$$\lim_{x \to c} f(x) = L, \qquad (1.1.17)$$

if and only if, whenever the difference $|x - c|$ is infinitesimal, the difference $|f(x) - L|$ is infinitesimal, as well. In formulas

$$\text{st}(x) = c \Rightarrow \text{st}(f(x)) = L. \qquad (1.1.18)$$

The standard part of x is sometimes referred to as its **shadow**. Therefore, the derivative of $f(x)$ is the shadow of the difference quotient.[10]

We can thus conclude that the standard part function is a form of projection from hyperreals to reals. As a consequence, using the algebraic formulation is somehow similar to performing non-standard calculus, the modern application of infinitesimals, in the sense of non-standard analysis, to differential and integral calculus. In effect, the extension of the real numbers, $\mathbb{R}$, is equivalent to providing the space of reals with a supplementary structure of infinitesimal lengths. This configures the hyperreal number system as an infinitesimal-enriched continuum, and the algebraic approach can be viewed as the algebraic version of non-standard calculus.

The great advantage of the infinitesimal-enrichment is that of successfully incorporating a large part of the technical difficulties at the foundational level of non-standard calculus. Similarly, in the algebraic formulation many numerical problems, mainly instability or convergence problems, are avoided by the presence of a supplementary structure of (finite) lengths both in $\mathbb{R}$, $\mathbb{R}^2$, and $\mathbb{R}^3$.

The use of an algebraic formulation instead of a differential one also has a justification based on the microstructure of matter. As previously said, when performing densities and rates, the intention is to formulate the field laws in an exact form. Nevertheless, the density finding process is carried out without considering whether a physical significance exists for the limit one is performing. In fact, since matter is discrete on a molecular scale, performing the limit process of the mean densities with the extent of the geometrical object going to zero makes no physical sense.

1.2 Field and Global Variables

We can classify the physical variables according to their nature, global or local.

Broadly speaking, the global variables are variables that are neither densities nor rates of other variables. In particular, we will call:

- **Global variable in space**, or space global variable, a variable that is not the line, surface, or volume density of another variable.
- **Global variable in time**, or time global variable, a variable that is not the rate or time derivative of another variable.

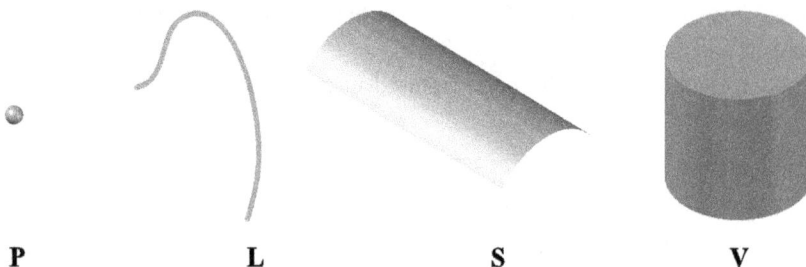

| P | L | S | V |

Figure 1.2. The four space elements and their notations.

The field variables are obtained from the global variables as densities of space global variables and rates of time global variables. Due to their point-wise nature, they are local variables.

The variables obtained by line, surface, or volume integration of field variables, the integral variables, are global variables, but there also exist global variables that are not integral variables.

One of the main consequences of using the left- rather than the right-hand side in Eq. (1.1.10) is that the nature of physical variables is different in the algebraic rather than the differential formulation, global in the first case and local in the second case.

In effect, in the differential formulation, some variables arise directly as functions of points and time instants, while the remaining variables are reduced to points and time instants functions by performing densities and rates and making use of the Cancelation Rule for limits. Thus, the physical variables of the differential formulation are point-wise and/or instant-wise field functions. Moreover, since we are ignoring approximations and accuracies, we must consider any infinitesimal region around a given point as a uniform region.

On the other hand, avoiding factorization and cancelation, the algebraic formulation uses global variables. Moreover, since, in doing so, the algebraic formulation preserves the length and time scales of the global physical variables, the physical variables, in spatial description,[24] turn out to be naturally associated with one of the four space elements[25] (point, line, surface, and volume, which are denoted with their initial capital letters in bold, **P**, **L**, **S**, and **V**, respectively, as shown in Fig. 1.2) and/or with one of the two time elements (time instant and time interval, which are denoted with **I** and **T**, respectively, as shown in Fig. 1.3).

---

[24] We speak about **spatial description**, or **Eulerian description**, when the subject of our observation is a space region and we want to know what happens inside it. Adopting this description, we can divide the space region into sub-regions, called control volumes, each of which is composed by points. Broadly speaking, we have a spatial description whenever we look at a space region as a whole, without focusing our attention on any specific object. The spatial description is typical of field theories, such as fluid flow, electromagnetism, gravitation, heat conduction, diffusion, and irreversible thermodynamics. For this reason, it is also called field description.

The Eulerian description, introduced by Jean-Baptiste le Rond d'Alembert (16 November 1717–29 October 1783), focuses on the current configuration, giving attention to what is occurring at a fixed point in space as time progresses, instead of giving attention to individual particles as they move through space and time. In this case the description of motion is made in terms of the spatial coordinates, that is, the current configuration is taken as the reference configuration.

[25] The existence of an association between the global physical variables of the spatial description of physical theories and the four space elements, **P**, **L**, **S**, and **V**, is established by the **Association Principle**. By way of example, in continuum mechanics the volume forces are associated with volumes. Analogously, the geometrical referents of surface forces are the surfaces, the geometrical referents of strains are the lines, and the geometrical referents of displacements are the points.

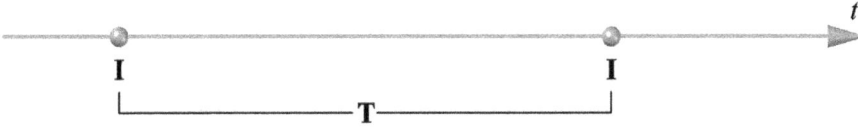

Figure 1.3. The two time elements and their notations.

The association between physical variables and space elements was discovered by Franklin Branin, an IBM engineer, in the late 1960s. Branin also pointed out the intimate relation between algebraic topology, network theory, and the vector calculus.

As an example of association between physical variables and space elements, the flux and the flow are associated with a surface (Fig. 1.4). The voltage, the magnetomotive force, the line integral of the fluid velocity, and the work of a force along a line in a force field are associated with a line. The mass content, the energy content, the entropy content, and the momentum content are associated with a volume. Moreover, displacements in solid mechanics, the kinetic potential in flow mechanics, the gauge function of electromagnetism, the iconal function in optics, and temperature are examples of variables associated with points in space (and time), without being densities or rates.

In the differential material description[26] of physics, the physical laws are ordinary differential equations, where the field variables are time derivatives and rates, while the algebraic material description makes use of global variables in space, which change with the two time elements (in material formulations, all the variations are time variations, while the system of which we are following the motion is prefixed).

In the differential spatial description of physics, the physical laws are partial differential equations, where the field variables are densities, while the algebraic spatial description makes use of global variables in time, which change with the four space elements (in spatial formulations, all the variations are geometrical variations, while the instant in which to compute the variations is prefixed).

In some physical theories, one may use both the material and the spatial descriptions. This is the case of fluid dynamics, electromagnetism, and quantum mechanics.

Note that there is a remarkable difference between global and field variables, when the domain of the physical problem is composed of more than one medium. In fact, while global variables are continuous through the interface of two different media, their variations can be discontinuous. Consequently, even field variables, which are densities and rates, are generally discontinuous. The

---

[26] When the subject of our observation is a physical system that evolves in time, for example a fluid body, we adopt the **material description**, or **Lagragian description**. The system under observation can be broken down into bodies, and bodies into particles. Broadly speaking, we have a material description whenever we concentrate our attention on one specific object of a space region, by ignoring what is happening to the other objects of the whole space region. The material description is typical of mechanics of rigid and deformable solids, analytical mechanics, and thermodynamics.

In the Lagrangian description the position and physical properties of the particles are described in terms of the material or referential coordinates and time. In this case the reference configuration is the configuration at $t = 0$. An observer standing in the referential frame of reference observes the changes in the position and physical properties as the material body moves in space as time progresses. The results obtained are independent of the choice of initial time and reference configuration.

Physical theory	Global variable	Referent geometrical object
Thermal conduction	**Temperature**	**[P]**
Thermal conduction	**Electrical potential**	**[P]**
Solid mechanics	**Displacement**	**[P]**
Fluid mechanics	**Velocity**	**[P]**
Electromagnetism	**Voltage**	**[L]**
Solid mechanics	**Stretching**	**[L]**
Acoustics	**Velocity circulation**	**[L]**
Electromagnetism	**Charge flow**	**[S]**
Fluid dynamics	**Discharge**	**[S]**
Thermal conduction	**Heat**	**[S]**
Solid mechanics	**Surface force**	**[S]**
Mechanics	**Mass content**	**[V]**
Mechanics	**Momentum content**	**[V]**
....		

Figure 1.4. Association between physical variables and points (**P**), lines (**L**), surfaces (**S**), and volumes (**V**) for several physical theories.

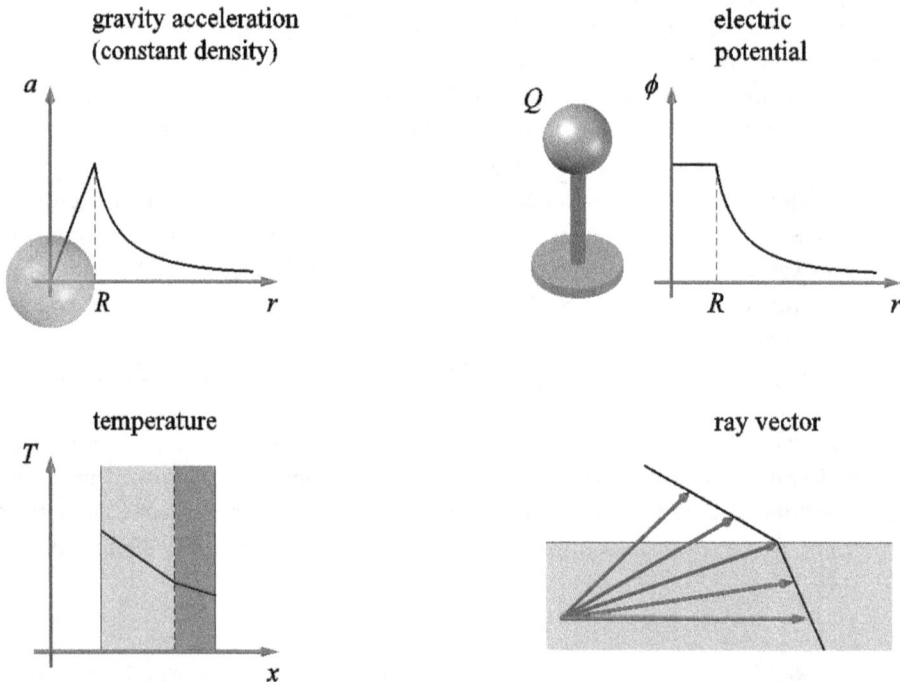

Figure 1.5. Continuity of the global variables associated with points in domains made of more than one medium.

same can be said for any other kind of discontinuities of the domain or the sources of the physical problem (some examples of continuity of the global variables are collected in Fig. 1.5).

Global variables are essential to the philosophy of the cell method, since, by using these variables, it is possible to obtain an algebraic formulation directly and, what is most important, the global variables involved in obtaining the formulation do not have to be differentiable functions. Moreover, by using the limit process on the mean densities and rates of the global variables, we can obtain the traditional field functions of the differential formulation.

The main difference between the two formulations – algebraic and differential – lies precisely in the fact that the limit process is used in the latter. In effect, since calculating the densities and rates of the domain variables is based on the assumption that global variables are continuous and differentiable, the range of applicability of differential formulation is restricted to regions without material discontinuities or concentrated sources, while that of the algebraic formulation is not restricted to such regions.

As will be better explained hereinafter (Chapter 6), the purpose of this book is not simply to classify the physical variables according to the geometrical and time elements on which they depend. The question takes on a much more deep meaning, since the ability of the CM to solve some of the problems affected by spurious solutions in the differential formulation lies, in part, just on the association between variables and space and/or time elements.

The comparison between global and field variables deserves a final comment, which will be essential for the comprehension of the difference between computational methods based on an algebraic rather than a differential formulation. Even in this last case, the comment arises from the association between physical variables and space elements in dimensions 0, 1, 2 and 3, which is ignored in the differential formulation, while it is emphasized in the algebraic formulation. In particular, the space distribution of the point-wise field functions of the differential formulation requires the introduction of coordinate systems, whose purpose is to create a correspondence between the points of the space and the numbers, that is, their coordinates. This allows us to describe geometry through mathematics. Now, in a differential formulation, the coordinate systems, together with a time axis, are the most appropriate frameworks to treat variables, since variables are point- and instant-wise functions, but we cannot state the same in an algebraic formulation, where variables also depend on geometrical and time objects of dimension greater than 0. Consequently, we must introduce some kind of generalization of the coordinate systems and time axes, in order to describe global variables in the algebraic formulation. In particular, we need some suitable reference structures, whose elements are endowed with spatial or time extents.

The generalization of the coordinate systems is achieved by introducing cell complexes[27] and by associating the global variables with the related space elements of the cell complexes, that is, the four space elements, $\mathbf{P}$, $\mathbf{L}$, $\mathbf{S}$, and $\mathbf{V}$, of the cell complexes. This allows us to describe global variables directly.

The branch of the mathematics that develops notions corresponding to those of the differential formulations, but based on global variables instead of field variables, is the algebraic

---

[27] In geometry, a **cell** is a polyhedral element (3-face, see Section 3.3) of a four-dimensional polytope (Section 3.2) or more than two-dimensional tessellation. Cells are facets (Section 3.3) for 4-polytopes and 3-honeycombs.

In topology, a **cell complex**, short for *CW* cell complex,[87] is a type of topological space,[92] introduced by John Henry Constantine Whitehead (11 November 1904–8 May 1960) to meet the needs of homotopy theory.[94] It is obtained by subdividing a space region into small disjoint blocks, or by topologically gluing together[95] basic building blocks. In algebraic topology, these blocks are called the **cells**, in agreement with the definition of a cell given in geometry, while the blocks are referred to as **elements** in some computational methods, like FEM and BEM. If the largest dimension of any of the cells is n, then the cell complex is said to have dimension n. The notion of cell complex is more general than those of mesh, grid, net, or lattice, commonly used in numerical analysis, and contains all these geometrical structures as particular cases.

topology. This is why the algebraic formulation uses notations of algebraic topology, in order to describe the four space elements of the cell complexes. Physical notions are therefore translated into mathematical notions through the intermediation of topology and geometry, and we are led to use algebra instead of differential calculus. Many properties of cell complexes were developed in algebraic topology, among them the notions of **orientation**, **duality**, and **incidence numbers**.

In this context, the role played by coordinate systems is substituted by a suitable labeling of the four space elements, which also takes into account some of the topological features of the four space elements, as their inner and outer orientations and duality (Section 2.2). We can thus conclude that the cell complexes and their labeling are the algebraic version of the points and their continuous mapping, established by the coordinate systems of the differential setting, respectively.

Also the generalization of the time axis is made by means of cell complexes, which, in this second case, are associated with the two time elements, that is, time instant, $\mathbf{I}$, and time intervals, $\mathbf{T}$. Algebraic topology is then employed even for the time cell complexes, in order to describe the time global variables.

In conclusion, the CM cell complexes do not have the same role that cell complexes have in the differential formulation. In fact, the CM cell complexes are not the result of a domain discretization, a process needed in differential formulations for managing the working region, as in the case of the FEM. They are required in the algebraic formulation in order to provide a suitable structure for describing global variables, since global variables are associated not only with points and time instants, as for the differential formulation, but also with lines, surfaces, volumes, and time intervals.

1.3 Set Functions in Physics

In mathematics, a set function is a function whose input is the **power set** (or **powerset**) of a set,[28] that is, the set of all subsets of a set. The power set of a set $\mathscr{A}$, written $\mathcal{P}(\mathscr{A})$, $P(\mathscr{A})$, $\mathbb{P}(\mathscr{A})$, $\wp(\mathscr{A})$, or $2^{\mathscr{A}}$, includes the empty set and $\mathscr{A}$ itself.

Symbolically, let $\mathscr{A}$ be a set and let V be a vector space, a set function is a map, ϕ, which associate an element of V with an element of $\mathcal{P}(\mathscr{A})$:

$$\phi : V \to \mathcal{P}(\mathscr{A}). \tag{1.3.1}$$

All physical quantities that refer to a system as a whole are set functions.[29] In physics, a set function always satisfies the requirement of additivity:[30]

$$\phi(A_1) + \phi(A_2) = \phi(A_1 \cup A_2), \tag{1.3.2}$$

where A_1 and A_2 are two disjoint sets of $\mathscr{A}$:

$$A_1 \cap A_2 = \varnothing. \tag{1.3.3}$$

---

[28] The existence of the power set of any set is postulated by the **axiom of power set**.

[29] In physics, the set functions are also called the **domain functions**.

[30] In mathematics, a set function does not necessarily satisfy additivity.

Any subset of $P(\mathscr{A})$ is called a **family of sets** over $\mathscr{A}$.

If $\mathscr{A}$ is a finite set with $n = |\mathscr{A}|$ elements, then the number of subsets of $\mathscr{A}$ is $|P(\mathscr{A})| = 2^n$. This is the motivation for the notation $2^{\mathscr{A}}$.

Examples of set functions are the length of a street, the altitude of an aircraft at a given instant, the maximum and minimum daily temperatures of a city, and the mass of a body.

Due to their association with elements provided with an extent, the global variables, defined in Section 1.2, are set functions.

1.4 A Comparison Between the Cell Method and the Discrete Methods

The cell method has often been compared to the direct or physical approach, initially used in the finite element method, or to the finite volume method and the finite difference method. In particular, the cell method may seem very similar to the vertex-based scheme of the FVM. However, on deeper analysis of the similarities and differences between the CM and the discrete methods, the CM is shown to be, for the moment, the only truly algebraic method.

In effect, the key point to bear in mind in building a truly algebraic formulation is that all operators must be discrete and use of the limit process must be avoided at each level of the formulation. The direct or physical approach is not suited to this, since it starts from the point-wise conservation equations of differential formulation (Fig. 1.6) and, for differential formulation,

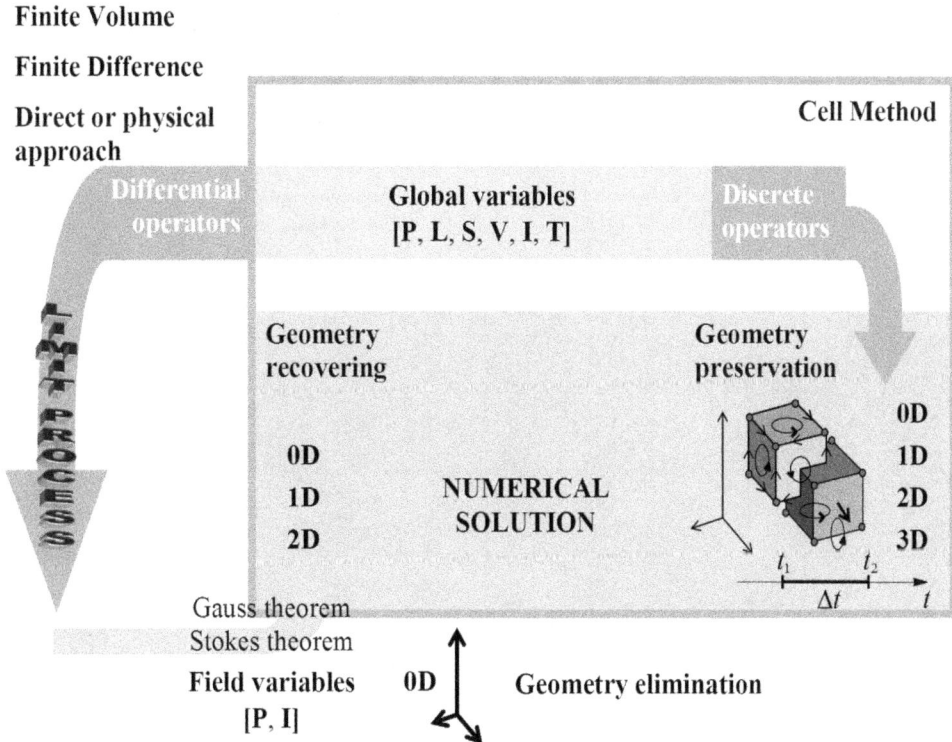

Figure 1.6. Building an algebraic formulation through the direct or physical approach, the finite volume method, the finite difference method, and the cell method.

there is the need for field functions, which depend on the point position and the instant value.

If the field functions are not described directly in terms of point position and instant values, they can be obtained by calculating the densities and rates of the global variables, as shown in Fig. 1.6.

The algebraic formulation is then derived from the differential formulation through an integration process (Fig. 1.6) that is needed because, while in differential formulation geometry must be eliminated from the physical laws, in the numerical solution geometry is essential.

The finite volume method and the finite difference method are also based on a differential formulation (Fig. 1.6). The cell method, on the contrary, uses global variables and formulates the governing equations in algebraic form directly.

CHAPTER 2

ALGEBRA AND THE GEOMETRIC INTERPRETATION OF VECTOR SPACES

In order to deepen the mathematical foundations of the cell method (CM), in this chapter we introduce some basics of exterior algebra (Section 2.1) and geometric algebra (Section 2.2), the two fundamental settings for the geometric study of vector spaces. Of special importance for the philosophy of the CM are the geometric interpretations of the operations on vectors, provided by both the exterior and geometric algebra, and the notions of extension of a vector by another vector, multivector (or p-vector), dual vector space, bialgebra, and covector. The geometric approach allows us to view the four space elements, **P**, **L**, **S**, and **V**, and the two time elements, **I** and **T**, as p-vectors of a geometric algebra, all inductively generated by the outer product of the geometric algebra. From the attitude and orientation of p-vectors, we then derive the two kinds of orientation for p-vectors, inner and outer orientations, which apply to both the space and the time elements. We also discuss how the orientation of a p-vector is induced by the orientation of the $(p-1)$-vectors on its boundary, and how the inner orientation of the attitude vector of a vector equals the outer orientation of its covector. This establishes an isomorphism between the orthogonal complement and the dual vector space of any subset of vectors. Finally, we can find a natural analog of this relationship in general Banach spaces. One of the most remarkable consequences of the relationship between inner and outer orientations is that the outer orientation depends on the dimension of the embedding space, while the inner orientation does not.

The notions of inner and outer orientations implicitly permeate the geometric algebra and are equivalent to the rules for the orientation of vectors in right-handed coordinate systems. We make them explicit in this book because they are at the basis of the CM description of physics, where the cell complexes for computational analysis are treated algebraically, as topological vector spaces.

2.1 The Exterior Algebra

The exterior algebra provides an algebraic setting in which to answer geometric questions and makes sense for spaces not just of geometric vectors, but of other vector-like objects such as vector fields or functions. The exterior algebra, or Grassmann algebra after Hermann Grassmann,[31] is the algebraic system whose product is the **exterior product**, or **wedge product**. It is the largest algebra that supports an alternating product on vectors.

2.1.1   The Exterior Product in Vector Spaces

The exterior product of vectors is an algebraic construction used in Euclidean geometry to study areas, volumes, and their higher-dimensional analogs. The exterior product of two vectors a and b, denoted by $a \wedge b$, is called a **bivector**, or **2-vector**,[32] and lives in a space called the **exterior square**, a geometrical vector space that differs from the original space of vectors. The magnitude of $a \wedge b$ can be interpreted as the area of the parallelogram with sides a and b, which, in three dimensions, can also be computed using the cross product[33] of the two vectors. The sum of any bivectors is a bivector.

More generally, the exterior product of any number k of vectors can be defined and is sometimes called a k-**blade**. It lives in a geometrical space known as the k-th exterior power. The magnitude of the resulting k-blade is the volume of the k-dimensional parallelotope[34] whose sides are the given vectors, just as the magnitude of the scalar triple product of vectors in three dimensions gives the volume of the parallelepiped spanned by those vectors. The k-blades, because they are the simple products of vectors, are called the simple elements of the algebra.

Since the exterior product is antisymmetric, $b \wedge a$ is the negation of the bivector $a \wedge b$, producing the opposite orientation. Consequently, the orientation (out of two) of the bivector is determined by the order in which the originating vectors are multiplied. The direction of the bivector is the plane determined by a and b, as long as they are linearly independent. The result of $a \wedge a$ is the zero bivector.

The exterior algebra contains objects that are not just k-blades, but sums of k-blades. Such a sum is called a k-vector.

The exterior product extends to the full exterior algebra, so that it makes sense to multiply any two elements of the algebra. Equipped with this product, the exterior algebra is an associative algebra, which means that,

---

[31] The exterior algebra was first introduced by Hermann Günther Grassmann (April 15, 1809–September 26, 1877) in 1844 under the blanket term of *Ausdehnungslehre*, or *Theory of Extension*. This referred more generally to an algebraic (or axiomatic) theory of extended quantities and was one of the early precursors to the modern notion of a vector space. Adhémar Jean Claude Barré de Saint-Venant (August 23, 1797, Villiers-en-Bière, Seine-et-Marne–January 1886, Saint-Ouen, Loir-et-Cher) also published similar ideas of exterior calculus for which he claimed priority over Grassmann.

[32] Not all bivectors can be generated as a single exterior product. Specifically, a bivector that can be expressed as one exterior product is called **simple**. In up to three dimensions all bivectors are simple, but in higher dimensions this is not the case.

[33] The **cross product**, **vector product**, or **Gibbs' vector product**, named after the American scientist Josiah Willard Gibbs (February 11, 1839–April 28, 1903), is a binary operation on two vectors in three-dimensional space. It results in a vector which is perpendicular to both vectors and therefore normal to the plane containing them. The direction of the third vector is established in accordance with the right-handed screw rule.[85]

[34] According to the definition given by Harold Scott MacDonald "Donald" Coxeter (February 9, 1907–March 31, 2003), a **parallelotope** is the generalization of a parallelepiped in higher dimensions.

$$\alpha \wedge (\beta \wedge \gamma) = (\alpha \wedge \beta) \wedge \gamma; \tag{2.1.1}$$

for any elements α, β, γ. The k-vectors have degree k, meaning that they are the sums of products of k vectors. When elements of different degrees are multiplied, the degrees add like multiplication of polynomials. This means that the exterior algebra is a graded algebra.[35]

In multilinear algebra[36] a **multivector** or **Clifford number** is an element of the (graded) exterior algebra on a vector space. "Multivector" may mean either homogeneous elements (all terms of the sum have the same grade or degree k), which are referred to as k-vectors or p-vectors, or may allow sums of terms in different degrees.

The k-th exterior power:

$$\Lambda^k V; \tag{2.1.2}$$

is the vector space of formal sums of k-multivectors. The product of a k-multivector and an ℓ-multivector is a $(k + \ell)$-multivector. So, the direct sum:

$$\bigoplus_k \Lambda^k V; \tag{2.1.3}$$

forms an associative algebra, which is closed with respect to the wedge product. This algebra, commonly denoted by ΛV, is called the exterior algebra of V.

The exterior algebra is one example of a bialgebra, meaning that its dual space also possesses a product, and this dual product is compatible with the exterior product. This dual algebra is precisely the algebra of alternating multilinear forms on V, and the pairing between the exterior algebra and its dual is given by the interior product.[37]

2.1.2 The Exterior Product in Dual Vector Spaces

In mathematics, any vector space, V, has a corresponding **dual vector space** (or just **dual space**, for short), V^*.

---

[35] An algebra A over a ring R is a graded algebra if it is graded as a ring.

[36] **Multilinear algebra** extends the methods of linear algebra. Just as linear algebra is built on the concept of a vector and develops the theory of vector spaces, multilinear algebra builds on the concepts of p-vectors and multivectors with Grassmann algebra (Section 2.2). Though Hermann Günther Grassmann (April 15, 1809–September 26, 1877) started the subject in 1844 with his *Ausdehnungslehre* (*Theory of Extension*)[31] and re-published in 1862, his work was slow to find acceptance as ordinary linear algebra provided sufficient challenges to comprehension. After some preliminary work by Elwin Bruno Christoffel (November 10, 1829–March 15, 1900), a major advance in multilinear algebra came in the work of Gregorio Ricci-Curbastro (12 January 1853–6 August 1925) and Tullio Levi-Civita (29 March 1873–29 December 1941).

[37] The **interior product** is a degree -1 antiderivation on the exterior algebra of differential forms on a smooth manifold.[51] The interior product, named in opposition to the exterior product, is also called interior or inner multiplication, or the inner derivative or derivation, but should not be confused with an inner product.[64]

There are two kinds of dual spaces: the **algebraic dual space** and the **continuous dual space**. The algebraic dual space is defined for all vector spaces.

Given any vector space V over a field F, the algebraic dual space V^*, also called the ordinary dual space, or simply the dual space, is defined as the set of all linear maps (linear functionals) from V to F:

$$\varphi : V \rightarrow F, \, v \mapsto \varphi(v); \tag{2.1.4}$$

where a linear map φ is a function from V to F, which is linear, that is, additive:[38]

$$\varphi(x + y) = \varphi(x) + \varphi(y) \text{ for all } x, y \in V; \tag{2.1.5}$$

and homogeneous of degree 1:[39]

$$\varphi(ax) = a(\varphi(x)) \text{ for all } x, y \in V, \, a \in F. \tag{2.1.6}$$

Equations (2.1.5) and (2.1.6) may be combined in the single relationship:

$$\varphi(ax + by) = a\varphi(x) + b\varphi(y) \text{ for all } x, y \in V, \, a, b \in F. \tag{2.1.7}$$

The dual space V^* itself becomes a vector space over F when equipped with the following addition and scalar multiplication, for all $\varphi, \psi \in V^*$, $x \in V$, and $\alpha \in F$:

$$(\varphi + \psi)(x) = \varphi(x) + \psi(x); \tag{2.1.8}$$

$$(\alpha\varphi)(x) = \alpha(\varphi(x)). \tag{2.1.9}$$

When defined for a topological vector space[40] X, subspace of V, there is a subspace X^* of V^*, corresponding to the continuous linear functionals, which constitutes a continuous dual space. Even the continuous dual space is often simply called the dual space.

---

[38] In algebra, an **additive function** (or **additive map**) is a function that preserves the addition operation.

[39] In mathematics, a **homogeneous function** is a function with multiplicative scaling behavior: if the argument is multiplied by a factor, then the result is multiplied by some power of this factor. Any linear function $f : V \rightarrow W$ is homogeneous of degree 1, where the degree is the exponent on the variable.

[40] In mathematics, a **topological vector space** (also called a **linear topological space**) blends a topological structure[92] (a uniform structure to be precise) with the algebraic concept of a vector space. A topological vector space, X, is a vector space over a topological field $\mathbf{K}$ (most often the real or complex numbers with their standard topologies) that is endowed with a topology, such that the vector addition $X \times X \rightarrow X$ and scalar multiplication $\mathbf{K} \times X \rightarrow X$ are continuous functions (where the domains of these functions are endowed with product topologies). $\mathbb{R}^n$ is a topological vector space, with the topology induced by the standard inner product. Every topological vector space has a continuous dual space – the set X^* of all continuous linear functionals, that is, continuous linear maps from the space into the base field $\mathbf{K}$.

For any finite-dimensional normed vector space or topological vector space, such as Euclidean n-space, the continuous dual and the algebraic dual coincide, since every linear functional is continuous.[41]

If V is finite-dimensional, then V^* has the same dimension as V.

Dual vector spaces for finite-dimensional vector spaces can be used for studying tensors.[42]

---

[41] In infinite dimensions, this is not true.

[42] **Tensors**, first conceived by Tullio Levi-Civita (29 March 1873–29 December 1941) and Gregorio Ricci-Curbastro (12 January 1853–6 August 1925), who continued the earlier work of Georg Friedrich Bernhard Riemann (September 17, 1826–July 20, 1866) and Elwin Bruno Christoffel (November 10, 1829–March 15, 1900) and others, as part of the absolute differential calculus, are geometric objects that describe linear relations between vectors, scalars, and other tensors. In particular, tensors are used to represent correspondences between the sets of geometric vectors. Because they express a relationship between vectors, tensors themselves must be independent of a particular choice of coordinate system. Taking a coordinate basis or frame of reference and applying the tensor to it results in an organized multi-dimensional array representing the tensor in that basis, or frame of reference. The coordinate independence of a tensor then takes the form of a "covariant" transformation law (Section 2.1.3) that relates the array computed in one coordinate system to that computed in another one. The **order** (also **degree** or **rank**) of a tensor is the dimensionality of the array needed to represent it, or equivalently, the number of indices needed to label a component of that array. In particular, a linear map can be represented by a matrix, a two-dimensional array, and therefore is a 2nd-order tensor. The numbers in the array are known as the **scalar components** of the tensor or simply its **components**.

In the finite-dimensional case, linear operators can be represented by matrices in the following way. Let K be a field, and U and V be finite-dimensional vector spaces over K. Let us select a basis $\mathbf{u}_1,...,\mathbf{u}_n$ in U and $\mathbf{v}_1,...,\mathbf{v}_m$ in V. Then let

$$\mathbf{x} = x^i \mathbf{u}_i$$

be an arbitrary vector in U (assuming Einstein convention[53]), and

$$A : U \to V$$

be a linear operator. It follows

$$A\mathbf{x} = x^i A\mathbf{u}_i = x^i \left(A\mathbf{u}_i \right)^j \mathbf{v}_j.$$

Then[9]

$$a_i^j := \left(A\mathbf{u}_i \right)^j \in K$$

is the matrix of the operator A, in fixed bases. a_i^j does not depend on the choice of $\mathbf{x}$, and

$$A\mathbf{x} = \mathbf{y}$$

if

$$a_i^j x^i = y^j.$$

(*continues next page*)

The **pairing** of a functional φ in the dual space V^* and an element x of V is sometimes denoted by a bracket:

$$\varphi(x) = [\varphi, x] = \langle \varphi, x \rangle. \tag{2.1.10}$$

The pairing defines a non-degenerate bilinear mapping:[43]

$$[\bullet, \bullet] : V^* \times V \to F. \tag{2.1.11}$$

Specifically, every non-degenerate bilinear form on a finite-dimensional vector space V gives rise to an isomorphism[44] from V to V^*, $\langle \bullet, \bullet \rangle$. Then, there is a natural isomorphism:

$$V \to V^*, \, v \mapsto v^*; \tag{2.1.12}$$

(*continues from previous page*)

Thus, in fixed bases, n-by-m matrices are in bijective correspondence to linear operators from U to V.

Just as the components of a vector change when we change the basis of the vector space, the entries of a tensor also change under such a transformation. If an index transforms, like a vector, with the inverse of the basis transformation, it is called **contravariant** and is traditionally denoted by an upper index, while an index that transforms with the basis transformation itself is called **covariant** and is denoted by a lower index.

[43] Let U, V, and W be three vector spaces over the same base field F; then a **bilinear map**, B, is an application from the Cartesian product of the first two vector spaces to the third vector space:

$$B : U \times V \to W;$$

which is linear in both arguments:

$$B(u + u', v) = B(u, v) + B(u', v);$$

$$B(u, v + v') = B(u, v) + B(u, v');$$

$$B(\lambda u, v) = B(u, \lambda v) = \lambda B(u, v).$$

A first immediate consequence of the definition is that $B(u, v) = 0$ whenever $x = 0$ or $y = 0$.

The case where W is F, we have a **bilinear form**, or **bilinear functional**. In particular, we have a bilinear form when U and V are vector spaces over the space of real numbers and W is $\mathbb{R}$:

$$B : U \times V \to \mathbb{R}.$$

In this case, the bilinear form is an extension of the scalar product, which is defined for vectors of the same vector space, to the elements of the two vector spaces U and V.

[44] In abstract algebra,[91] two mathematical structures are said to be **isomorphic** if there is an one-to-one mapping of mathematical elements between them. Therefore, an **isomorphism** is a bijective homomorphism.[115] The **isomorphism theorems** are three theorems that describe the relationship between quotients, homomorphisms, and subobjects. Versions of the theorems exist for groups, rings, vector spaces, modules, Lie algebras, and various other

(*continues next page*)

given by:[9]

$$v^*(w) := \langle v, w \rangle; \tag{2.1.13}$$

where $v^* \in V^*$ is said to be the **dual vector** of $v \in V$. The inverse isomorphism is given by

$$V^* \to V, \, f \mapsto f^*; \tag{2.1.14}$$

where f^* is the unique element of V for which, for all $w \in V$:

$$\langle f^*, w \rangle = f(w). \tag{2.1.15}$$

(*continues from previous page*)

algebraic structures. In universal algebra, the isomorphism theorems can be generalized to the context of algebras and congruences.

First isomorphism theorem

Let G and H be groups, and let $\varphi : G \to H$ be a homomorphism. Then:

1. The kernel of φ is a normal subgroup of G.
2. The image of φ is a subgroup of H.
3. The image of φ is isomorphic to the quotient group[76] $G/\ker(\varphi)$.

Second isomorphism theorem

Let G be a group. Let S be a subgroup of G, and let N be a normal subgroup of G. Then:

1. The product SN is a subgroup of G.
2. The intersection $S \cap N$ is a normal subgroup of S.
3. The quotient groups $(SN)/N$ and $S/(S \cap N)$ are isomorphic.

Technically, it is not necessary for N to be a normal subgroup, as long as S is a subgroup of the normalizer of N. In this case, the intersection $S \cap N$ is not a normal subgroup of G, but it is still a normal subgroup of S.

Third isomorphism theorem

Let G be a group. Let N and K be normal subgroups of G, with

$$K \subseteq N \subseteq G;$$

then

1. the quotient N/K is a normal subgroup of the quotient G/K,
2. the quotient group $(G/K)/(N/K)$ is isomorphic to G/N,

In an infinite-dimensional Hilbert space,[45] analogous results hold by the Riesz representation theorem.[46] Let H be a Hilbert space, and let H^* denote its dual space, consisting of all continuous linear functionals from H into the field $\mathbb{R}$ or $\mathbb{C}$. If x is an element of H, then there is a mapping $H \to H^*$ into the continuous dual space H^*. However, this mapping is antilinear,[47] rather than linear. The function φ_x, defined by:[9]

$$\varphi_x(y) := \langle y, x \rangle, \ \forall y \in H; \tag{2.1.16}$$

---

[45] The mathematical concept of a **Hilbert space**, named after the German mathematician David Hilbert (January 23, 1862–February 14, 1943), generalizes the notion of Euclidean space. It extends the methods of vector algebra and calculus from the two-dimensional Euclidean plane and three-dimensional space to spaces with any finite or infinite number of dimensions. A Hilbert space is an abstract vector space possessing the structure of an inner product that allows length and angle to be measured. Furthermore, Hilbert spaces must be complete, a property that stipulates the existence of enough limits in the space to allow the techniques of calculus to be used.

In particular, H is a real or complex inner product space that is also a complete metric space with respect to the distance function induced by the inner product. To say that H is a real inner product space means that H is a real vector space on which there is an inner product $\langle x, y \rangle$ associating a real number with each pair of elements x, y of H. Such an inner product will be bilinear, that is, linear in each argument:

$$\langle ax_1 + bx_2, y \rangle = a\langle x_1, y \rangle + b\langle x_2, y \rangle;$$
$$\langle x, ay_1 + by_2 \rangle = a\langle x, y_1 \rangle + b\langle x, y_2 \rangle.$$

The norm is the real-valued function:

$$\|x\| = \sqrt{\langle x, x \rangle};$$

and the distance d between two points x, y in H is defined in terms of the norm by

$$d(x, y) = \|x - y\| = \sqrt{\langle x - y, x - y \rangle}.$$

That this function is a distance function means

1. that it is symmetric in x and y;
2. that the distance between x and itself is zero, and otherwise the distance between x and y must be positive;
3. that the triangle inequality holds, meaning that the length of one leg of a triangle xyz cannot exceed the sum of the lengths of the other two legs:

$$d(x, z) \leq d(x, y) + d(y, z).$$

This last property is ultimately a consequence of the more fundamental Cauchy–Schwarz inequality, which asserts

$$|\langle x, y \rangle| \leq \|x\| \cdot \|y\|$$

with equality if and only if x and y are linearly dependent.

[46] In functional analysis the **Riesz representation theorem** describes the dual of a Hilbert space.[45] It is named in honour of Frigyes Riesz (January 22, 1880–February 28, 1956), the older brother of the mathematician Marcel Riesz (16 November 1886–4 September 1969).

[47] A mapping

$$f : V \to W$$

(continues next page)

where $\langle \bullet, \bullet \rangle$ denotes the inner product of the Hilbert space, is an element of H^* and every element of H^* can be written uniquely in this form.

A topology on the dual space, X^*, of a topological vector space,[40] X, over a topological field, $\mathbf{K}$, can be defined as the coarsest topology (the topology with the fewest open sets) such that the dual pairing $X^* \times X \to \mathbf{K}$ is continuous. This turns the dual space into a locally convex topological vector space.[48] This topology is called the **weak* topology**, that is, a weak topology defined on the dual space X^*. In order to distinguish the weak topology from the original topology on X, the original topology is often called the **strong topology**. If X is equipped with the weak topology, then addition and scalar multiplication remain continuous operations, and X is a locally convex topological vector space.

Elements of the algebraic dual space V^* are sometimes called **covectors**, or **1-forms**, and are denoted by bold, lowercase Greek. They are linear maps from V to its field of scalars.

If V consists of the space of geometrical vectors (arrows) in the plane, then the level curves of an element of V^* form a family of parallel lines in V. So, an element of V^* can be intuitively thought of as a particular family of parallel lines covering the plane. To compute the value of a functional on a given vector, one needs only to determine which of the lines the vector lies on. Or, informally, one "counts" how many lines the vector crosses.

More generally, if V is a vector space of any (finite) dimension, then the level sets of a linear functional in V^* are parallel **hyperplanes** in V (see Fig. 2.1 for the three-dimensional vector space), and the action of a linear functional on a vector can be visualized in terms of these hyperplanes, or **p-planes**,[49] in the sense that the number of (1-form) hyperplanes intersected by a vector equals the interior product between the covector and the vector (Fig. 2.2).

Fig. 2.2 also provides the geometrical interpretation of the sum between covectors, defined by Eq. (2.1.8).

(*continues from previous page*)

from a complex vector space to another is said to be **antilinear** (or **conjugate-linear** or **semilinear**, though the latter term is more general) if

$$f(ax + by) = \bar{a}f(x) + \bar{b}f(y)$$

for all a, b in $\mathbb{C}$ and all x, y in V, where $\bar{a}$ and $\bar{b}$ are the complex conjugates of a and b, respectively

[48] In functional analysis and related areas of mathematics, **locally convex topological vector spaces** or **locally convex spaces** are examples of topological vector spaces (TVS) which generalize normed spaces. They can be defined as topological vector spaces whose topology is generated by translations of balanced, absorbent, and convex sets. Alternatively they can be defined as a vector space with a family of seminorms, and a topology can be defined in terms of that family. Although in general such spaces are not necessarily normable, the existence of a convex local base for the zero vector is strong enough for the Hahn–Banach theorem to hold, yielding a sufficiently rich theory of continuous linear functionals.

[49] This method of visualizing linear functionals is sometimes introduced in general relativity texts, such as Gravitation by Misner, Thorne & Wheeler (1973).

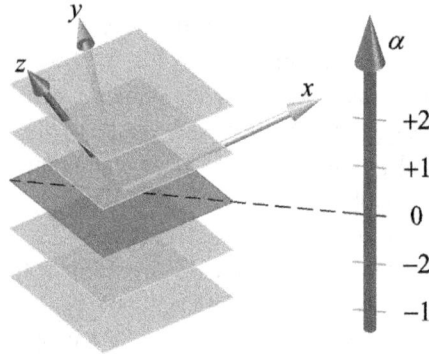

Figure 2.1. Geometric interpretation of a 1-form α as a stack of hyperplanes of constant value, each corresponding to those vectors that α maps to a given scalar value shown next to it along with the "sense" of increase. The zero plane (dark gray) is through the origin.

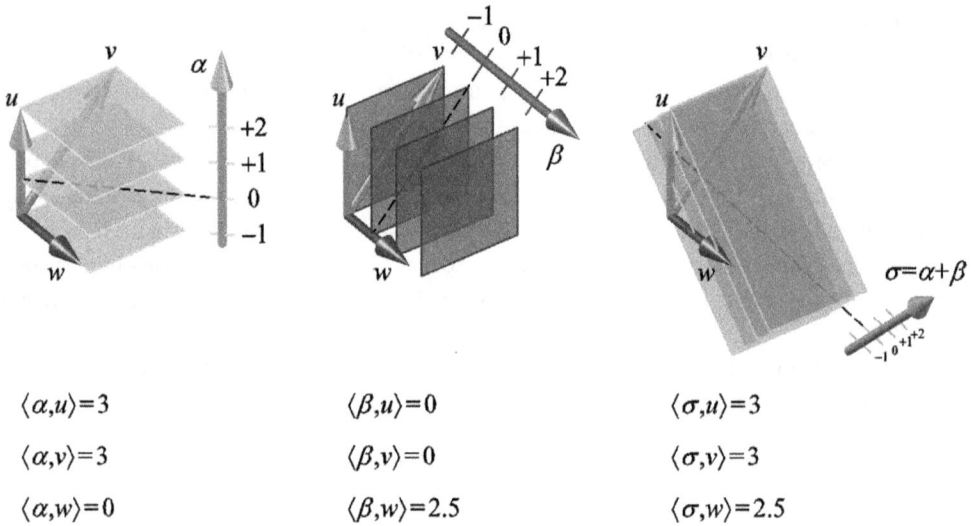

$$\langle\alpha,u\rangle=3 \qquad \langle\beta,u\rangle=0 \qquad \langle\sigma,u\rangle=3$$

$$\langle\alpha,v\rangle=3 \qquad \langle\beta,v\rangle=0 \qquad \langle\sigma,v\rangle=3$$

$$\langle\alpha,w\rangle=0 \qquad \langle\beta,w\rangle=2.5 \qquad \langle\sigma,w\rangle=2.5$$

Figure 2.2. Linear functionals (1-forms) α, β, their sum σ, and vectors u, v, w, in 3d Euclidean space.

With reference to Fig. 2.3, where we have represented only the first two parallel planes of the linear functional $\mathbf{a}^*$ (in the three-dimensional vector space), the second plane is given by the locus:

$$\alpha\left(\mathbf{u}\right)=\mathbf{a}^*\left(\mathbf{u}\right)=\langle\mathbf{a},\mathbf{u}\rangle=1; \tag{2.1.17}$$

where $\mathbf{a}$ is orthogonal to the level sets of the linear functional $\mathbf{a}^*$. Moreover, due to the Riesz representation theorem, the covector $\mathbf{a}^*$ can be uniquely represented by $\mathbf{a}$, its associated vector.

In the special case, where the vector space is the Euclidean space, the interior product is substituted by the Euclidean inner product (scalar product):

$$\mathbf{a}^*\left(\mathbf{u}\right)=\mathbf{a}\cdot\mathbf{u}=1. \tag{2.1.18}$$

From Eq. (2.1.17) it follows that, denoted by d the distance between the two planes (Fig. 2.3), the norm of the vector $\mathbf{a}$ is equal to

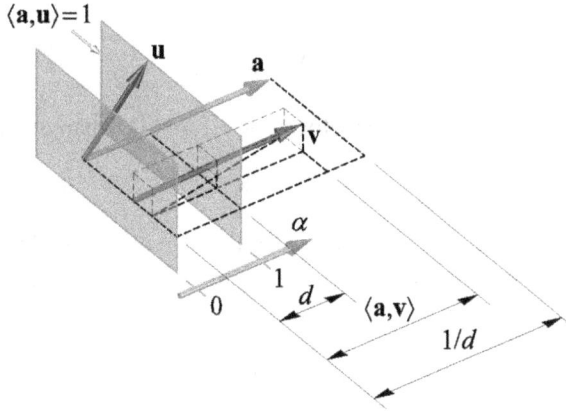

Figure 2.3. Representation of the covector α^* as a pair of parallel planes.

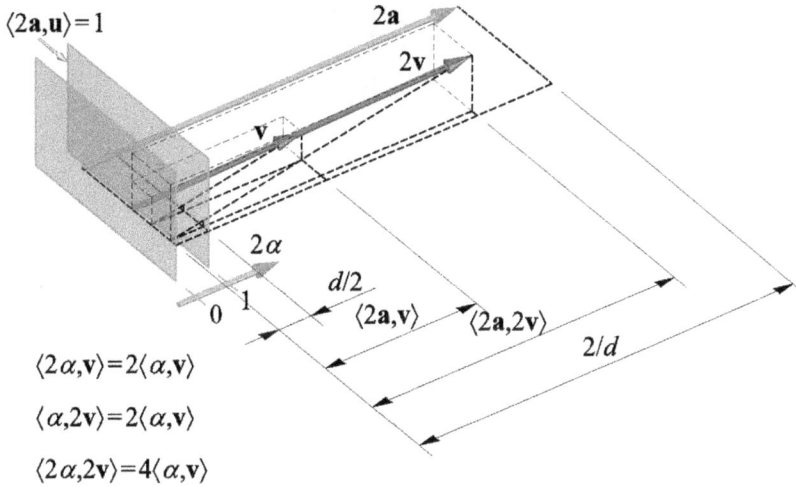

Figure 2.4. Inner products of a 1-form α and 1-vector $\mathbf{v}$, each scalar multiplied by 1 and 2.

$$\|\mathbf{a}\| = \frac{1}{d}. \tag{2.1.19}$$

Consequently, the distance between the planes of the covector $2\mathbf{a}^*$, whose norm is $2/d$:

$$\|2\mathbf{a}\| = 2\|\mathbf{a}\| = \frac{2}{d}; \tag{2.1.20}$$

is half the distance between the planes of the covector $\mathbf{a}^*$ (Fig. 2.4).

Since V^* is a vector space, there exist in V^* the null vector, which is called the **trivial** linear functional:

$$\mathbf{0}^*(\mathbf{x}) = 0; \tag{2.1.21}$$

and the opposite element:

$$(-\varphi)(\mathbf{x}) = -\varphi(\mathbf{x}).\qquad(2.1.22)$$

In multilinear algebra,[36] a **multilinear form**, or k-form, is a map of the type:

$$f : V^k \to \mathbf{K};\qquad(2.1.23)$$

where V is a vector space over the field $\mathbf{K}$, which is separately linear in each its k variables. The k-forms are generated by the exterior product on covectors (Fig. 2.5). This allows us to define the k-th exterior power:

$$\Lambda^k\left(V^*\right);\qquad(2.1.24)$$

as the vector space of formal sums of k-forms.

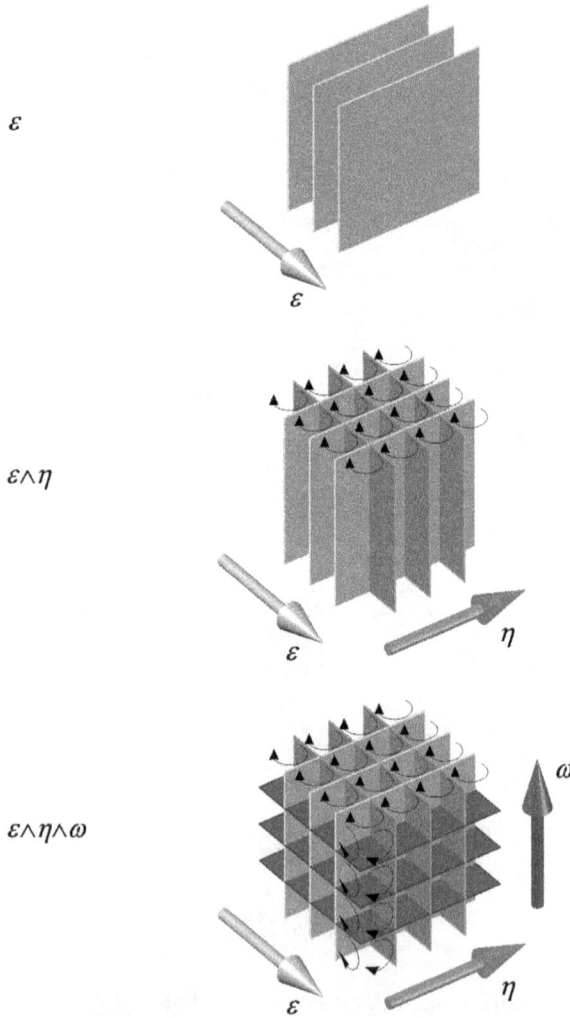

Figure 2.5. Geometric interpretation for the exterior product of k 1-forms (ε, η, ω) to obtain a k-form ("mesh" of coordinate surfaces, here planes), for $k = 1, 2, 3$. The "circulations" show orientation.

An important type of multilinear forms is alternating multilinear forms which have the additional property of changing their sign under exchange of two arguments. The set of all alternating multilinear forms is a vector space, as the sum of two such maps, or the product of such a map with a scalar, is again alternating. By the universal property of the exterior power, the space of alternating forms of degree k on V is naturally isomorphic with the dual vector space $\left(\Lambda^k V\right)^*$. If V is finite-dimensional, then the latter is naturally isomorphic to $\Lambda^k\left(V^*\right)$.

In formal terms, there is a correspondence between the graded dual of the graded algebra ΛV and alternating multilinear forms on V. The exterior product of multilinear forms defined above is dual to a coproduct defined on ΛV, giving the structure of a coalgebra.

The **counit** is the homomorphism:

$$\varepsilon : \Lambda\left(V\right) \to K; \tag{2.1.25}$$

which returns the 0-graded component of its argument. The coproduct and counit, along with the exterior product, define the structure of a bialgebra on the exterior algebra.

2.1.3 Covariant and Contravariant Components

Covariance and **contravariance**[50] describe how the quantitative description of certain geometric or physical entities changes with a change of basis, when the basis is not an orthogonal basis. For holonomic bases,[51] this is determined by a change from one coordinate system to another.

For a vector to be basis-independent, the components of the vector must contravary with a change of basis to compensate. That is, the components must vary with the inverse transformation to that of the change of basis, and transform in the same way as the coordinates. In other words, if the reference axes were rotated in one direction, the component representation of the vector would rotate in exactly the opposite way. The components of vectors (as opposed to

---

[50] The terms covariant and contravariant were introduced by James Joseph Sylvester (3 September 1814–15 March 1897) in 1853, in order to study algebraic **invariant theory**, which is a branch of abstract algebra[91] dealing with actions of groups on algebraic varieties from the point of view of their effect on functions. The theory dealt with the question of explicit description of polynomial functions that do not change, or are invariant, under the transformations from a given linear group. Classically, the term "invariant theory" refers to the study of invariant algebraic forms (equivalently, symmetric tensors[42]) for the action of linear transformations. This was a major field of study in the latter part of the 19th century.

[51] In mathematics, a **manifold** is a topological space[92] that near each point resembles Euclidean space. More precisely, each point of an n-dimensional manifold has a neighborhood that is homeomorphic[89] to the Euclidean space of dimension n. A **differentiable manifold** is a type of manifold that is locally similar enough to a linear space to allow one to do calculus.

A **holonomic basis** or **coordinate basis** for a differentiable manifold is a set of basis vector fields $\left\{e_k\right\}$ such that some coordinate system $\left\{x^k\right\}$ exists for which

$$\left\{e_k\right\} = \frac{\partial}{\partial x^k}.$$

those of dual vectors) are said to be **contravariant**[52] (raised indices, written v^i). The vector itself is said to be cotravariant.

Mathematically, if the coordinate system undergoes a transformation described by an invertible matrix M, so that a coordinate vector $\mathbf{x}$ is transformed to $\hat{\mathbf{x}} = M\mathbf{x}$:

$$\hat{x}_i = \sum_j M_i{}^j x_j = M_i{}^j x_j; \qquad (2.1.26)$$

where in the second expression, the summation sign was suppressed in accordance with Einstein notation,[53] then a contravariant vector $\mathbf{v}$ must be similarly transformed via $\hat{\mathbf{v}} = M\mathbf{v}$, and the components, v^i, of $\mathbf{v}$, transform with the inverse of the matrix M:

$$\hat{v}^i = \left(M^{-1}\right)^i{}_j v^j. \qquad (2.1.27)$$

Given a basis $B = \{\mathbf{e}_1,...,\mathbf{e}_n\}$ in V, the contravariant components in Einstein notation are denoted by upper indices:

$$\mathbf{v} = v^i \mathbf{e}_i = \begin{bmatrix} \mathbf{e}_1 & \mathbf{e}_2 & ... & \mathbf{e}_n \end{bmatrix} \begin{bmatrix} v^1 \\ v^2 \\ ... \\ v^n \end{bmatrix}. \qquad (2.1.28)$$

Consequently, vectors are represented as $n \times 1$ matrices (column vectors).

---

[52] Examples of vectors with contravariant components (contravariant vectors) include the position of an object relative to an observer, or any derivative of position with respect to time, including velocity, acceleration, and jerk.

[53] The **Einstein notation**, or **Einstein summation convention**, was introduced to physics by Albert Einstein (14 March 1879–18 April 1955) in 1916. It is a notational convention that implies summation over a set of indexed terms in a formula, thus achieving notational brevity. According to this convention, when an index variable appears twice in a single term it implies summation of that term over all the values of the index. So

$$y = c_i x^i,$$

where the indices can range over the set $\{1, 2, 3\}$, is the Einstein notation for

$$y = \sum_{i=1}^3 c_i x^i = c_1 x^1 + c_2 x^2 + c_3 x^3.$$

An index that is summed over is a **summation index**, in this case i. It is also called a **dummy index** since any symbol can replace i without changing the meaning of the expression, provided that it does not collide with index symbols in the same term.

An index that is not summed over is a **free index** and should be found in each term of the equation or formula if it appears in any term.

Einstein notation can be applied in slightly different ways. When dealing with covariant and contravariant vectors, each index occurs once in an upper (superscript) and once in a lower (subscript) position in a term. A covariant vector can only be contracted with a contravariant vector, corresponding to summation of the products of coefficients. On the other hand, when there is a fixed coordinate basis (or when not considering coordinate vectors), one may choose to use only subscripts.

Whatever be the basis, B, of V, it is always possible to construct a specific basis in V^*, $B^* = \{e^1, ..., e^n\}$, called the **dual basis**.

For a dual vector to be (dual) basis-independent, the components of the dual vector must covary with a change of dual basis to remain representing the same covector. That is, the components must vary by the same transformation as the change of dual basis (or, equivalently, change oppositely to the coordinates). The components of dual vectors (as opposed to those of vectors) are said to be **covariant**[54] (lowered indices, written v_i). The dual vector itself is said to be covariant.

Mathematically, the components, v_i, of a covector $\mathbf{v}$, transform with the transformation matrix, M, itself:

$$\hat{v}_i = M_i{}^j v_j. \tag{2.1.29}$$

In Einstein notation, covariant components are denoted by lower indices:

$$\mathbf{v} = v_i \mathbf{e}^i = \begin{bmatrix} v_1 & v_2 & \cdots & v_n \end{bmatrix} \begin{bmatrix} \mathbf{e}^1 \\ \mathbf{e}^2 \\ \cdots \\ \mathbf{e}^n \end{bmatrix}. \tag{2.1.30}$$

Covectors are represented as $1 \times n$ matrices (row covectors).

In conclusion, contravariant vectors are represented with covariant basis vectors, and covariant vectors are represented with contravariant basis vectors.

Fig. 2.6 shows how to represent a vector both in tangent basis vectors to the coordinate curves and in dual basis, covector basis, or cobasis, normal vectors to coordinate surfaces, in 3D general curvilinear coordinates (q^1, q^2, q^3). Note that the basis and cobasis do not coincide since the basis is not orthogonal.

The dual basis, B^*, is a set of linear functionals on V, defined by the relation:

$$e^i (c_1 \mathbf{e}_1 + ... + c_n \mathbf{e}_n) = c_i, \quad i = 1, ..., n; \tag{2.1.31}$$

for any choice of coefficients $c_i \in F$. In particular, letting in turn each one of those coefficients be equal to one and the other coefficients zero, gives the system of equations:

$$e^i (\mathbf{e}_j) = \delta^i{}_j; \tag{2.1.32}$$

where $\delta^i{}_j$ is the Kronecker delta symbol:

$$\delta^i{}_j = \begin{cases} 0, & \text{if } i \neq j; \\ 1, & \text{if } i = j. \end{cases} \tag{2.1.33}$$

---

[54] Examples of vectors with covariant components (covariant vectors) generally appear when taking a gradient of a function.

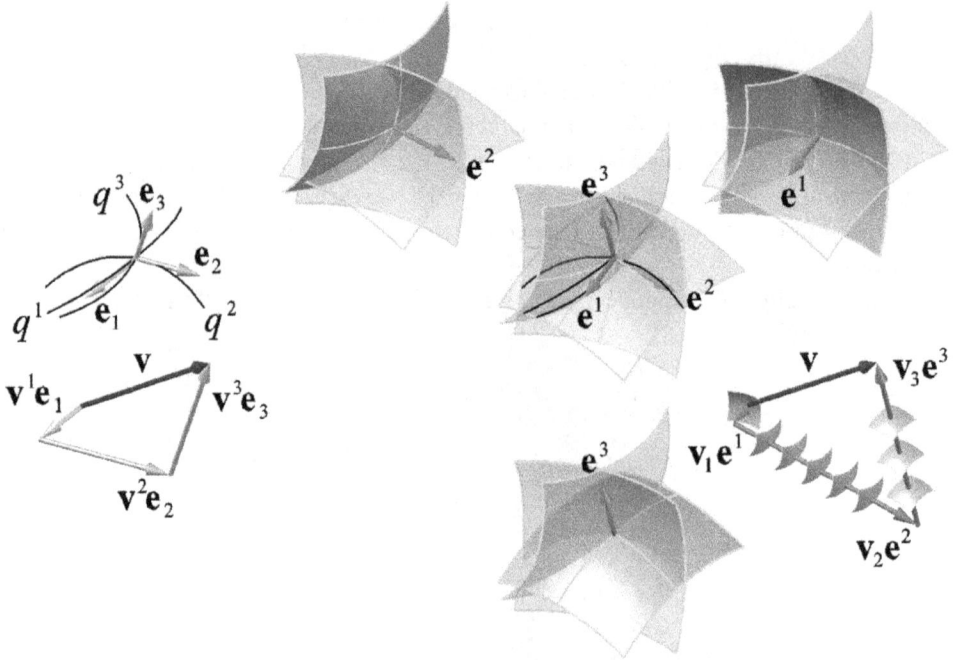

Figure 2.6. A vector **v** (dark gray) in tangent basis vectors $\mathbf{e}_1$, $\mathbf{e}_2$, $\mathbf{e}_3$ (left) and in dual basis $\mathbf{e}^1$, $\mathbf{e}^2$, $\mathbf{e}^3$ (right).

The basis vectors are biorthogonal if these pair to 1 if and only if the indexes are equal, and to zero otherwise. Consequently, the two sets B and $B*$ form, by definition, a **biorthogonal system**. Since the evaluation of a covector on a vector is a pairing, the biorthogonality condition becomes

$$\left\langle \mathbf{e}^i, \mathbf{e}_j \right\rangle = \delta^i{}_j. \tag{2.1.34}$$

This allows one to introduce a **dot product** (or **scalar product**, or **inner product**) between two vectors, by defining "pairing with a vector" to be "evaluation by its corresponding covector." For basis vectors, this means:[9]

$$\mathbf{e}_i \cdot v := \left\langle \mathbf{e}^i, v \right\rangle; \tag{2.1.35}$$

and the basis vectors satisfy

$$\mathbf{e}_i \cdot \mathbf{e}_j = \delta_{ij}. \tag{2.1.36}$$

The notation for the superscripts and subscripts on the delta switches to match whether one is using a covector and a vector or two vectors. Formally one is considering $\delta^i{}_j$ as the **contravariant metric tensor** and δ_{ij} as the **covariant metric tensor**.

In finite dimensions one can alternatively see that the biorthogonality conditions imposes n independent linear conditions (because there are n basis vectors and they are linearly independent) on each element of the dual, so each element of the dual set is determined uniquely, since the dual space has dimension n. The action of dual vectors on an n-dimensional vector space, V, can be interpreted by viewing elements of V as $n \times 1$ column matrices and elements of the dual space, $V*$, as $1 \times n$ row matrices that act as linear functionals by left matrix multiplication.

In particular, if we interpret $\mathbb{R}^n$ as the space of columns of n real numbers, its dual space is typically written as the space of rows of n real numbers. Such a row acts on $\mathbb{R}^n$ as a linear functional by ordinary matrix multiplication. Since the bases are orthonormal with respect to their bilinear forms,[43] the linear operator A^*, whose rows are the elements of the dual basis, $\mathbf{e}^i$, is the adjoint[55] of the linear operator A, whose columns are the elements of the basis of V, $\mathbf{e}_i$:

$$\left(A^*\mathbf{e}_i \right) \cdot \mathbf{e}^j = \left(A\mathbf{e}^j \right) \cdot \mathbf{e}_i; \tag{2.1.37}$$

where A^* is the linear operator from V to V^*:

$$A^* : V \to V^*; \tag{2.1.38}$$

---

[55] The **adjoint operator** generalizes the notion of transpose of a matrix to a linear operator. Let L be a linear operator between two isomorphic[44] vector spaces, V and W, placed in duality by a non-degenerate bilinear form:

$$L : V \to W .$$

The adjoint of the operator L, denoted by L^*, is the linear operator from V to W:

$$L^* : V \to W;$$

such that[9]

$$\langle L\mathbf{v}, \mathbf{v}' \rangle \triangleq \langle L^*\mathbf{v}', \mathbf{v} \rangle;$$

where $\mathbf{v}$ belongs to the domain of the operator L and $\mathbf{v}'$ belongs to the domain of the adjoint operator, L^*:

$$\mathbf{v} \in \mathrm{D}(L) \subseteq V, \ \mathbf{v}' \in \mathrm{D}(L^*) \subseteq V .$$

In components, the definition of the adjoint operator can be written as

$$\sum_{h,k=1}^{n} v'_k L_{kh} v_h \triangleq \sum_{h,k=1}^{n} v_h L^*_{hk} v'_k;$$

from which it follows that the i-th row of L^* is the i-th column of L:

$$L^*_{hk} = L_{kh} .$$

If L is a differential operator with formal part $\mathscr{L}$, and $\mathscr{L}^*$ is the formal part of the adjoint operator L^*, the two formal differential operators $\mathscr{L}$ and $\mathscr{L}^*$ are **formally adjoint operators**.

When W is equal to V^*, the dual vector space of V, the two operators L and L^* coincide and the operator L is called a **self-adjoint** operator. Moreover, if the domain of the adjoint contains the domain of the operator:

$$\mathrm{D}(L) \subseteq \mathrm{D}(L^*);$$

then the operator L is **symmetric**.

and A is the linear operator from V^* to V:

$$A : V^* \to V. \tag{2.1.39}$$

For a vector $\mathbf{v}$ of the vector space V and a covector ε of the dual vector space V^*, we obtain

$$\left(A^* \mathbf{v} \right) \cdot \varepsilon = \left(A \varepsilon \right) \cdot \mathbf{v}; \tag{2.1.40}$$

or, in components:

$$\sum_{h,k=1}^{n} \varepsilon_k a^{*k}_{\ h} v^h = \sum_{h,k=1}^{n} v^h a_h^{\ k} \varepsilon_k. \tag{2.1.41}$$

It follows the relation of transposition between the matrices:

$$a^{*k}_{\ h} = a_h^{\ k}. \tag{2.1.42}$$

Note that, being a linear operator between different vector spaces, speaking of symmetry of A makes no sense, since the adjoint of A cannot coincide with A. Analogously, also speaking of symmetry of A^* makes no sense.

2.2   The Geometric Algebra

Vector algebra and geometric algebra (GA) are alternative approaches to providing additional algebraic structures on vector spaces, with geometric interpretations, particularly vector fields in multivariable calculus and applications in mathematical physics. Vector algebra is specific to Euclidean three-space, while geometric algebra uses multilinear algebra and applies in all dimensions and signatures, notably $3 + 1$ space-time as well as 2 dimensions. They are mathematically equivalent in 3 dimensions, though the approaches differ. Vector algebra is more widely used in elementary multivariable calculus, while geometric algebra is used in some more advanced treatments and is proposed for elementary use as well. In advanced mathematics, particularly differential geometry, neither is widely used, with differential forms being far more widely used.

Geometric algebra gives emphasis on geometric interpretations and physical applications.[56] A geometric algebra is the Clifford algebra[57] of a vector space over the field of real

---

[56] Specific examples of geometric algebras applied in physics include the algebra of physical space, the spacetime algebra, and the conformal geometric algebra. Geometric calculus, an extension of GA that includes differentiation and integration can be further shown to incorporate other theories such as complex analysis, differential geometry, and differential forms. David Orlin Hestenes (born May 21, 1933, in Chicago Illinois) and Chris J. L. Doran argue that it provides compact and intuitive descriptions in many areas, including classical and quantum mechanics, electromagnetic theory and relativity.

[57] **Clifford algebras** are a type of associative algebra. They are named after the English geometer William Kingdon Clifford (4 May 1845–3 March 1879). As K-algebras (an algebra over a field K, where K is called the base field of A), Clifford algebras generalize the real numbers, complex numbers, quaternions and several other hypercomplex number systems. The theory of Clifford algebras is intimately connected with the theory of quadratic forms and orthogonal transformations.

numbers endowed with a quadratic form.[58] Clifford algebras are closely related to exterior algebras (Section 2.1). The distinguishing multiplication operation that defines the geometric algebra as a unital ring[59] is the **geometric product**.[60]

A wide range of spaces can be studied using geometric algebra, including Euclidean and non-Euclidean spaces of any dimension. Two non-Euclidean examples include that of the

---

[58] A **quadratic form** is a homogeneous polynomial of degree 2 in a number of variables. For example:

$$2x^2 + 4xy - 3y^2$$

is a quadratic form in the variables x and y.

[59] A **ring** is a set R equipped with two binary operations, $+$ and $\cdot$, called *addition* and *multiplication* (the symbol $\cdot$ is often omitted and multiplication is just denoted by juxtaposition), that map every pair of elements of R to a unique element of R. These operations must satisfy some properties, called **ring axioms**, which must be true for all a, b, c in R. Many authors do not require rings to have a multiplicative identity, since many algebras of functions considered in analysis are not unital, as the algebra of functions decreasing to zero at infinity, especially those with compact support on some (non-compact) space. Such a structure is also called **non-unital ring**, or **pseudo-ring**, or a **rng** (pronounced *rung*), where the term "rng" is meant to suggest that it is a "ring" without an "identity element", i. The axioms that are required in the definition of a rng are seven:

R is an Abelian group[113] under the addition, meaning that:
1. $(a+b)+c = a+(b+c)$: the addition is associative.
2. $0+a = a$: there is an element 0 in R that is the zero element of the addition.
3. $a+b = b+a$: the addition is commutative.
4. $a+(-a) = (-a)+a = 0$: for each a in R there exists $-a$, the inverse element of a in R.

Multiplication is associative:
5. $(a \cdot b) \cdot c = a \cdot (b \cdot c)$.

Multiplication distributes over addition, meaning that:
6. $a \cdot (b+c) = (a \cdot b) + (a \cdot c)$: multiplication is left distributive over addition.
7. $(b+c) \cdot a = (b \cdot a) + (c \cdot a)$: multiplication is right distributive over addition.

Some other authors assume as one of the ring axioms the existence of a multiplicative identity:
8. $a \cdot 1 = 1 \cdot a = a$: there is an element 1 in R that is the multiplicative identity.

Rings that satisfy all eight of the above axioms are sometimes, for emphasis, referred to as **unital rings** (also called **unitary rings**, **rings with unity**, **rings with identity** or **rings with 1**). For example, the set of even integers satisfies the first seven axioms, but it does not have a multiplicative identity, and therefore is not a unital ring. Conversely, a Clifford algebra[57] is a unital associative algebra, which contains and is generated by a vector space V equipped with a quadratic form Q.[58] Specifically, The Clifford algebra $C\ell(V,Q)$ is the "freest" algebra generated by V, subject to the condition:

$$v^2 = Q(v)1 \quad \text{for all } v \in V \; ;$$

where the product on the left is that of the algebra, and the 1 is its multiplicative identity.

Although ring addition is commutative, ring multiplication is not required to be commutative, that is, $a \cdot b$ need not equal $b \cdot a$. Rings that also satisfy commutativity for multiplication (such as the ring of integers) are called **commutative rings**.

[60] The geometric product was first briefly mentioned by Hermann Günther Grassmann (April 15, 1809–September 26, 1877), who was chiefly interested in developing the closely related but more limited exterior algebra. In 1878, William Kingdom Clifford (4 May 1845–3 March 1879) greatly expanded on Grassmann's work to form what are now usually called Clifford algebras in his honor.

spacetime algebra for Minkowski spacetime[61] and the universal geometric algebra.[62] Although the algebra is created using real number scalars, the complex numbers and quaternions can still appear as other elements of the geometric algebra. For example, there can exist elements that have square -1, as with imaginary numbers. In fact, the real numbers and the complex numbers can both be considered as special cases of geometric algebras. Other physically important

---

[61] In mathematical physics, **Minkowski space** or **Minkowski spacetime**, named after the German mathematician Hermann Minkowski (June 22, 1864–January 12, 1909), is the mathematical space setting in which Einstein's theory of special relativity is most conveniently formulated. In this setting, the three ordinary dimensions of space are combined with a single dimension of time to form a four-dimensional manifold[51] for representing a spacetime. The symmetry group of a Euclidean space is the Euclidean group, while for a Minkowski space it is the Poincaré group. In 1905 (published 1906) it was noted by Jules Henri Poincaré (29 April 1854–17 July 1912) that, by taking time to be the imaginary part of the fourth spacetime coordinate $\sqrt{-1}\,ct$, a Lorentz transformation can be regarded as a rotation of coordinates in a four-dimensional Euclidean space with three real coordinates representing space, and one imaginary coordinate, representing time, as the fourth dimension. Since the space is then a pseudo-Euclidean space, the rotation is a representation of a hyperbolic rotation. This idea was elaborated by Minkowski, who used it to restate the Maxwell equations in four dimensions. Minkowski showed in 1907 that his former student Albert Einstein's special theory of relativity (1905), presented algebraically by Einstein, could also be understood geometrically as a theory of four-dimensional space-time. He concluded that time and space should be treated equally, and so arose his concept of events taking place in a unified four-dimensional space-time continuum. The beginning part of his address delivered at the 80th *Assembly of German Natural Scientists and Physicians* (September 21, 1908) is now famous:

> *"The views of space and time which I wish to lay before you have sprung from the soil of experimental physics, and therein lies their strength. They are radical. Henceforth space by itself, and time by itself, are doomed to fade away into mere shadows, and only a kind of union of the two will preserve an independent reality."*

Einstein, at first, viewed Minkowski's treatment as a mere mathematical trick, before eventually realizing that a geometrical view of space-time would be necessary in order to complete his own later work in general relativity (1915). Formally, Minkowski space is a four-dimensional real vector space equipped with a nondegenerate, symmetric bilinear form with signature $(-,+,+,+)$. Some may also prefer the alternative signature $(+,-,-,-)$. In general, mathematicians and general relativists prefer the former, while particle physicists tend to use the latter. In a further development, Minkowski gave an alternative formulation that did not use the imaginary time coordinate, but represented the four variables (x, y, z, t) of space and time in coordinate form, in a four-dimensional affine space. Points in this space correspond to events in space-time, and there is a defined light-cone associated with each point. Events not on the light-cone are classified by their relation to the apex as *space-like* or *time-like*, while the light-cone of an event is the set of all *light-like* vectors (null vectors) at that event. In the presence of gravity, spacetime is described by a curved four-dimensional manifold for which the tangent space to any point is a four-dimensional Minkowski space. Thus, the structure of Minkowski space is still essential in the description of general relativity. In the realm of weak gravity, spacetime becomes flat and looks globally, not just locally, like Minkowski space. For this reason, Minkowski space is often referred to as *flat spacetime*.

[62] A **universal geometric algebra** (UGA) is a type of geometric algebra generated by real vector spaces endowed with an indefinite quadratic form. The universal geometric algebra $\mathcal{G}(n,n)$, of order 2^{2n}, is defined as the Clifford algebra[57] of $2n$-dimensional pseudo-Euclidean space $\mathbb{R}^{n,n}$. This algebra is also called the "mother algebra". The vectors in this space generate the algebra through the geometric product. UGA contains all finite-dimensional geometric algebras (GA). The elements of UGA are called multivectors. Every multivector can be written as the sum of several r-vectors. Some r-vectors are scalars ($r = 0$), vectors ($r = 1$) and bivectors ($r = 2$). Scalars are identical to the real numbers. Complex numbers are not used as scalars because there already exist structures in UGA that are equivalent to the complex numbers. The geometric product for multivectors is called the generalized geometric product.

structures appear in geometric algebras: for example, the gamma matrices introduced by the Dirac equation appear as elements in a certain geometric algebra.

2.2.1 Inner and Outer Products Originated by the Geometric Product

Taking the geometric product among vectors can yield bivectors, trivectors, or general p-vectors. The addition operation combines these into general multivectors, which are the elements of the ring. This includes, among other possibilities, a well-defined sum of a scalar and a vector, an operation that is impossible by the traditional vector addition.

Given a finite-dimensional real quadratic space[63] $V = \mathbb{R}^n$ with quadratic form:

$$Q = V \rightarrow \mathbb{R}; \tag{2.2.1}$$

the geometric algebra for this quadratic space is the Clifford algebra $\mathcal{C}\ell(V, Q)$. If a, b, and c are vectors, then the geometric product is associative:

$$a(bc) = (ab)c; \tag{2.2.2}$$

and distributive over addition:

$$a(b + c) = ab + ac. \tag{2.2.3}$$

Moreover, the geometric product has the following property:

$$a^2 \in \mathbb{R}; \tag{2.2.4}$$

where the square is not necessarily positive. A further important property of the geometric product is the existence of elements with multiplicative inverse, also known as units. If $a^2 \neq 0$ for some vector a, then a^{-1} exists and is equal to

---

[63] A quadratic form q in n variables over K induces a map from the n-dimensional coordinate space K^n into K:

$$Q(v) = q(v), \quad v = \left[v_1, ..., v_n\right]^{\mathrm{T}} \in K^n.$$

The map Q is a **quadratic map**, which means that:

- For all a in K and v in V:

$$Q(av) = a^2 Q(v).$$

- When the characteristic of K is not two, the map:

$$B_Q : V \times V \rightarrow K, \quad B_Q(v, w) = \frac{1}{2}\left(Q(v + w) - Q(v) - Q(w)\right);$$

is bilinear over K.

The pair (V, Q) consisting of a finite-dimensional vector space V over K and a quadratic map from V to K is called a **quadratic space** and B_Q is the associated bilinear form of Q. The notion of a quadratic space is a coordinate-free version of the notion of quadratic form. Sometimes, Q is also called a quadratic form.

$$a^{-1} = \frac{a}{a^2}. \tag{2.2.5}$$

Not all the elements of the algebra are necessarily units. For example, if u is a vector in V such that $u^2 = 1$, the elements $1 \pm u$ have no inverse since they are zero divisors:

$$(1-u)(1+u) = 1 - uu = 1 - 1 = 0. \tag{2.2.6}$$

From the axioms above, we find that we may write the geometric product of any two vectors a and b as the sum of a symmetric product and an antisymmetric product:

$$ab = \frac{1}{2}(ab + ba) + \frac{1}{2}(ab - ba). \tag{2.2.7}$$

The symmetric product defines the inner product of vectors a and b:[9]

$$a \cdot b := \frac{1}{2}(ab + ba) = \frac{1}{2}\left((a+b)^2 - a^2 - b^2\right); \tag{2.2.8}$$

which is a real number, because it is a sum of squares, and is not required to be positive definite. It is not specifically the inner product on a normed vector space.[64]

The antisymmetric product in Eq. (2.2.7) is equal to the exterior product of the contained exterior algebra and defines the outer product of vectors a and b:

$$a \wedge b := \frac{1}{2}(ab - ba). \tag{2.2.9}$$

---

[64] An inner product of the vector space V over the field F is a map:

$$\langle \bullet, \bullet \rangle : V \times V \to F \; ;$$

where the field of scalars denoted by F is either the field of real numbers, $\mathbb{R}$, or the field of complex numbers, $\mathbb{C}$, which satisfies the following three axioms for all vectors $x, y, z \in V$ and all scalars $a \in F$:

- Conjugate symmetry (in $\mathbb{R}$, it is symmetric):

$$\langle x, y \rangle = \overline{\langle y, x \rangle}.$$

- Linearity in the first argument:

$$\langle ax, y \rangle = a \langle x, y \rangle \;;$$

$$\langle x + y, z \rangle = \langle x, z \rangle + \langle y, z \rangle.$$

- Positive-definiteness:

$$\langle x + x \rangle \geq 0 \;;$$

with equality only for $x = 0$.

For Euclidean vector spaces,[73] the inner product $\langle \mathbf{u}, \mathbf{v} \rangle = \mathbf{u}^\mathrm{T} \mathbf{v}$, better known as the dot product, is the trace of the outer product $\mathbf{u} \otimes \mathbf{v}$,[65] provided that $\mathbf{u}$ is represented as an $m \times 1$ column vector and $\mathbf{v}$ as an $n \times 1$ column vector with $m = n$.

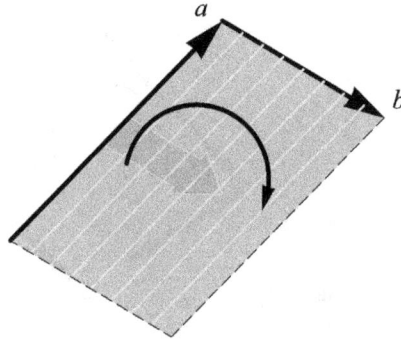

Figure 2.7. The extension of vector a along vector b provides the geometric interpretation of the outer product $a \wedge b$.

Note that the outer product of GA is not the outer product of linear algebra.[65] Geometrically, the outer product $a \wedge b$ can be viewed by placing the tail of the arrow b at the head of the arrow a and extending vector a along vector b (Fig. 2.7).

The resulting entity is a two-dimensional subspace, and we call it a bivector. It has an area equal to the size of the parallelogram spanned by a and b. The senses of a and b orientate the sides of the parallelogram and define a sense of traversal of its boundary. In the case of Fig. 2.7, the traversal sense is a clockwise sense, which can be depicted by a clockwise arc.

Note also that a bivector has no shape. Using a parallelogram to visualize the area provides an intuitive way of understanding, but a bivector is just an oriented area, in the same way a vector is just an oriented length (Fig. 2.8).

Being the extension of a vector along another vector, the outer product can be considered, in some ways, the opposite of the dot product, which projects a vector onto another vector.

The geometric interpretation of the outer product $b \wedge a$ is achieved by placing the tail of the arrow a at the head of the arrow b and extending vector b along vector a. This reverses the circulation of the boundary (Fig. 2.9), while it does not change the area of the parallelogram spanned by a and b.

The cross product[33] of two vectors in three dimensions with positive-definite quadratic form is closely related to their outer product.

In conclusion, the geometric product in Eq. (2.2.7) can be written as the sum between a scalar and a bivector:

$$ab = a \cdot b + a \wedge b. \tag{2.2.10}$$

---

[65] In linear algebra, the outer product typically refers to the tensor product of two vectors. The result of applying the outer product to a pair of vectors is a matrix. In particular, The outer product $\mathbf{u} \otimes \mathbf{v}$ is equivalent to a matrix multiplication $\mathbf{u}\mathbf{v}^{\mathrm{T}}$:

$$\mathbf{u} \otimes \mathbf{v} = \mathbf{u}\mathbf{v}^{\mathrm{T}};$$

provided that $\mathbf{u}$ is represented as an $m \times 1$ column vector and $\mathbf{v}$ as an $n \times 1$ column vector (which makes $\mathbf{v}^{\mathrm{T}}$ a row vector). For complex vectors, it is customary to use the conjugate transpose of $\mathbf{v}$ (denoted by $\mathbf{v}^{\mathrm{H}}$):

$$\mathbf{u} \otimes \mathbf{v} = \mathbf{u}\mathbf{v}^{\mathrm{H}}.$$

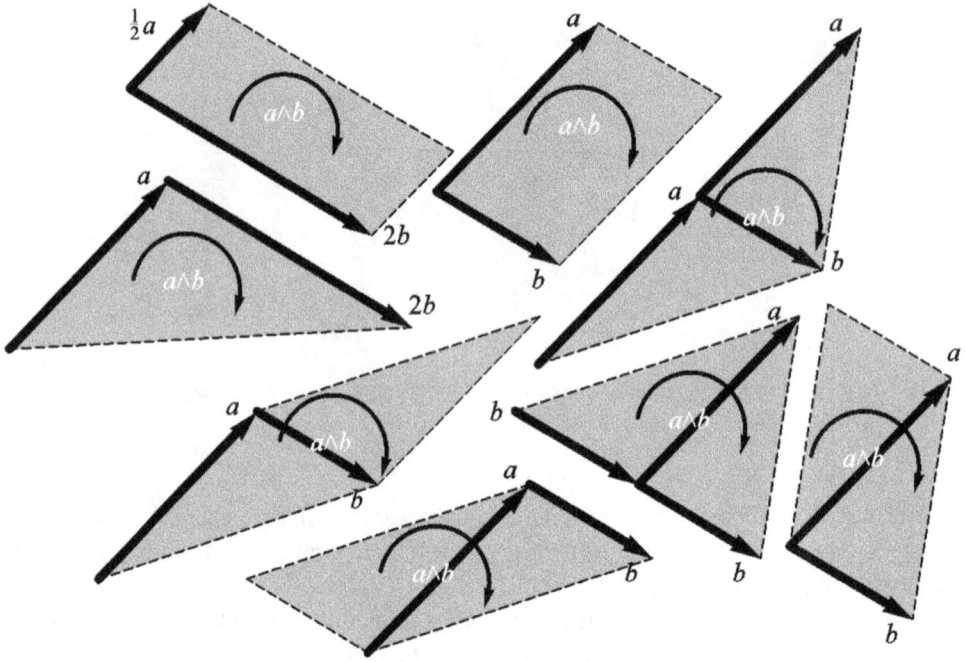

Figure 2.8. Different shapes for the same bivector, $a \wedge b$.

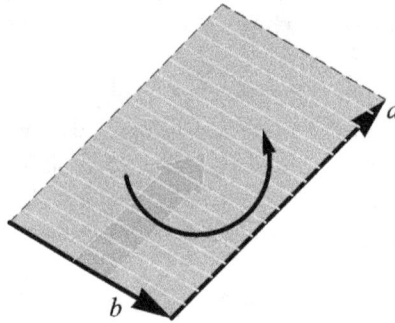

Figure 2.9. The extension of vector b along vector a provides the geometric interpretation of the outer product $b \wedge a$.

The scalar and the bivector are added by keeping the two entities separated, in the same way in which, in complex numbers, we keep the real and imaginary parts separated. Being the sum of a scalar and a bivector, the geometric product is a particular combination of blades, that is, is a multivector.[66]

a and b are **parallel** if all their geometric product is equal to their inner product, whereas a and b are **perpendicular** if their geometric product is equal to their outer product (Fig. 2.10).

---

[66] A multivector is a linear combination of different k-blades. In $\mathbb{R}^2$ it will contain, in general, a scalar part, a vector part and a bivector part.

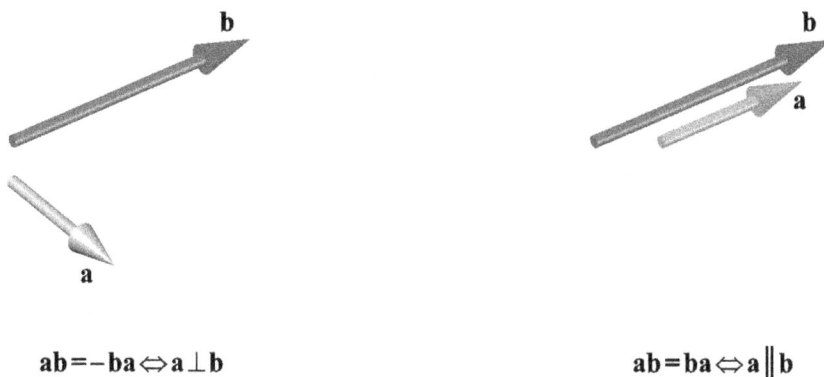

$$ab = -ba \Leftrightarrow a \perp b \qquad\qquad ab = ba \Leftrightarrow a \parallel b$$

Figure 2.10. Given two vectors a and b, if the geometric product ab is anticommutative they are perpendicular (left), if it is commutative they are parallel (right).

In a geometric algebra for which the square of any nonzero vector is positive, the inner product of two vectors can be identified with the dot product of standard vector algebra.

Most instances of geometric algebras of interest have a nondegenerate quadratic form. If the quadratic form is fully degenerate:[67]

$$Q = 0; \qquad\qquad (2.2.11)$$

---

[67] Given a quadratic form in n variables, $q_A(x_1, ..., x_n)$, its coefficients can be arranged into an $n \times n$ real symmetric matrix. The coefficients a_{ij} of the matrix and the variables x_i of the quadratic form are related by the formula:

$$q_A(x_1, ..., x_n) = \sum_{i,j=1}^{n} a_{ij} x_i x_j.$$

One of the most important questions in the theory of quadratic forms is how much can one simplify a quadratic form q_A by a homogeneous linear change of variables. A fundamental theorem due to Carl Gustav Jacob Jacobi (10 December 1804–18 February 1851) asserts that q_A can be brought to a **diagonal form**:

$$q_A = \lambda_1 \tilde{x}_1^2 + \lambda_2 \tilde{x}_2^2 + ... + \lambda_n \tilde{x}_n^2,$$

so that the corresponding symmetric matrix is diagonal, and this is even possible to accomplish with a change of variables given by an orthogonal matrix – in this case the coefficients $\lambda_1, \lambda_2, ..., \lambda_n$ are in fact determined uniquely up to a permutation. If the change of variables is given by an invertible matrix, not necessarily orthogonal, then the coefficients λ_i can be made to be 0, 1, and −1. Sylvester's law of inertia, named after the English mathematician James Joseph Sylvester (3 September 1814–15 March 1897), states that the numbers of 1 and −1 are invariants of the quadratic form, in the sense that any other diagonalization will contain the same number of each. The case when all λ_i have the same sign is especially important. The quadratic form is called **positive definite** when all λ_i are equal to 1, and **negative definite** when all λ_i are equal to −1. If none of the terms are equal to 0, then the form is called **nondegenerate**. This case includes positive definite, negative definite, and indefinite (a mix of 1 and −1) quadratic form. Equivalently, a nondegenerate quadratic form is one whose associated symmetric form is a nondegenerate bilinear form. If all the terms are equal to 0, then the form is called **fully degenerate**.

the inner product of any two vectors is always zero, and the geometric algebra $C\ell(V, Q)$ is then simply the exterior algebra $\Lambda(V)$. One can thus consider the Clifford algebra $C\ell(V, Q)$ as an enrichment (or more precisely, a quantization[68]) of the exterior algebra on V with a multiplication that depends on Q (one can still define the exterior product independent of Q). For nonzero Q there exists a canonical linear isomorphism between $\Lambda(V)$ and $C\ell(V, Q)$, whenever the ground field K does not have characteristic two.[69] That is, they are naturally isomorphic as vector spaces, but with different multiplications (in the case of characteristic two, they are still isomorphic as vector spaces, just not naturally). Clifford multiplication together with the privileged subspace is strictly richer than the exterior product since it makes use of the extra information provided by Q.

Let $\{e_i\}$ be a frame, that is, a set of n basis vectors that span an n-dimensional vector space. The frame dual to the $\{e_i\}$ are the elements of the geometric algebra that are the multiplicative inverses of the basis vectors, which are the elements denoted by $\{e^i\}$ satisfying

$$e^i \cdot e_j = \delta^i{}_j; \tag{2.2.12}$$

where $\delta^i{}_j$ is the Kronecker delta symbol.

Let:

$$\varepsilon = e_1 \wedge \dots \wedge e_n; \tag{2.2.13}$$

be a pseudoscalar formed the basis (this pseudoscalar is not necessarily unit). The dual basis vectors may be built constructively according to the formula:

$$e^i = (-1)^{i-1} e_1 \wedge \dots \wedge \check{e}_i \wedge \dots \wedge e_n \varepsilon^{-1}; \tag{2.2.14}$$

where in the long wedge product the check symbol on the i-th term means that the product is taken with respect to all of the basis vectors except that one, which is removed from the product. The long wedge product is therefore of grade $n-1$ so that, when multiplied by the grade n

---

[68] **Quantization** is a procedure for constructing a quantum field theory starting from a classical field theory. This is a generalization of the procedure for building quantum mechanics from classical mechanics. One also speaks of **field quantization**, as in the "quantization of the electromagnetic field", where one refers to photons as field "quanta" (for instance as light quanta). This procedure is basic to theories of particle physics, nuclear physics, condensed matter physics, and quantum optics. The fundamental notion that a physical property may be "quantized" is referred to as "the hypothesis of quantization". This means that the magnitude can take on only certain discrete values.

Clifford algebras may be thought of as quantizations of the exterior algebra, in the same way that the Weyl algebra is a quantization of the symmetric algebra. Weyl algebras and Clifford algebras admit a further structure of a *-algebra and can be unified as even and odd terms of a superalgebra.

[69] The **characteristic** of a ring R, often denoted by $\mathrm{char}(R)$, is defined to be the smallest number of times one must use the ring's multiplicative identity element (1) in a sum to get the additive identity element (0). That is, $\mathrm{char}(R)$ is the smallest positive number n such that

$$\underbrace{1 + \dots + 1}_{n \text{ summands}} = 0.$$

The ring is said to have characteristic zero if this repeated sum never reaches the additive identity.

inverse pseudoscalar, returns a dual basis vector that is grade 1. Note that, unlike the situation in traditional linear algebra, the dual frame of geometric algebra lies within the same space as the original basis, not in an abstract dual space.

Due to the presence of an inner product, the Riesz representation theorem in Eq. (2.1.16), restricted to real numbers, takes the form in Eq. (2.1.18):

$$\alpha^*(\mathbf{u}) = \alpha \cdot \mathbf{u} = 1; \tag{2.2.15}$$

where α is the unique vector that represents the covector α^*.

As we can express an n-dimensional vector as a linear combination of the basis vectors, e_i, writing it as a n-tuple of real numbers, likewise bivectors can be expressed as linear combinations of basis bivectors. By way of example, consider the real number decomposition $\mathbf{a} = (\alpha_1, \alpha_2)$ and $\mathbf{b} = (\beta_1, \beta_2)$, of the two vectors $\mathbf{a}$ and $\mathbf{b}$ of the Euclidian Plane $\mathbb{R}^2$, onto the basis vectors $\mathbf{e}_1$ and $\mathbf{e}_2$ (Fig. 2.11):

$$\mathbf{a} = \alpha_1 \mathbf{e}_1 + \alpha_2 \mathbf{e}_2; \tag{2.2.16}$$

$$\mathbf{b} = \beta_1 \mathbf{e}_1 + \beta_2 \mathbf{e}_2. \tag{2.2.17}$$

The outer product of $\mathbf{a}$ and $\mathbf{b}$ provides

$$\begin{aligned}\mathbf{a} \wedge \mathbf{b} &= (\alpha_1 \mathbf{e}_1 + \alpha_2 \mathbf{e}_2) \wedge (\beta_1 \mathbf{e}_1 + \beta_2 \mathbf{e}_2) \\ &= \alpha_1 \mathbf{e}_1 \wedge \beta_1 \mathbf{e}_1 + \alpha_1 \mathbf{e}_1 \wedge \beta_2 \mathbf{e}_2 + \alpha_2 \mathbf{e}_2 \wedge \beta_1 \mathbf{e}_1 + \alpha_2 \mathbf{e}_2 \wedge \beta_2 \mathbf{e}_2.\end{aligned} \tag{2.2.18}$$

By reordering the scalar multiplications, we obtain

$$\mathbf{a} \wedge \mathbf{b} = \alpha_1 \beta_1 \mathbf{e}_1 \wedge \mathbf{e}_1 + \alpha_1 \beta_2 \mathbf{e}_1 \wedge \mathbf{e}_2 + \alpha_2 \beta_1 \mathbf{e}_2 \wedge \mathbf{e}_1 + \alpha_2 \beta_2 \mathbf{e}_2 \wedge \mathbf{e}_2. \tag{2.2.19}$$

Now, since the outer product of a vector by itself equals zero, the outer product $\mathbf{a} \wedge \mathbf{b}$ can be written as

$$\mathbf{a} \wedge \mathbf{b} = \alpha_1 \beta_2 \mathbf{e}_1 \wedge \mathbf{e}_2 + \alpha_2 \beta_1 \mathbf{e}_2 \wedge \mathbf{e}_1; \tag{2.2.20}$$

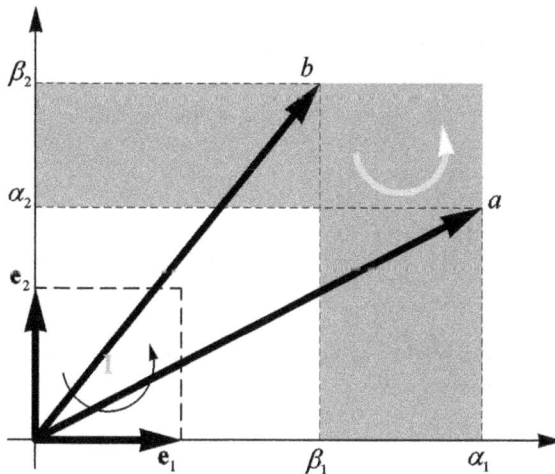

Figure 2.11. Vector decomposition in two-dimensional basis.

where, since (Fig. 2.11):

$$\mathbf{e}_1 \wedge \mathbf{e}_2 = \mathbf{I}; \tag{2.2.21}$$

$$\mathbf{e}_2 \wedge \mathbf{e}_1 = -\mathbf{I}; \tag{2.2.22}$$

we can finally express the bivector $\mathbf{a} \wedge \mathbf{b}$ in terms of the basis bivector, $\mathbf{I}$, which, in Euclidean plane represents $\mathbf{e}_{12} = \mathbf{e}_1 \wedge \mathbf{e}_2$:

$$\mathbf{a} \wedge \mathbf{b} = (\alpha_1 \beta_2 - \alpha_2 \beta_1)\mathbf{I}. \tag{2.2.23}$$

Since α_1, α_2, β_1, and β_2 denote the vector components of $\mathbf{a}$ and $\mathbf{b}$, the round bracketed term can be geometrically interpreted as the norm of a sum of bivectors, that is, an algebraic sum of areas:

$$\mathbf{a} \wedge \mathbf{b} = \begin{bmatrix} +1 & -1 \end{bmatrix} \begin{bmatrix} \alpha_1 \beta_2 \\ \alpha_2 \beta_1 \end{bmatrix} \mathbf{I} = \alpha_1 \wedge \beta_2 + \alpha_2 \wedge \beta_1; \tag{2.2.24}$$

where $\alpha_1 \beta_2$ is multiplied by +1 because $\alpha_1 \wedge \beta_2$ has the same sense as $\mathbf{I}$, while $\alpha_2 \beta_1$ is multiplied by −1 because $\alpha_2 \wedge \beta_1$ has the opposite sense.

The difference between $\alpha_1 \beta_2$ and $\alpha_2 \beta_1$ equals the size of the colored area in Fig. 2.11.

Since all spaces $\mathbb{R}^n$ generate a set of basis blades that make up a geometric algebra of subspaces, denoted by Cl_n, a possible basis for Cl_2, generated by $\mathbb{R}^2$, is

$$\left\{ \underbrace{\mathbf{1}}_{\text{basis scalar}} , \underbrace{\mathbf{e}_1, \mathbf{e}_2}_{\text{basis vectors}} , \underbrace{\mathbf{I}}_{\text{basis bivector}} \right\}, \tag{2.2.25}$$

where $\mathbf{1}$ is used to denote the basis 0-blade, or scalar-basis.

Analogously, in three-dimensional space $\mathbb{R}^3$, there are three basis vectors, three basis bivectors, and one basis trivector, besides than the scalar-basis (Fig. 2.12). Therefore, a possible basis for Cl_3 is

$$\left\{ \underbrace{\mathbf{1}}_{\text{basis scalar}} , \underbrace{\mathbf{e}_1, \mathbf{e}_2, \mathbf{e}_3}_{\text{basis vectors}} , \underbrace{\mathbf{e}_{12}, \mathbf{e}_{13}, \mathbf{e}_{23}}_{\text{basis bivectors}} , \underbrace{\mathbf{e}_{123}}_{\text{basis trivector}} \right\}. \tag{2.2.26}$$

The total number of basis blades for an algebra can be calculated by adding the numbers required for all basis k-blades:

$$\sum_{k=0}^{n} \binom{n}{k} = 2^n; \tag{2.2.27}$$

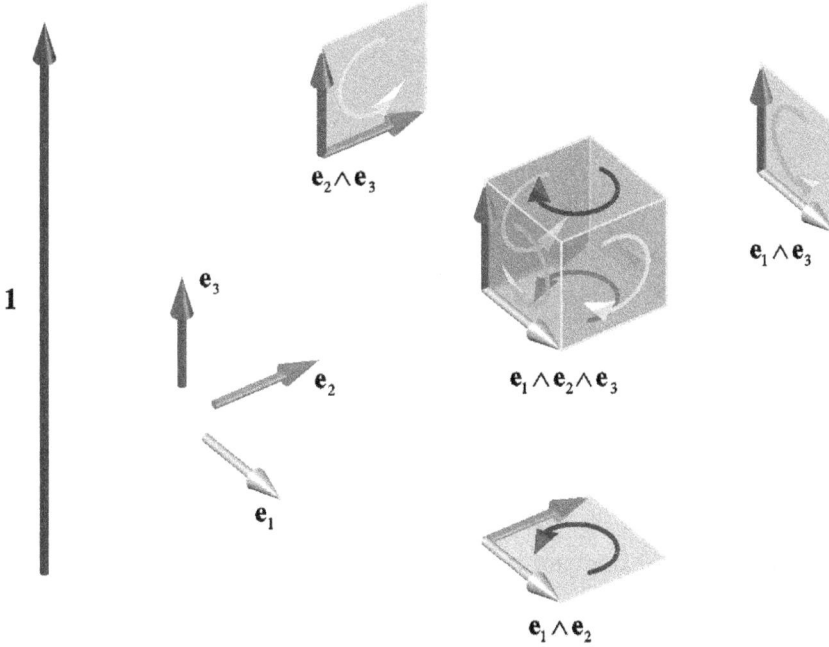

Figure 2.12. Standard n-vector basis in 3D: unit scalar 1 (represented by a black number line), unit vectors, unit bivectors, and a unit trivector.

where the number of basis k-blades needed in an n-dimensional space to represent arbitrary k-blades is provided by the binomial coefficient:[70]

$$\binom{n}{k} = \frac{n!}{k!(n-k)!}.$$

(2.2.28)

This is because a basis k-blade is uniquely determined by the k basis vectors from which it is constructed. There are n different basis vectors in total and $\binom{n}{k}$ is the number of ways to choose k elements from a set of n elements.

---

[70] **Binomial coefficients** are a family of positive integers that are indexed by two non-negative integers. The binomial coefficient indexed by n and k is usually written as $\binom{n}{k}$. It is the coefficient of the x^k term in the polynomial expansion of the binomial power $(1+x)^n$. Under suitable circumstances the value of the coefficient is given by the expression $\frac{n!}{k!(n-k)!}$. Arranging binomial coefficients into rows for successive values of n, and in which k ranges from 0 to n, gives a triangular array called Pascal's triangle.[71]

$$
\begin{array}{ccccccc}
 & & & & 1 & & & & & n=0 \\
 & & & 1 & & 1 & & & & n=1 \\
 & & 1 & & 2 & & 1 & & & n=2 \\
 & 1 & & 3 & & 3 & & 1 & & n=3 \\
1 & & 4 & & 6 & & 4 & & 1 & n=4 \\
1 & 5 & & 10 & & 10 & & 5 & 1 & n=5
\end{array}
$$

$$d=0 \quad d=1 \quad d=2 \quad d=3 \quad d=4 \quad d=5$$

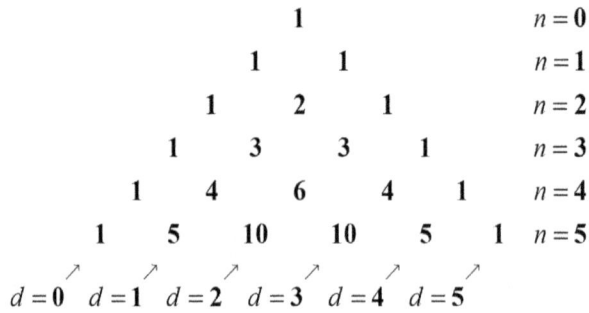

Figure 2.13. The first six rows of Pascal's triangle, from row 0 to row 5.[71]

Therefore, the number of basis k-blades in an n-dimensional space is the k-th entry in row n of Pascal's triangle[71] (Fig. 2.13).

2.2.2 The Features of *p*-vectors and the Orientations of Space Elements

The various objects of geometric algebra, the p-vectors, are charged with three attributes, or features: **attitude**, **orientation**, and **magnitude**. The second feature, taken singularly and combined with the first feature, gives rise to the two kinds of orientation in space, inner and outer orientations.

2.2.2.1 Inner Orientation of Space Elements

The second feature of p-vectors, the orientation, is, more properly, an **inner orientation**, because it does not depend on the embedding space. For example, a vector **u** in three dimensions

---

[71] **Pascal's triangle** is a triangular array of the binomial coefficients.[70] It is named after the French mathematician Blaise Pascal (19 June 1623–19 August 1662) in much of the Western world, although other mathematicians studied it centuries before him in South Asia (the earliest explicit depictions of a triangle of binomial coefficients occur in the 10th century in an Ancient Indian book written by Pingala, in or before the 2nd century BC, and, in 1068, four columns of the first sixteen rows were given by the mathematician Bhattotpala), Greece (where the numbers of the triangle arose from the study of figurate numbers[100]), Iran (discussed by the Persian mathematician, Abū Bakr ibn Muhammad ibn al Ḥusayn al-Karajī, or al-Karkhī, c. 953 in Karaj or Karkh—c. 1029, the **Khayyam-Pascal triangle**, or **Khayyam triangle**, was later repeated by the Persian poet-astronomer-mathematician Ghiyāth ad-Dīn Abu'l-Fath 'Umar ibn Ibrāhīm al-Khayyām Nīshāpūrī, 18 May 1048–4 December 1131), China (known in the early 11th century through the work of the Chinese mathematician Jia Xian, approximately 1010–1070 AD, the triangle is called **Yang Hui's triangle** in China, after Yang Hui, ca. 1238–1298, who presented it in 13th century), Germany (Petrus Apianus, 16 April 1495–21 April 1552, published the triangle on the frontispiece of his book on business calculations in the 16th century), and Italy (where it is referred to as **Tartaglia's triangle**, after the Italian algebraist Niccolò Fontana Tartaglia, 1499/1500, Brescia–13 December 1557, Venice). The rows of Pascal's triangle are conventionally enumerated starting with row $n=0$ at the top (Fig. 2.13).

The entries in each row are numbered from the left, beginning with $d=0$ (Fig. 2.13). Rows and entries are numbered in this way since, in doing so, the coefficient of the x^d term in the polynomial expansion of the binomial power $(1+x)^n$ is given by the d-th entry in row n. The n-th row contains $n+1$ entries. The shallow diagonals of the triangle sum to the Fibonacci numbers. The entries in each row are staggered relative to the numbers in the adjacent rows. To build the triangle, write the number 1 on row 0. Each element of following rows is the sum of the number above and to the left and the number above and to the right. If either the number to the right or left is not present, substitute a zero in its place (the first and last entries are always equal to 1).

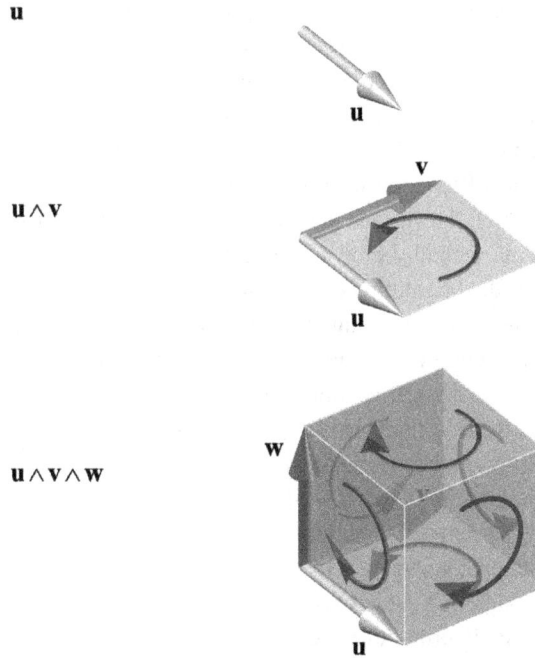

Figure 2.14. Geometric interpretation for the exterior product of p vectors to obtain a p-vector (parallelotope elements[34]), where $p = 1, 2, 3$. The "circulations" show the inner orientation.

has an orientation given by the sense of a straight line parallel to it (often indicated by an arrowhead, as in Fig. 2.14).

Similarly, a bivector $\mathbf{u} \wedge \mathbf{v}$ in three dimensions,[32] generated by the exterior product between $\mathbf{u}$ and $\mathbf{v}$ (Section 2.1), has an orientation (sometimes denoted by a curved arrow in the plane containing $\mathbf{u}$ and $\mathbf{v}$, as in Fig. 2.14) indicating a choice of sense of traversal of its boundary (its **circulation**). Finally, a trivector $\mathbf{u} \wedge \mathbf{v} \wedge \mathbf{w}$ has an orientation (sometimes denoted by curved arrows in the planes of its two-faces (see Section 3.3), as in Fig. 2.14) indicating a choice of sense of circulation of the faces on its boundary.[72] All the former orientations are inner orientations.

---

[72] The choice is given for one face of the boundary and extended to the other faces, in accordance with **Möbius law for edges**, named after the German mathematician and theoretical astronomer August Ferdinand Möbius (November 17, 1790–September 26, 1868): the orientations of two adjacent faces induce an opposite orientation on the common edge. Möbius law for edges may be directly derived from the definition of exterior product. In fact, on the three planes that contain the pairs of vectors $(\mathbf{u}, \mathbf{v})$, $(\mathbf{v}, \mathbf{w})$, and $(\mathbf{w}, \mathbf{u})$ of the trivector $\mathbf{u} \wedge \mathbf{v} \wedge \mathbf{w}$, where the pairs are ordered in way to preserve the order of the vectors in the exterior product, the orientations of the planes are given by the directions of the three exterior products $\mathbf{u} \wedge \mathbf{v}$, $\mathbf{v} \wedge \mathbf{w}$, and $\mathbf{w} \wedge \mathbf{u}$, respectively. In the remaining three planes, where the direction of the normal edge is reversed, the orientations are given by the directions of the three exterior products $-(\mathbf{u} \wedge \mathbf{v})$, $-(\mathbf{v} \wedge \mathbf{w})$, and $-(\mathbf{w} \wedge \mathbf{u})$. The negative sign comes from the fact that the common vertex of the remaining planes is the origin of the right-handed triple of vectors $(-\mathbf{u}, -\mathbf{v}, -\mathbf{w})$ and the exterior product $-\mathbf{u} \wedge -\mathbf{v} \wedge -\mathbf{w}$ is equal to $-(\mathbf{u} \wedge \mathbf{v} \wedge \mathbf{w})$. If we rotate two opposite faces of a cube in the senses of their own orientations, one face moves toward the other with a clockwise rotation. Approaching its opposite face, each face moves along the direction followed by a right-handed screw positioned on the face, when the right-handed screw is clockwise rotated, that is, screwed. Conversely, the cube deforms by assuming the shape of a left-handed screw. Therefore, for each pair of faces, we can associate the

(*continues next page*)

The term "inner" refers to the fact that the circulations are defined for the boundaries of the elements, by choosing an order for the vertexes. Therefore, we move and stay on the boundaries of the elements, without going out from the elements themselves.

In geometric algebra, the inner orientation is the geometric interpretation of the exterior geometric product among vectors. In particular, the inner orientation of a plane surface can be viewed as the orientation of the exterior product between two vectors $\mathbf{u}$ and $\mathbf{v}$ (the bivector $\mathbf{u} \wedge \mathbf{v}$) of the plane on which the surface lies, that is, the sense of the rotation that would align the first vector, $\mathbf{u}$, with the second vector, $\mathbf{v}$ (Fig. 2.14). Analogously, the inner orientation of a volume can be viewed as the orientation of the exterior product between three vectors $\mathbf{u}$, $\mathbf{v}$, and $\mathbf{w}$ (the trivector $\mathbf{u} \wedge \mathbf{v} \wedge \mathbf{w}$) of the three-dimensional space containing the volume (Fig. 2.14).

It is worth noting that, contrarily to the definition of an inner orientation in a narrow sense, the definition of a positive or a negative inner orientation implicitly implies the notion of external observer. In fact, as far as a plane surface is concerned, the result of the exterior product $\mathbf{u} \wedge \mathbf{v}$ is positive if and only if the sense of the rotation that would align $\mathbf{u}$ with $\mathbf{v}$ is anticlockwise for the observer that watches the plane in which $\mathbf{u}$ with $\mathbf{v}$ lie. In other words, since speaking of clockwise or anticlockwise sense of the rotation makes no sense in a two-dimensional space, an observer that lives in dimension 2 cannot discriminate whether the clockwise or anticlockwise is the positive sense of rotation. A sense of rotation in a plane is clockwise or anticlockwise only if it is related to the oriented normal direction.

By choosing the anticlockwise as the positive sense of rotation, the observer can be represented as a vector $\mathbf{w}$ that forms an ordered right-handed triple of vectors $(\mathbf{u}, \mathbf{v}, \mathbf{w})$, together with $\mathbf{u}$ and $\mathbf{v}$, where $\mathbf{w}$ is normal to the plane in which $\mathbf{u}$ and $\mathbf{v}$ lie. If the magnitude of $\mathbf{w}$ is the scalar given by the exterior product $\mathbf{u} \wedge \mathbf{v}$, $\mathbf{w}$ is the result of the vector product[33] $\mathbf{u} \times \mathbf{v}$. Consequently, the exterior product is provided with a sign only if "watched" by an observer living in the three-dimensional space, thus giving rise to a vector product. In other words, the exterior product lives in a geometrical vector space of dimension 2, while the vector product that provides its sign can be defined only in three-dimensional (or more than three-dimensional) spaces.

In conclusion, the inner orientation of a surface is not positive or negative in itself. Neither choosing the sign of the inner orientation can be considered an arbitrary convention. Providing the inner orientation of a surface with a sign makes sense only when the surface is "watched" by an external observer, that is, only when the surface is studied in an embedding space of dimension greater than 2, the dimension of the surface. If this is the case, a plane surface has a positive or negative inner orientation when the associated (in the three-dimensional space) vector product is positive or negative, respectively. For example, the surface originated by the bivector $\mathbf{u} \wedge \mathbf{v}$ in Fig. 2.14 has a positive inner orientation (the surface is watched form above). The same surface would have a negative inner orientation if watched from below.

(*continues from previous page*)

motion direction of a face over the other with the motion direction of a right-handed screw, which is screwed on one face or the other. Since the pairs of faces are three and they are mutually orthogonal, we finally obtain three mutually orthogonal right-handed screws that, in turn, are associated with right-handed coordinates in three-dimensional space. In conclusion, the passage from right-handed coordinates to left-handed coordinates is equivalent to the change of inner orientation of a volume. It is worth noting that, in what observed, the kind of screw, right-handed rather than left-handed, is associated with the relative motion of one face over its opposite face, not with the shape assumed by the cube after the deformation, which is of the opposite type.

As can be seen in Fig. 2.14, all the six faces of the positive trivector $\mathbf{u} \wedge \mathbf{v} \wedge \mathbf{w}$ have a negative inner orientation when they are watched by an external observer, while they have a positive inner orientation when they are watched by a local observer that is inside the volume. This happens since the inner volume of the trivector is the intersection of the six positive half-spaces, that is, the half-spaces of the six observers that watch the positive surfaces originated by the trivector. By relating the sign of the inner orientation to the external observer also in this second case, the positive inner orientation of a volume is the one watched by the external observer. As a consequence, the inner orientation of a volume is positive when the inner orientations of all its faces are negative. In the case of Fig. 2.14, therefore, the volume generated by the trivector $\mathbf{u} \wedge \mathbf{v} \wedge \mathbf{w}$ has a positive inner orientation.

The concept of inner orientation defined above did not apply to zero-dimensional vector spaces (points). However, it is useful to be able to assign different inner orientations to a point. Since every zero-dimensional vector space is naturally identified with $\mathbb{R}^0$, the operation on one point that is equivalent to an exterior product produces elements in $\mathbb{R}^1$, which is a vector space provided with an orientation (given by the standard basis). These one-dimensional elements have a positive inner orientation if they are oriented as the basis of $\mathbb{R}^1$, a negative inner orientation otherwise. In other words, also this latter time we fix the sign of the inner orientation of the element by choosing an order of traversal for the elements of its boundary, that is, the points that compose it. The extension of the outer product to zero-dimensional vectors (points) provides:[9]

$$\mathbf{P} \wedge \mathbf{Q} \triangleq \mathbf{u} ; \tag{2.2.29}$$

which has the geometrical meaning of point $\mathbf{P}$ extended toward point $\mathbf{Q}$. The extension of the outer product preserves the antisymmetric property of the product, since $\mathbf{Q} \wedge \mathbf{P}$ (point $\mathbf{Q}$ extended toward point $\mathbf{P}$) is the negation of $\mathbf{P} \wedge \mathbf{Q}$:

$$\mathbf{Q} \wedge \mathbf{P} = -\mathbf{u}. \tag{2.2.30}$$

What if we extend a zero-dimensional subspace along a one-dimensional one? A point extended by a vector results in an oriented length, which can be represented by the vector itself (Fig. 2.15). Consequently, a bound vector[73] with origin in $\mathbf{P}$, $(\mathbf{P}, \mathbf{u})$, can be seen as the outer product between $\mathbf{P}$ and the free vector[73] $\mathbf{u}$:

$$\mathbf{P} \wedge \mathbf{u} \triangleq (\mathbf{P}, \mathbf{u}). \tag{2.2.31}$$

Then, since the bound vector $(\mathbf{P}, \mathbf{u})$ is often denoted by simply $\mathbf{u}$, as its free vector, we can also write

$$\mathbf{P} \wedge \mathbf{u} = \mathbf{u}. \tag{2.2.32}$$

For consistency, we must therefore define the outer product between the vector $\mathbf{u}$ and the point $\mathbf{P}$ as the negation of $\mathbf{u}$:

$$\mathbf{u} \wedge \mathbf{P} \triangleq -\mathbf{u}. \tag{2.2.33}$$

---

[73] A **Euclidean vector** (sometimes called a **geometric** or **spatial vector**, or simply a **vector**) is a geometric object that has magnitude (or length) and direction. A Euclidean vector with fixed initial and terminal point is called a **bound vector**. When only the magnitude and direction of the vector matter, then the particular initial point is of no importance, and the vector is called a **free vector**. Thus, two arrows $\overline{AB}$ and $\overline{A'B'}$ in space represent the same free vector if they have the same magnitude and direction.

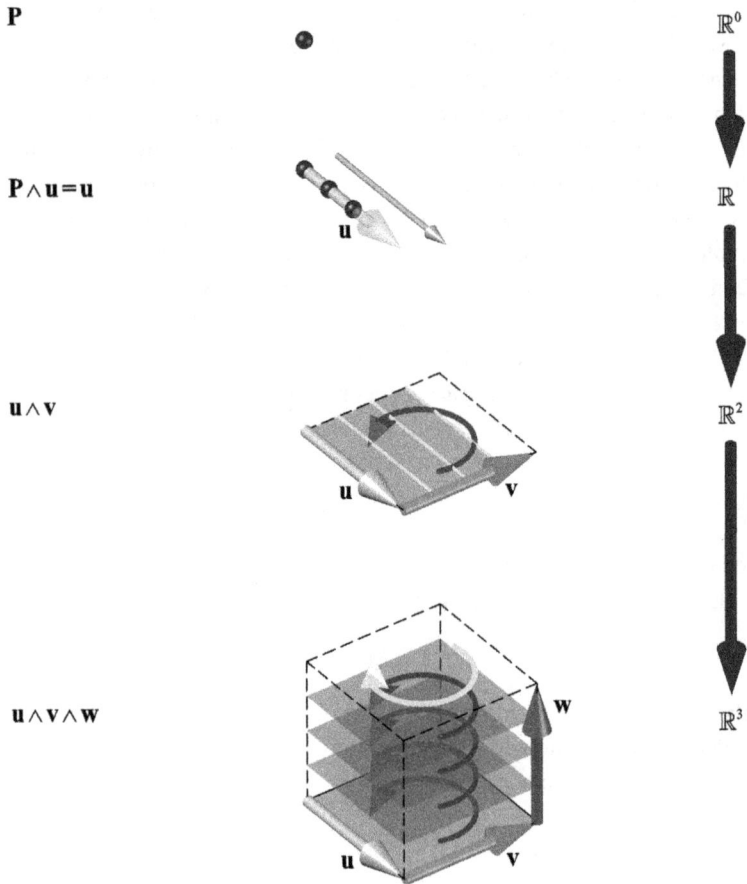

Figure 2.15. The inner orientation of a p-space element is induced by the $(p-1)$-space elements on its boundary.

Note that, as for the bivector, whether or not the one-dimensional element is in the same sense of the basis can be evaluated only by an observer that does not lie on the one-dimensional space. Since there always exists a plane that contains a given line (the one-dimensional space) and a given point that does not belong to the line (the point in which the observer is), the observer belongs to a two-dimensional space, at least. Thus, the evaluation of the sign of the one-dimensional element is possible only in dimension 2.

Moreover, in analogy to the direction of the vector product $\mathbf{u} \times \mathbf{v}$, which is orthogonal both to $\mathbf{u}$ and $\mathbf{v}$, the result of the operation $\mathbf{P} \times \mathbf{Q}$, defined in $\mathbb{R}^1$ on elements of $\mathbb{R}^0$, has the direction of a line that is orthogonal both to $\mathbf{P}$ and $\mathbf{Q}$. In three-dimensional space, where we can define infinite sub-spaces of dimension 1, each provided with its own basis, this operation produces elements in the direction of any line of the three-dimensional space. Being orthogonal to each direction of the three-dimensional space, the point is orthogonal to the three-dimensional space itself and to each volume of the space.

In conclusion, we can define two inner orientations of a point, the outward and the inward orientations. The first orientation is given by the outgoing lines, while the second orientation is given by the incoming lines (Fig. 2.16). In the first case, the point is called a **source**, while, in the second case, is called a **sink**.

As far as the sign of the inner orientation of the point is concerned, by making use of the notion of observer in this latter case too, each incoming line can be viewed as the sense along which the external observer watches the point. In this sense, a sink is a point with a positive inner orientation, while a source is a point with a negative inner orientation (Fig. 2.16).

Note also that even the trivector can be viewed as an extension. In fact, it is originated by a bivector extended by a third vector (Fig. 2.15). Like bivectors (Fig. 2.8), a trivector has no shape, but only volume and sign. Even though a box helps to understand the nature of trivectors intuitively, it could have been any shape (Fig. 2.17).

In conclusion, since the positive or negative inner orientation of a p-space element is induced by the positive or negative inner orientation of the $(p-1)$-space elements on its boundary, we derive the inner orientations and their signs inductively (Fig. 2.15). This allows us to extend the procedure for finding the inner orientation of the space elements to spaces of any dimension.

In the following, a point, a line, a surface, and a volume endowed with inner orientations will be denoted by putting bars over their symbols (Figs. 2.16, 2.19, 2.20).

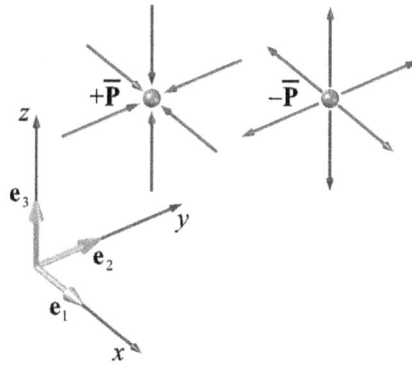

Figure 2.16. Positive and negative inner orientations of a point.

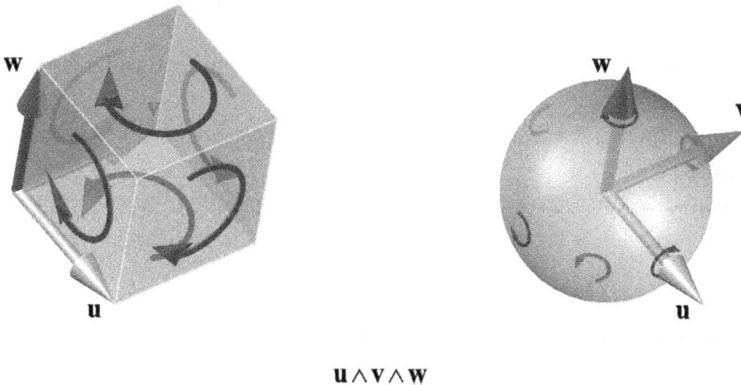

$$\mathbf{u} \wedge \mathbf{v} \wedge \mathbf{w}$$

Figure 2.17. The geometric interpretation of the exterior product of three vectors $\mathbf{u}$, $\mathbf{v}$, $\mathbf{w}$ as an oriented volume. Two possible shapes are shown: a parallelepiped and a sphere. The actual shape is irrelevant to the exterior product.

Figure 2.18. The two inner orientations of a curve.

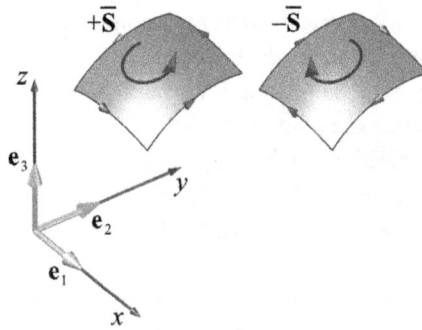

Figure 2.19. Positive and negative inner orientations of a surface.

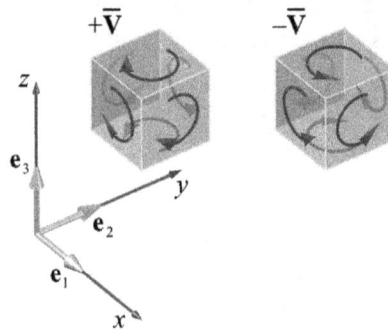

Figure 2.20. Positive and negative inner orientations of a volume.

The inner orientations of a curve,[74] a surface (not necessarily plane), and a volume are given by the inner orientations of the p-vectors of greater dimension in their tangent spaces (Figs. 2.18–2.20).

Electric potential and velocity potential are examples of physical variables associated with the inner orientations of points.

Strains, in continuum mechanics, are examples of physical variables associated with the inner orientations of lines.

The work involved in a thermodynamic cycle is an example of a physical variable whose sign depends on the inner orientation of the area that the cycle includes.

2.2.2.2 Outer Orientation of Space Elements

As far as the first feature of p-vectors is concerned, the attitude is part of the description of how p-vectors are placed in the space they are in. Thus, the notion of attitude is related to the notion of embedding of a p-vector in its space, or space immersion.

---

[74] A **curve** (also called a **curved line** in older texts) is, generally speaking, an object similar to a line but which is not required to be straight.

In particular, a vector in three dimensions has an attitude given by the family of straight lines parallel to it (possibly specified by an unoriented ring around the vector), a bivector in three dimensions has an attitude given by the family of planes associated with it (possibly specified by one of the normal lines common to these planes), and a trivector in three dimensions has an attitude that depends on the arbitrary choice of which ordered bases are positively oriented and which are negatively oriented.[75]

Between a p-vector and its attitude there exists the same kind of relationship that exists between an element a of a set X and the equivalence class[76] of a in the quotient set of X by a given equivalence relationship. In the special case of the attitude of a vector in the three-dimensional space, the set is that of the straight lines and the equivalence relationship is that of parallelism between lines. In this sense, the family of lines mentioned in the definition of an attitude of a vector is one equivalence class of the quotient set of straight lines by the equivalence relation of parallelism. One of the invariants of the equivalence relation of parallelism is the family of planes that are normal to the lines in a given equivalence class. Since we can choose any of the parallel planes for representing the invariant, we can speak both in terms of family of parallel planes and in terms of one single plane.

Similar considerations may also be applied to the relationship between bivectors and their attitudes, or trivectors and their attitudes. Thus, we can describe the attitude of a p-vector either in terms of its equivalence class (the family of parallel lines, when the p-vector is a vector), or in terms of its class invariant (the family of normal planes, when the p-vector is a vector), that is, the equivalence class of its orthogonal complement.[77] In particular, as far as the description in terms of class invariants is concerned, the attitude of a vector $\mathbf{u}$ can be viewed as a family of

---

[75] In the three-dimensional Euclidean space, right-handed bases are typically declared to be positively oriented, but the choice is arbitrary, as they may also be assigned a negative orientation. A vector space with an orientation is called an **oriented** vector space, while one without a choice of orientation is called **unoriented**.

[76] Given a set X and an equivalence relation ~ on X, the **equivalence class** of an element n in X is the subset of all elements in X which are equivalent to n. Equivalence classes among elements of a structure are often used to produce a smaller structure whose elements are the classes, distilling a relationship every element of the class shares with at least one other element of another class. This is known as **modding out** by the class. Two elements of the set are considered equivalent (with respect to the equivalence relation) if and only if they are elements of the same class. The class may assume the identity of one of the original elements. The equivalence class of an element a is denoted by $[a]$ and may be defined as the set of elements that are related to a by ~ :

$$[a] = \{x \in X : a \sim x\} \cdot$$

The intersection of any two different classes is empty; the union of all the classes equals the original set. A set of nonempty equivalence classes in X given an equivalence relation ~ is a partition of X, where, in general, a **partition** of a set G is a set P of nonempty subsets of G, such that every element of G is an element of a single element of P. The set of all equivalence classes in X given an equivalence relation ~ is denoted as $X/\sim$ and called the **quotient set** of X by ~. If X is a topological space,[92] $X/\sim$ is called a **quotient space**.[95]

If ~ is an equivalence relation on X, and $P(x)$ is a property of elements of X such that whenever $x \sim y$, $P(x)$ is true if $P(y)$ is true, then the property P is said to be a **class invariant** under the relation ~, or an **invariant** of ~, or **well-defined** under the relation ~. More generally, an invariant with respect to an equivalence relation is a property that is constant on each equivalence class.

[77] In the mathematical fields of linear algebra and functional analysis, the **orthogonal complement** of a subspace W of a vector space V equipped with a bilinear form B is the set $W^\perp$ of all vectors in V that are orthogonal to every vector in W. Informally, it is called the **perp**, short for **perpendicular complement**. It is a subspace of V and its dimension,

(continues next page)

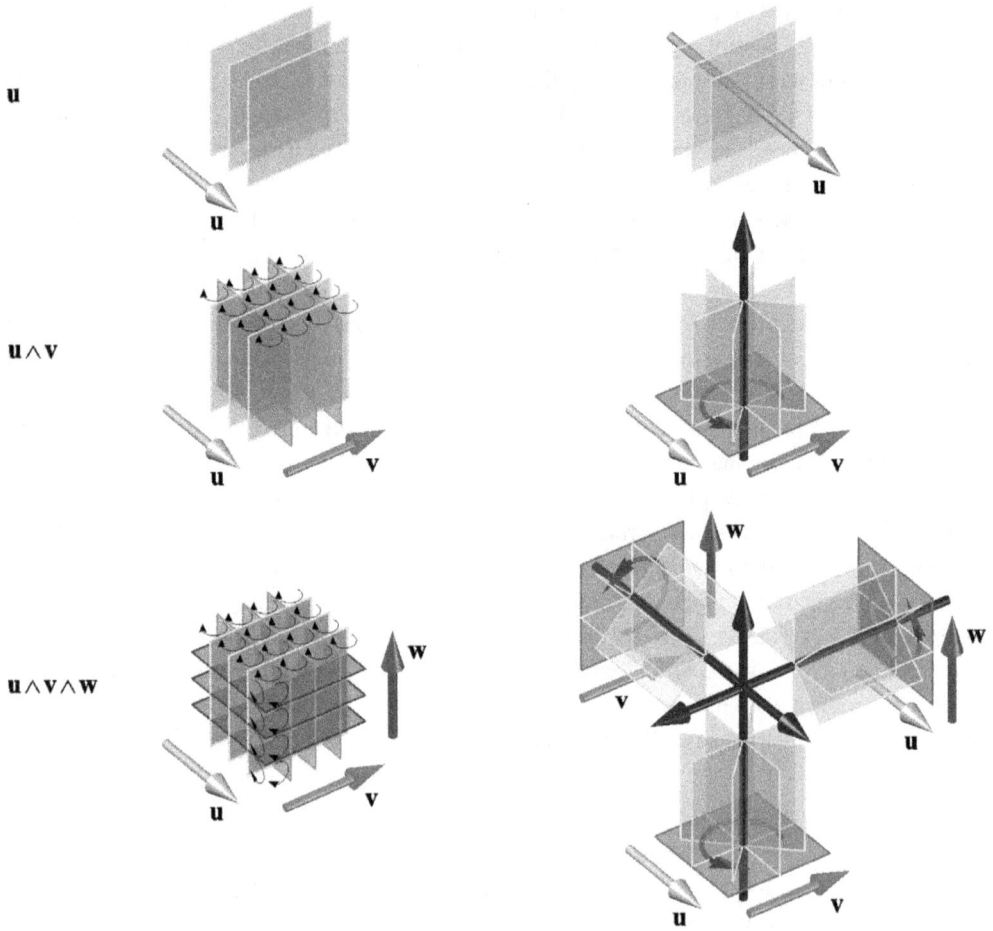

Figure 2.21. Geometric interpretation of the attitude of a p-vector in terms of class invariants.

normal planes (Fig. 2.21), each one originated by the translation of a plane normal to **u**, along the direction of **u** (the planes span the direction of **u**). Equivalently, the attitude of **u** can be represented by an arbitrary plane of the family of normal planes.

Analogously, the attitude of a bivector $\mathbf{u} \wedge \mathbf{v}$ can be viewed as two families of parallel planes (Fig. 2.21), the first family normal to **u** and the second family normal to **v** (the planes span both the directions of **u** and **v**). Since **u** and **v** are linearly independent in their common plane, the planes that span both the directions of **u** and **v** originate all the planes normal to $\mathbf{u} \wedge \mathbf{v}$, that is, all the planes parallel to $\mathbf{u} \times \mathbf{v}$. These planes can be represented by the line of intersection between an arbitrary plane of the first family and an arbitrary plane of the second family. The intersection line is parallel to all planes of the two families and to vector $\mathbf{u} \times \mathbf{v}$.

(*continues from previous page*)

$\dim W^\perp$, is the complement of $\dim W$ with respect to $\dim V$, that is, the result of subtracting $\dim W$ from $\dim V$. It follows that the sum of $\dim W^\perp$ and $\dim W$ always equals $\dim V$:

$$\dim W^\perp + \dim W = \dim V.$$

Finally, the attitude of a trivector $\mathbf{u} \wedge \mathbf{v} \wedge \mathbf{w}$ can be viewed as three families of parallel planes (Fig. 2.21), provided that the three families are normal to $\mathbf{u}$, $\mathbf{v}$, and $\mathbf{w}$, respectively (the planes span the three directions of $\mathbf{u}$, $\mathbf{v}$, and $\mathbf{w}$). If $\mathbf{u}$, $\mathbf{v}$, and $\mathbf{w}$ are linearly independent, then the three families originate all the plane of the three-dimensional space. A possible representation of all the planes of the three-dimensional space, under the equivalence relation of parallelism, is achieved by choosing a point of the space and considering the set of all the planes that contain the point. Being common to all the planes, the point can be used for representing the whole set of planes, which, in turn, represents all the planes of the three-dimensional space.

In conclusion, as for the inner orientation of the p-vectors, also the attitude of the p-vectors is defined inductively, starting from the 1-vector. This allows us to define the attitude of the p-vectors even in dimension greater than 3.

By comparison between Figs. 2.21 and 2.3, it follows that the same family of parallel planes represents both the set of planes that are normal to $\mathbf{u}$ and the set of hyperplanes of $\mathbf{u}^*$, the dual vector of $\mathbf{u}$. Consequently, the attitude of the class invariant of a vector $\mathbf{u}$ equals the attitude of the covector $\mathbf{u}^*$. This is ultimately a consequence of the Riesz representation theorem (Eq. (2.1.16)), which allows us to represent a covector by its related vector. Said in terms of equivalence relation of parallelism, the class invariant of a vector equals the equivalence class of its covector. This establishes a bijective correspondence between the attitude of the orthogonal complement of a vector $\mathbf{u}$ and the attitude of the covector, $\mathbf{u}^*$, of $\mathbf{u}$.

Due to the Riesz representation theorem, the bijective correspondence extends also to the second feature, that is, the orientations of a vector and its covector, since the order of the hyperplanes is determined by the sense of $\mathbf{u}$ (Fig. 2.3). This allows us to define a second type of orientation for the covector $\mathbf{u}^*$, which we call the **outer orientation** since it is induced by the (inner) orientation of $\mathbf{u}$ and has the geometrical meaning of sense of traversal of the hyperplanes of $\mathbf{u}^*$. In doing so, we have established a bijective correspondence between the inner orientation of a vector and the outer orientation of its covector. On the other hand, since it is always possible to define an inner orientation for $\mathbf{u}^*$ (by choosing a basis bivector for $\mathbf{u}^*$), the duality between vectors and covectors will result in an outer orientation for $\mathbf{u}$, induced by the inner orientation of $\mathbf{u}^*$. Therefore, the inner orientation of a covector induces an outer orientation on its vector.

Moreover, since the equivalence classes of $\mathbf{u}^*$ are in bijective correspondence with the attitude of $\mathbf{u}$, to fix the inner orientation of $\mathbf{u}^*$ is also equivalent to fixing an orientation, which is an inner orientation, for the attitude of $\mathbf{u}$. In doing so, the attitude of $\mathbf{u}$ becomes an attitude vector,[78] and its inner orientation equals the outer orientation of $\mathbf{u}$. Therefore, by providing the attitude with an orientation, we establish an isomorphism between the orthogonal complement and the dual vector space of any subset of vectors. There is a natural analog of this relationship in general Banach spaces.[79]

---

[78] Leonhard Euler (15 April 1707–18 September 1783) introduced a vectorial way to describe any rotation, with a vector on the rotation axis and module equal to the value of the angle. Therefore any orientation can be represented by a rotation vector (also called the Euler vector) that leads to it from the reference frame. When used to represent an orientation, the rotation vector is commonly called the orientation vector, or **attitude vector**.

[79] A **Banach space**, named after the Polish mathematician Stefan Banach (March 30, 1892–August 31, 1945), is a vector space X over the field $\mathbb{R}$ of real numbers, or over the field $\mathbb{C}$ of complex numbers, which is equipped with a norm $\| \bullet \|$ and which is complete with respect to that norm, that is to say, for every Cauchy sequence:

$$\{x_n\}_{n=1}^{\infty} ;$$

(continues next page)

In this case one defines $W^\perp$, the orthogonal complement of W, to be a subspace, of the dual of V, defined similarly as the annihilator:[80]

$$W^\perp = \left\{ x \in V^* : \forall y \in W, x(y) = 0 \right\} ; \tag{2.2.34}$$

where W is a closed linear subspace in V.

This means that the pairing between the geometric algebra and its dual can be described by the invariants of the equivalence relation of parallelism.[76]

In conclusion, we can define the orientation of a vector by providing either its inner orientation or the inner orientation of its attitude vector (which is also the outer orientation of the vector). The latter, in turn, is equal to the inner orientation of the covector.

It remains to be determined how to relate the inner orientation of a vector with its outer orientation. In other words, it remains to be put in bijective correspondence the inner orientation of **u** with the inner orientation of the hyperplanes of **u*** or, which is the same, with the inner orientation of the attitude vector of **u**. We can do this in two ways, by establishing two different criteria. The starting point is, once again, the equivalence class of **u** by the equivalence relation of parallelism. We have said that each equivalence class is a family of (unoriented) parallel straight lines. Therefore, we can represent an equivalence class by ideally holding together all the straight lines, or a representative set of straight lines, of the class. In other words, holding

---

(continues from previous page)

in X, there exists an element x in X such that:

$$\lim_{n \to \infty} x_n = x, \text{ that is, } \lim_{n \to \infty} \left\| x_n - x \right\|_X = 0.$$

Completeness of a normed space is preserved if the given norm is replaced by an equivalent one. All norms on a finite-dimensional vector space are equivalent. Every finite-dimensional normed space is a Banach space.

[80] Let S be a subset of V. The **annihilator** of S in V^*, denoted here by S^O, is the collection of linear functionals $f \in V^*$ such that:

$$[f, s] = \langle f, s \rangle = 0 ;$$

for all $s \in S$. That is, S^O consists of all linear functionals $f : V \to F$ such that the restriction to S vanishes:

$$f|_S = 0.$$

The annihilator of a subset is itself a vector space. In particular:

$$\varnothing^O = V^*$$

is all of V^* (vacuously), whereas

$$V^O = 0$$

is the zero subspace.

If W is a subspace of V then the quotient space[76] V/W is a vector space in its own right, and so has a dual. By the first isomorphism theorem,[44] a functional $f : V \to F$ factors through V/W if and only if W is in the kernel of f. There is thus an isomorphism:

$$(V/W)^* \cong W^O.$$

As a particular consequence, if V is a direct sum of two subspaces A and B, then V^* is a direct sum of A^O and B^O.

Figure 2.22. Relationship between the right-hand grip rule and the two orientations of a surface: inner orientation (on the plane on which the right fist is) and outer orientation (along the direction of the thumb). The figure also provides the two orientations of a vector: inner orientation (along the direction of the thumb) and outer orientation (on the plane on which the right fist is).

together a family of lines is equivalent to partitioning the straight lines of the three-dimensional space in subsets of lines that are parallel to a given vector.

Since a curved arrow is often used, both in mathematics and in physics, for indicating the sense of fingers curling when closing a hand around something, we can use a curved arrow for denoting the closed hand that holds the straight lines, thus considering the entire family (the equivalence class) of straight lines. This means that we can use a curved arrow for denoting the attitude vector of a vector. In particular, we can establish the desired duality relation by extending the thumb in the sense of the vector and using the sense of closure of the hand for providing the attitude vector with an inner orientation. Since we have two hands, the use of the right hand rather than the left hand gives rise to a criterion, the right-hand grip rule,[81] rather than another criterion, the left-hand grip rule. It is usual to use right hands in right-handed coordinate systems (Fig. 2.22) and left hands in left-handed coordinate systems.

This bijective correspondence allows us to represent the orientation of $\mathbf{u}$ as the inner orientation of $\mathbf{u}^*$, rather than the inner orientation of $\mathbf{u}$ (Fig. 2.23). As a consequence, the orientation of $\mathbf{u}$ can be described either in terms of inner or outer orientations, being the two descriptions absolutely interchangeable.

As discussed in Section 2.2.2.1, in the three-dimensional space a point is orthogonal to any trivector and a trivector is orthogonal to any point. This allows us to extend the bijective correspondence between orthogonal complements and dual vectors to the relationship between points and volumes (including the relationship between their inner and outer orientations). As a consequence, the outer orientation of a point is defined by the inner orientation of a surrounding volume of arbitrary shape. This, in turn, is determined by the inner orientations of the planes

---

[81] The **right-hand grip rule** is a different form of the right-hand rule.[84] If the coordinates are right-handed and you place your right fist on the plane, then your fingers will curl from the first axis to the second axis and your thumb will point along the third axis. The rule is used either when a vector must be defined to represent the rotation of a body, a magnetic field or a fluid, or, vice versa, when it is necessary to decode the rotation vector, to understand how the corresponding rotation occurs. The first form of the rule is used to determine the direction of the exterior product of two vectors. In the special case where the rule is used to determine the direction of the torque vector, if you grip the imaginary axis of rotation of the rotational force so that your fingers point in the direction of the force, then the extended thumb points in the direction of the torque vector.

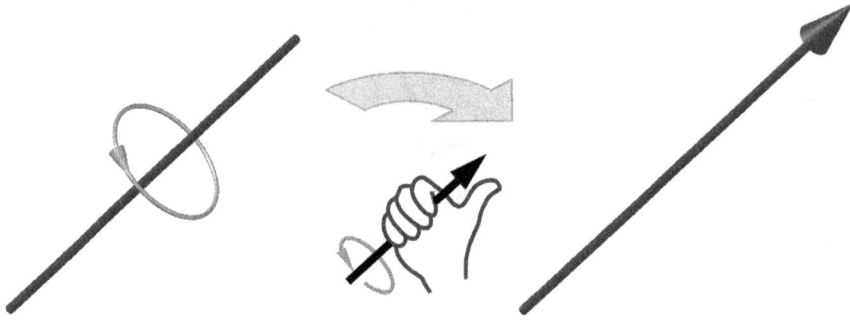

Figure 2.23 Equivalence between outer and inner orientations of a vector, in right-handed coordinates.

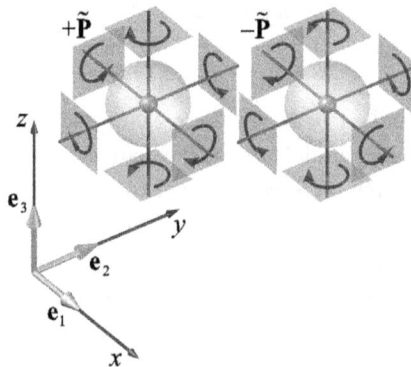

Figure 2.24. Relationship between outer orientation of a point and inner orientation of a volume.

that are tangent to the volume surface (Fig. 2.17), that is, by the outer orientations of the normal straight lines that cross the volume surface. Since it is always possible to choose a spherical volume around the point, the outer orientation of the point is ultimately given by the half-lines that have the given point in common (Fig. 2.24).

In conclusion, the correspondence between space elements provided with outer orientation and their orthogonal complements,[77] which turn out to be provided with inner orientation, leads us to define a relation of duality between space elements of different dimensions. In the three-dimensional space, the **dual space element**, or just **dual**[82] for short, of a space element of dimension p is the normal space element of dimension $3 - p$. That is, the sum of the dimensions of a space element and its dual element always equals three, the dimension of the space they are in. In other words, if we regard any space element equipped with an orientation as the result of an exterior product on vectors, we can define the dual of the p-dimensional space element as the orthogonal complement[77] of the p-dimensional space element.

In particular, the dual element of a point is a volume, the dual element of a straight line is a surface, the dual element of a surface is a straight line, and the dual element of a volume is a point. The outer orientation of a space element is given by the inner orientation of its dual space element (Fig. 2.25).

---

[82] With this notation, the term "dual" is used as a substantive, not as an adjective.

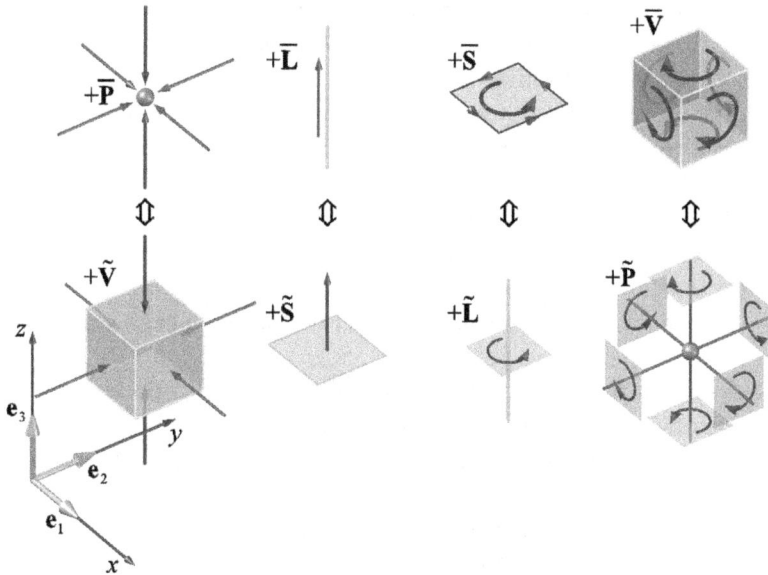

Figure 2.25. Relationship between the inner orientation of a p-space element and its dual element, of dimension $p-1$.

Note that the relationship between space elements and their duals extends to the signs of inner and outer orientations. Consequently, a positive or negative inner orientation of a p-space element induces a positive or negative outer orientation in its dual element, respectively. In particular, since the point with a positive inner orientation is a sink, a volume has a positive outer orientation when the normal lines enter the volume, as in the old scientific literature.[83] For accuracy, according to what observed on the relationship between the sign of the orientation and the observer, the positive outer orientation of a volume is indicated by inward normals and the point with a positive inner orientation is a sink whenever the observer is external, while the positive outer orientation of a volume is indicated by outward normals and the point with a positive inner orientation is a source whenever the observer is a local observer, ideally positioned inside the volume. Therefore, the signs of the outer orientations of points and volumes depend on the kind of description, material[26] rather than spatial.[24] Since different physical disciplines use different descriptions, this explains why the physical disciplines use different notations for the positive orientation of volumes.[83]

Note also that the choice of the inward orientation as the positive outer orientation of the volume is compatible with the inner orientation of its faces. In fact, the positive inner

---

[83] In James Clerk Maxwell's time (13 June 1831–5 November 1879), inward arrows indicated the positive orientation of a volume. This led Maxwell to introduce the notion of convergence, changed to divergence when the outward normals were taken as positive. The inward convention is still used in hydraulics and geotechnics, despite the outward convention is the most used today.

orientation of the faces of the volumes is clockwise for an external observer, thus fixing in the inward direction the positive direction for crossing the faces themselves.

Finally, a surface has a positive outer orientation when the line crossing it has a positive inner orientation, represented by a vector that points toward the external observer. A line has a positive outer orientation when the normal plane has a positive inner orientation for the observer that is watching the plane.

A point, a line, a surface, and a volume endowed with outer orientations will be denoted by putting tildes over their symbols (Figs. 2.24–2.28).

The outer orientations of a curve, a (not necessarily plane) surface and a volume are given by the outer orientations of the p-vectors of greater dimension in their tangent spaces (Figs. 2.26–2.28).

Scalar magnetic potential and stream function in fluid dynamics are examples of physical variables associated with points with outer orientations.

The outer orientation of a line is useful for describing, for example, the rotatory polarization of a light beam.

The heat that crosses a surface, the surface forces in continuum mechanics, the mass current, the energy current, and the charge current are examples of surface variables whose sign

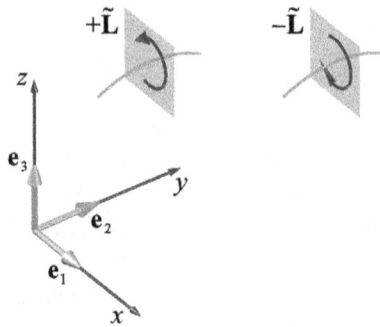

Figure 2.26. Positive and negative outer orientations of a curve.

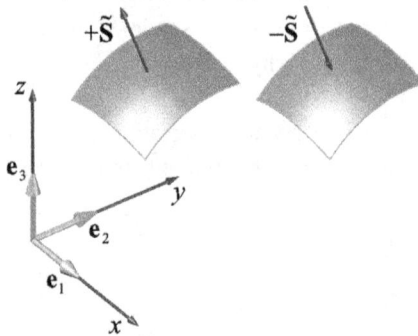

Figure 2.27. Positive and negative outer orientations of a surface.

depends on the crossing direction of the surface, that is, on the outer orientation of the surface to which they are related.

The notion of divergence is associated with the idea that outward normals indicate a positive volume[83] (observer inside the volume).

Being an orthogonal complement, the dual of a p-dimensional space element has dimension $n - p$, in the n-dimensional space. This means that the outer orientation depends on the dimension of the embedding space, while the inner orientation does not. As an example, the outer orientation of a line in the three-dimensional space is the orientation of a curved arrow on a normal plane (Fig. 2.29a), while it is the direction of a crossing arrow in the two-dimensional space (Fig. 2.29b), and two outgoing or incoming arrows, denoting expansion or contraction of the line, respectively, when it is embedded in the one-dimensional space (Fig. 2.29c). In Fig. 2.30, we have collected the positive outer orientations of the space elements in three-dimensional, two-dimensional, and one-dimensional spaces.

The dependence of the outer orientation on the dimension of the embedding space also allows us to define the class invariant by the equivalence relation of parallelism between p-vectors in the n-dimensional space.

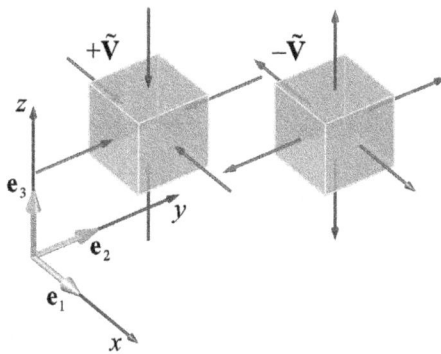

Figure 2.28. Positive and negative outer orientations of a volume.

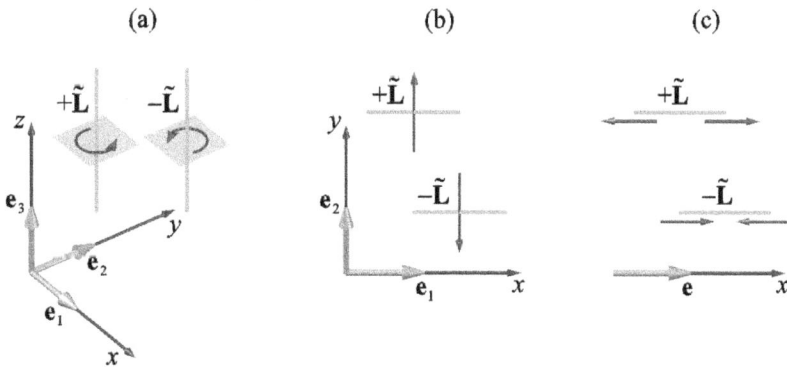

Figure 2.29. The two outer orientations (positive and negative orientations) of a line in three-dimensional, two-dimensional, and one-dimensional spaces.

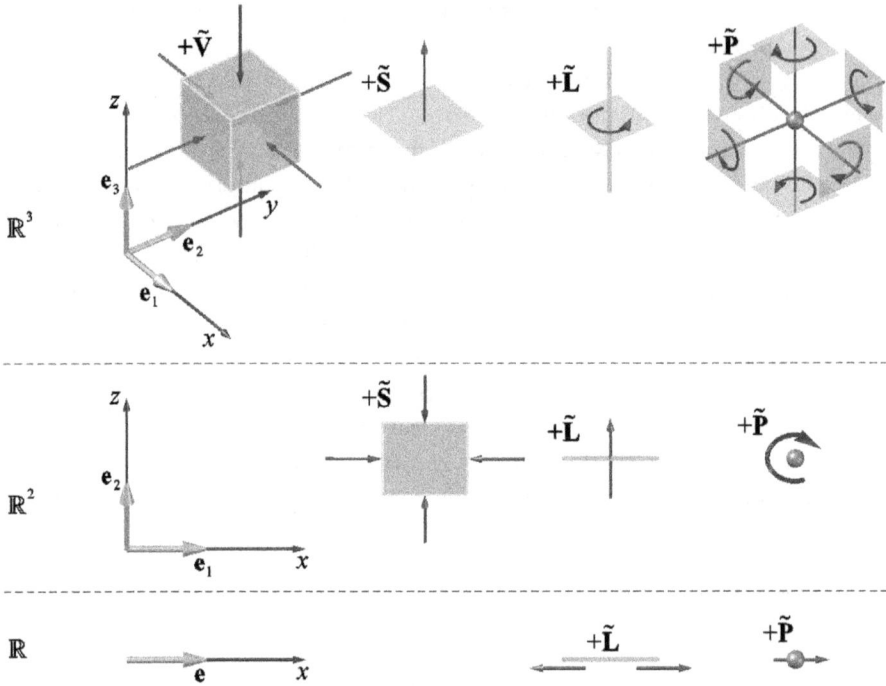

Figure 2.30. How the positive outer orientation depends on the dimension n of the embedding space.

In conclusion, the notions of inner and outer orientations are implicit in geometric algebra, despite they are not formalized there.

Even the relationship between the inner and outer orientations and the related notion of orthogonal complement (or dual element) are implicit, both in mathematics and physics. They are given by the right-hand rule,[84] which is equivalent to the right-hand grip rule[81] and the right-handed screw rule.[85] As far as the right-hand grip rule is concerned, in fact, we have already seen that the sense of fingers curling gives both the inner orientation of the plane on which the right fist is and the outer orientation of its dual, that is, the line that is directed as the thumb (Fig. 2.22). Analogously, the direction of the thumb gives both the inner orientation of the line that is directed as the thumb and the outer orientation of its dual, that is, the plane on which the right fist is. Thus, broadly speaking, the thumb (with its sense) and the fist (with the curling sense of its fingers) are each the orthogonal complement of the other and their orientations stand in relation of duality.

---

[84] The **right-hand rule** was invented for use in electromagnetism by British physicist John Ambrose Fleming (29 November 1849–18 April 1945) in the late 19th century. When applied to the cross product $\mathbf{a} \times \mathbf{b}$, with the thumb, index, and middle fingers at right angles to each other (with the index finger pointed straight), the middle finger points in the direction of the cross product when the thumb represents $\mathbf{a}$ and the index finger represents $\mathbf{b}$.

[85] According to the **right-handed screw rule**, also called the **corkscrew-rule**, curl the fingers of your right hand around a right-handed screw and hold it with your thumb pointing down. Turn the screw clockwise, as viewed looking down in the direction of your extended thumb. This clockwise rotation will cause a right-handed screw to move in the downward direction. A right-handed screw (or a corkscrew) that is turned in the sense of the rotation that would align the first vector, $\mathbf{a}$, with the second vector, $\mathbf{b}$, moves in the direction of the cross product $\mathbf{a} \times \mathbf{b}$.

Moreover, the idea that the line in the direction of the thumb is one arbitrary element of the set of lines that are normal to the plane of the fist is implicit in the definition of cross product, which is a free vector.[73] This makes the position of the cross product indeterminate, allowing us to draw the vector everywhere in the space. Analogously, the four fingers of the fist are associated with the four parallel planes, that is, with the common attitude of the four planes.

Since a rotation of a right-handed screw in the curling sense of the right hand fingers will cause the right-handed screw to move in the direction of the right thumb, we can identify the right fist with a right-handed screw. This allows us to extend the relationships established above to the right-handed screw rule. This second time, however, we can also establish relations of dual orientation between the volume and its orthogonal complement, the point. In fact, in a right-handed coordinate system the three pairs of opposite faces of a positive cube rotate as three right-handed screws (Fig. 2.31), which are screwed on the faces of the cube.[72] Therefore, the clockwise rotation of the right-handed screw on the cube faces gives the positive inner orientation of a volume in a right-handed coordinate system. On the other hand, the clockwise rotations of the three right-handed screws define the outer orientations of the normals to the faces of the cube, that is, the positive outer orientation of the point that is the dual of the volume. Thus, the rotation of the right-handed screw gives both the inner orientation of a volume and the outer orientation of its dual, that is, the point.

Finally, the motion directions of the right-handed screws define the outer orientation of the volume originated by the trivector $\mathbf{u} \wedge \mathbf{v} \wedge \mathbf{w}$, where $\mathbf{u}$, $\mathbf{v}$, and $\mathbf{w}$ are directed as the three right-handed coordinate axes. This latter time, in fact, the clockwise rotation of the right-handed screws on the cube faces will cause the screws to move inward, thus providing also the positive inner orientation of the point that is the dual of the volume. Therefore, the motion direction of the right-handed screw gives both the inner orientation of a point and the outer orientation of its dual, that is, the volume. This means that the right-handed screw rule implicitly establishes also a relation of duality between the rotation sense and the motion direction of a right-handed screw. Thus, broadly speaking, rotation sense and motion direction are each the orthogonal complement of the other.

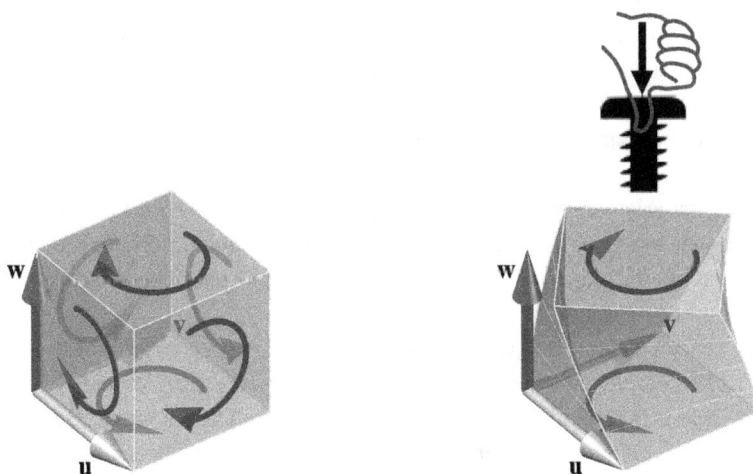

Figure 2.31. Relationship between the inner orientation of a cube and the orientation of a screw.

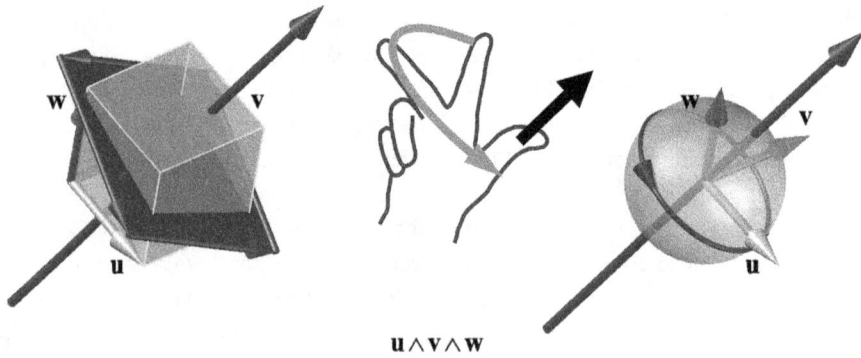

$$\mathbf{u} \wedge \mathbf{v} \wedge \mathbf{w}$$

Figure 2.32. The geometric interpretation of the exterior product of three vectors **u**, **v**, **w** as an oriented volume compared with the right-hand rule. The actual shape is irrelevant.

As far as the third rule is concerned, the right-hand rule, the finger disposition in the right hand naturally provides both the inner orientation of the point that is the origin of a right-handed coordinate system, with axes in the directions of the fingers, and the outer orientation of the volume that surrounds the hand.

At the same time, by rotating one finger at a time toward the other, we obtain both the (equatorial) inner orientation of the volume that surrounds the hand and the outer orientation of the "origin" of the fingers, given by the rotation sense around the third finger, which represents one of the half-lines outgoing the "origin". This is the same relationship provided by the spinning-top analogy of the exterior product $\mathbf{u} \wedge \mathbf{v} \wedge \mathbf{w}$ between three vectors **u**, **v**, and **w** directed as three right-handed axes, where the orientation on a solid ball representing the trivector can be thought of as giving the ball a counterclockwise spin around any given axis (Fig. 2.32).

Finally, by rotating the first finger toward the second finger, the rotation sense gives both the inner orientation of the plane in which the two fingers lie and the outer orientation of the line in the direction of the third finger, while the third finger gives both the inner orientation of the line along it and the outer orientation of the plane in which first and second fingers lie.

In conclusion, we could say that the right-hand grip rule, the right-hand rule, and the right-handed screw rule (or the equivalent corkscrew-rule) establish the pairing (Section 2.1.2) between the orientations of the p-vectors and their duals, the p-covectors, for $p = 1, 2, 3$. This allows us to describe space elements in dimension $p = 1, 2, 3$ either in terms of inner or outer orientations.

Inner and outer orientations of space elements assume a great importance in the description of physics, because the signs of global physical variables (Section 1.2) depend on the orientation. This is why we formalize their definitions in this book, making their notions and their relationships explicit.

Once the space elements and their orientations have been viewed in these terms, the elements of the cell complexes[27] used in computational physics are elements of a topological vector space[40] and can be studied in the context of the geometric algebra. Being a p-vector, each cell has a dual cell and the set of dual cells can be viewed as the set of multilinear forms (Section 2.1.2) on the topological vector space. In particular, if X is a normed space equipped with the weak topology (Section 2.1.2), then the dual space X^* is itself a normed vector space and its norm gives rise to a topology, called the strong topology, on X^*.[86] Consequently, the cell complexes are, more properly, CW cell complexes.[87] Moreover, since, different from the weak topology, the strong topology is not a coarsest topology (Section 2.1.2), having a strong topology on the dual space means that there is not a unique way for defining the dual space elements. Thus, different definitions for the dual space elements may lead them to overlap.

---

[86] This is the topology of **uniform convergence**.

[87] In a **CW cell complexes**, the "C" stands for "closure-finite" and the "W" for "weak topology", where **closure-finite** means that each closed cell is covered by a finite union of open cells. Specifically, in mathematics, a **cover** of a set X is a collection of sets whose union contains X as a subset. Formally, if

$$C = \left\{ U_\alpha : \alpha \in A \right\}$$

is an indexed family of sets U_α, then C is a cover of X if

$$X \subseteq \bigcup_{\alpha \in A} U_\alpha.$$

In topology, if the set X is a topological space,[92] then a cover C of X is a collection of subsets U_α of X whose union is the whole space X. In this case we say that "C covers X", or that "the sets U_α cover X".

CHAPTER 3

ALGEBRAIC TOPOLOGY AS A TOOL FOR TREATING GLOBAL VARIABLES WITH THE CM

After a brief introduction on the basic nomenclature of algebraic topology (Section 3.1), in Section 3.2 we discuss the topological notions of simplices and simplicial complexes, which form a special class of topological spaces.

The definitions of faces, facets, and cofaces of a k-polytope are given in Section 3.3. The same definitions apply to a k-simplex, with the particularity that the faces of a k-simplex are simplices themselves. These faces are called the m-faces of the k-simplex. It is also discussed how, given a k-simplex, the number of m-faces of that k-simplex is a figurate number. Consequently, this number is a binomial coefficient and can be found along the diagonals of Pascal's triangle. It is also equal to the numbers of basis $(m+1)$-blades needed in a $(k+1)$-dimensional space to represent arbitrary $(m+1)$-blades.

The graph theory, briefly exposed in Section 3.4, allows us to introduce the main tools of the incidence geometry, that is, the incidence numbers and the incidence matrices. These tools can be applied even to cell complexes, since an oriented cell complex is the generalization of an oriented graph to directed simplices of dimension greater to 1. One of the main consequences of this generalization is that we can extend to cell complexes the notion of dual graph of a plane graph, attaining a dual cell complex. In particular, since the dual graph depends on a particular embedding, the dual cell complex, as the dual graph, is not unique.

The boundary and the coboundary of a p-cell are defined in Section 3.5, and put in relationship with the notions of incidence relation and incidence matrix. In the special case of the three-dimensional space, the CM uses the incidence matrices for describing the incidence relations between the space elements and their coboundaries, and between the time elements and their coboundaries. We thus obtain three incidence matrices for the four space elements and one incidence matrix for the two time elements. The former three incidence matrices are denoted by $\mathbf{G}$, $\mathbf{C}$, and $\mathbf{D}$. Their incidence numbers are arranged in the matrices in way that $\mathbf{G}$, $\mathbf{C}$, and $\mathbf{D}$ turn out to be not simply the generalization of the incidence matrices of the graph theory, but their transposes. In this section, it is also shown how each cell of a plane cell complex can be viewed as a two-dimensional space, where the points of the cell, with their labeling and inner orientation, play the role of a basis scalar, the edges of the cell, with their labeling and inner orientation, play the role of basis vectors, and the cell itself, with its inner orientation, plays the role of basis bivector. This clarifies in which sense the cell complexes and their labeling are equivalent, in the algebraic setting, to the coordinate systems and their continuous maps in $\mathbb{R}^n$.

The chain complex and cochain complex are introduced in Section 3.6. These algebraic means are useful for the definition of the boundary operators, the boundary process, the coboundary operators, and the coboundary process. Coboundary operators and coboundary process, in particular, play an essential role in defining the mathematical structure of the CM governing equations, as will be shown in Chapter 5.

The discrete p-forms, treated in Section 3.7, are further essential tools in the algebraic formulation of the CM. More specifically, the discrete p-forms generalize the notion of field functions and are the algebraic version of the exterior differential forms. The discrete p-forms also allow us to find the relation between the boundary operator and the coboundary operator, leading to the algebraic form of the generalized theorem of Stokes.

Finally, in Section 3.8, we discuss the geometrical structure of a four-dimensional cell complex, suitable for a unified description of space and time global variables. The basic cell of this four-dimensional cell complex is the tesseract. Some p-cells of the CM space/time tesseract are associated with a variation of the space variables, some other p-cells are associated with a variation of the time variables, and some other p-cells are associated with a variation of both the space and time variables. This is only one of the many similarities between the CM space/time tesseract and the four-dimensional Minkowski spacetime.

3.1 Some Notions of Algebraic Topology

Topology[88] is the mathematical study of shapes and spaces. It is a major area of mathematics, concerned with the most basic properties of space, such as connectedness, continuity, and boundary. Topology studies the properties of geometric figures that are preserved under continuous deformations, including stretching and bending, but not tearing or overlapping (gluing). Since these kind of deformations are homeomorphisms,[89] that is, continuous, invertible, and with continuous inverses transformations, topology studies the properties of geometric figures that are invariant under homeomorphisms.

---

[88] The word topology comes from the Greek τόπος, "place", and λόγος, "study". Ideas that are now classified as topological were expressed as early as 1736, in Leonhard Euler's paper on the Seven Bridges of Königsberg.

[89] A **homeomorphism** (from the ancient Greek language: ὅμοιος (homoios) meaning "similar" and μορφή (morphē) meaning "shape", "form", not to be confused with homomorphism[115]) or **topological isomorphism** or **bicontinuous function** is a continuous function between topological spaces[92] that has a continuous inverse function. Homeomorphisms are the isomorphisms in the category of topological spaces – that is, they are the mappings that preserve all the topological properties of a given space. Two spaces with a homeomorphism between them are called **homeomorphic**, and from a topological viewpoint they are the same.

Topology has three subfields:

- **Algebraic topology**[90] uses tools from abstract algebra[91] to study topological spaces.[92]
- **Point-set topology** establishes the foundational aspects of topology and investigates concepts inherent to topological spaces.[92] Examples include compactness and connectedness.
- **Geometric topology** primarily studies manifolds[51] and their embeddings (placements) in other manifolds. A particularly active area is **low-dimensional topology**, which studies manifolds of four or fewer dimensions. This includes **knot theory**, the study of mathematical knots.

The basic goal of algebraic topology is that to measure degrees of connectivity, using algebraic constructs such as homology[93] and homotopy groups.[94] This is achieved by looking for invariants, which are used in order to classify the topological spaces.

An n-dimensional closed cell is the image of an n-dimensional closed ball, or closed n-ball, under an attaching map,[95] where

- an **image** is the subset of a function's codomain, which is the output of the function on a subset of its domain;

---

[90] An older name for algebraic topology was **combinatorial topology**. This older name dates from the time when topological invariants of spaces (for example, the Betti numbers) were regarded as derived from combinatorial decompositions of spaces, such as decomposition into simplicial complexes (see Section 3.2).

[91] In algebra, which is a broad division of mathematics, **abstract algebra** is a common name for the sub-area that studies algebraic structures in their own right.

[92] A **topological space** is a set of points, along with a set of neighborhoods for each point, that satisfy a set of axioms relating points and neighborhoods. The definition of a topological space relies only upon set theory. Manifolds[51] and metric spaces are specializations of topological spaces with extra structures or constraints.

[93] In algebraic topology, **homology** (in part from Greek ὁμός, "identical") is a certain general procedure to associate a sequence of abelian groups[113] or modules[114] with a given topological space.[92] It is a rigorous mathematical method for defining and categorizing holes in a shape.

[94] **Homotopy groups** are used in algebraic topology to classify topological spaces.[92] The first and simplest homotopy group is the **fundamental group**, which records the information about loops in a space.

[95] Suppose that X is a topological space[92] and A is a subspace of X. One can identify all points in A to a single equivalence class[76] and leave points outside of A equivalent only to themselves. The resulting quotient space is a topological space, denoted by X/A. This quotient space is the adjunct space of the two classes, via the map that identifies the points of the boundary of A with themselves. In mathematics, an **adjunction space** (or **attaching space**) is a common construction in topology where one topological space is attached or "glued" onto another. Specifically, let C_1 and C_2 be topological spaces with S a subspace of C_2. Let

$$f : S \to C_1$$

be a continuous map (called the **attaching map**). One forms the adjunction space $C_1 \cup_f C_2$ by taking the disjoint union of C_1 and C_2 and identifying x with $f(x)$ for all x in S. Schematically, $C_1 \cup_f C_2$ is the quotient space:

$$C_1 \cup_f C_2 = (C_1 \amalg C_2)/\{f(S) \sim S\}.$$

A common example of an adjunction space is given when C_2 is a closed n-ball (or cell) and S is the boundary of the ball, the $(n-1)$-sphere. Inductively attaching cells along their spherical boundaries to this space results in an example of a CW complex.[87] If S is a space with one point, then the adjunction is the **wedge sum** of C_1 and C_2, that is, a "one-point union" of C_1 and C_2.

- in three-dimensional Euclidean space, a **ball** is the space inside a sphere. It may be a **closed ball** (including the boundary points) or an **open ball** (excluding them). The definition extends to lower and higher-dimensional Euclidean spaces and to metric spaces in general. In particular, the ball is a bounded interval for $n = 1$ and the interior of a circle (a **disk**) for $n = 2$.

Let (M, d) be a metric space, namely a set M with a metric (distance function) d. The **open (metric) ball of radius $r > 0$ centered at a point** p in M, usually denoted by $B_r(p)$ or $B(p, r)$, is defined by[9]

$$B_r(p) \triangleq \{x \in M : d(x, p) < r\} \quad \forall r > 0. \tag{3.1.1}$$

The **closed (metric) ball of radius $r > 0$ centered at a point** p in M, which may be denoted by $B_r[p]$ or $B[p, r]$, is defined by

$$B_r[p] \triangleq \{x \in M : d(x, p) \leq r\} \quad \forall r > 0. \tag{3.1.2}$$

Since the definition of a ball (open or closed) requires $r > 0$, a ball centered at a point p always includes p itself.

In Cartesian space $\mathbb{R}^n$ with the p-norm L_p, an open ball is the set:

$$B(r) = \left\{ x \in \mathbb{R}^n : \sum_{i=1}^{n} |x_i|^p < r^p \right\} \quad \forall p \geq 1; \tag{3.1.3}$$

while a closed ball is the set:

$$B[r] = \left\{ x \in \mathbb{R}^n : \sum_{i=1}^{n} |x_i|^p \leq r^p \right\} \quad \forall p \geq 1. \tag{3.1.4}$$

In a topological space,[92] X, balls are not necessarily induced by a metric. An (open or closed) **n-dimensional topological ball** of X is any subset of X which is homeomorphic[89] to an (open or closed) Euclidean n-ball.

- Let X and Y be topological spaces[92] with A a subspace of Y. The **attaching map**[95] is any continuous map, f, from A to X:

$$f : A \to X. \tag{3.1.5}$$

In algebraic topology, it is usual to introduce cell complexes,[27] mainly in the restricted form of simplicial complexes (Section 3.2), and to consider the vertices, edges, surfaces, and volumes of a cell complex as p-cells, that is, cells of different dimensions.

The **n-skeleton** (pl. n-skeleta) of a cell complex is the union of the cells whose dimension is at most n. If the union of a set of cells is closed, then this union is itself a cell complex, called a **subcomplex**. Thus the n-skeleton is the largest subcomplex of dimension n or less.

A cell complex is often constructed by defining its skeleta inductively. Begin by taking the 0-skeleton to be a discrete space. Next, attach 1-cells to the 0-skeleton. Here, the 1-cells are attached to points of the 0-skeleton via some continuous map from unit 0-sphere, that is, S_0. Define the 1-skeleton to be the identification space obtained from the union of the 0-skeleton, 1-cells, and the identification of points of boundary (Section 3.5) of 1-cells by assigning an identification mapping from the boundary of the 1-cells into the 1-cells. In general, given the $(n-1)$-skeleton

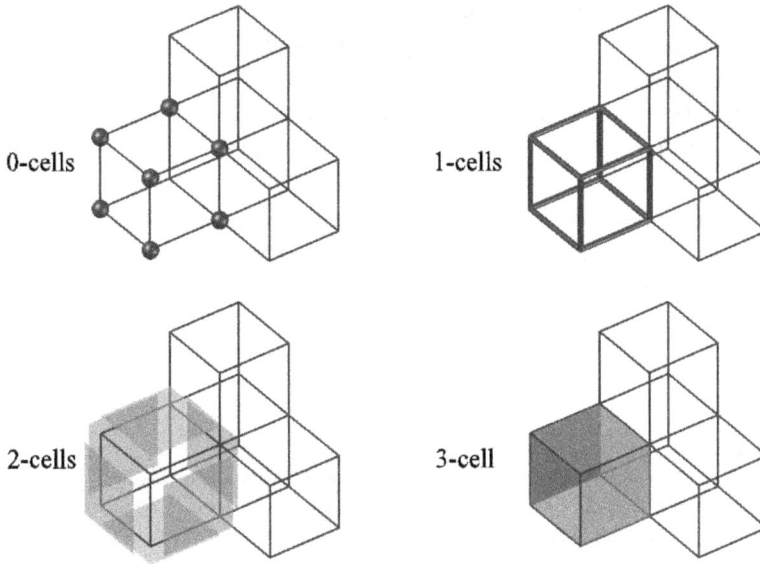

Figure 3.1. The four space elements in algebraic topology.

and a collection of (abstract) closed n-cells, as above, the n-cells are attached to the $(n-1)$-skeleton by some continuous mapping from S_{n-1}, and making an identification (equivalence relation) by specifying maps from the boundary of each n-cell into the $(n-1)$-skeleton. The n-skeleton is the identification space obtained from the union of the $(n-1)$-skeleton and the closed n-cells by identifying each point in the boundary of an n-cell with its image.

Following the terminology of algebraic topology, we will denote the vertices of a cell complex as 0-cells, the edges as 1-cells, the surfaces as 2-cells, and the volumes as 3-cells (Fig. 3.1).

Note the difference between the terminologies of algebraic topology and geometry, where the points are 1-cells, the edges are 2-cells, the surfaces are 3-cells, and the three-dimensional volumes are 4-cells.

3.2 Simplices and Simplicial Complexes

A **simplex** (plural **simplexes** or **simplices**) is the smallest convex set[96] containing the given vertices.

---

[96] In Euclidean space, an object is **convex** if, for every pair of points within the object, every point on the straight line segment that joins them is also within the object.

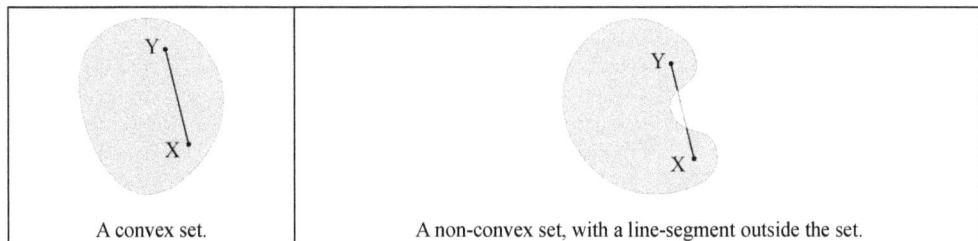

| A convex set. | A non-convex set, with a line-segment outside the set. |

The formal definition of simplex makes use of the notion of **polytope** (Fig. 3.2), a geometric object with flat sides that exists in any general number of dimensions.[97] Specifically, a *k*-**simplex** is a *k*-dimensional polytope that is the convex hull[98] of its $k + 1$ vertices.

A 2-simplex is a triangle, a 3-simplex is a tetrahedron, a 4-simplex is a 4-cell,[99] and so on. A single point may be considered a 0-simplex, and a line segment may be considered a 1-simplex.

In algebraic topology, simplices are used as building blocks to construct an interesting class of topological spaces,[92] called **simplicial complexes**. These spaces are built from simplices glued together[95] in a combinatorial fashion.

---

[97] A polygon is a polytope in two dimensions, or 2-polytope, a polyhedron is a polytope in three dimensions, or 3-polytope, a polychoron is a polytope in four dimensions, or 4-polytope, and so on in higher dimensions. The original definition of polytope, broadly followed by Ludwig Schläfli (15 January 1814–20 March 1895), Thorold Gosset (1869–1962), and others, begins with the zero-dimensional point as a 0-polytope (vertex). A one-dimensional polytope, the 1-polytope (edge), is constructed by bounding a line segment with two 0-polytopes. Then, 2-polytopes (polygons) are defined as plane objects whose bounding facets (Section 3.3) are 1-polytopes, 3-polytopes (polyhedra) are defined as solids whose facets are 2-polytopes, and so forth.

[98] The **convex hull**, or **convex envelope**, of a set X of points in the Euclidean plane or Euclidean space may be defined as
1. the (unique) minimal convex set[96] containing X,
2. the intersection of all convex sets containing X,
3. the set of all convex combinations of points in X, or
4. the union of all simplices with vertices in X.

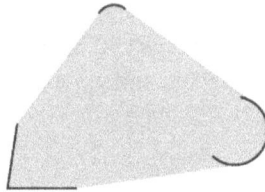

The convex hull of the set in dark gray is the convex in light gray.

[99] The **4-cell** in algebraic topology, 5-cell in geometry, is a four-dimensional object bounded by five tetrahedral cells. It is also known as the **pentachoron**, or **pentatope**, or **tetrahedral hyperpyramid**, or **hypercell**. It is a 4-simplex, the simplest possible convex regular 4-polytope (four-dimensional analogue of a polyhedron) and is analogous to the tetrahedron in three dimensions and the triangle in two dimensions. Its maximal intersection with three-dimensional space is the triangular prism and its vertex figure is a tetrahedron, where the **vertex figure** is, broadly speaking, the figure exposed when a corner of a polyhedron or polytope is sliced off. The pentachoron can be constructed from a tetrahedron, by adding a 5th vertex such that it is equidistant from all the other vertices of the tetrahedron (it is essentially a four-dimensional pyramid with a tetrahedral base).

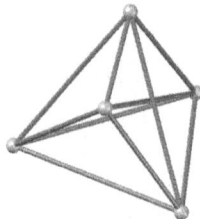

The pentachoron

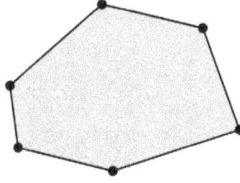

Figure 3.2. A two-dimensional polytope.

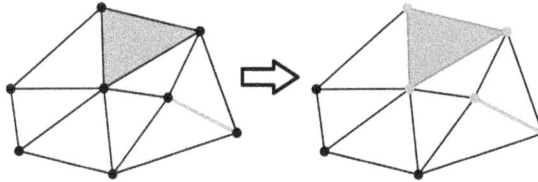

Figure 3.3. Two simplices (left) and their closure (right).

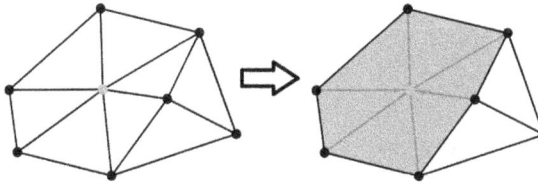

Figure 3.4. A simplex (left) and its star (right).

A simplicial complex, K, satisfies the following two conditions:

1. Any face (see Section 3.3) of a simplex from K is also in K.
2. The intersection of any two simplices $\sigma_1, \sigma_2 \in K$ is a face of both σ_1 and σ_2.

A **simplicial k-complex** is a simplicial complex, where the largest dimension of any simplex equals k.

Let K be a simplicial complex and let S be a collection of simplices in K.

The **closure** of S (denoted by $\text{Cl}\,S$) is the smallest simplicial subcomplex of K that contains each simplex in S (Fig. 3.3). $\text{Cl}\,S$ is obtained by repeatedly adding to S each face (Section 3.3) of every simplex in S.

The **star** of S (denoted by $\text{St}\,S$) is the set of all simplices in K that have any faces in S (Fig. 3.4). The star is generally not a simplicial complex itself.

The **link** of S (denoted by $\text{Lk}\,S$) is the closed star of S minus the stars of all faces of S (Fig. 3.5):

$$\text{Lk}\,S = \text{Cl}\,\text{St}\,S - \text{St}\,\text{Cl}\,S \qquad (3.2.1)$$

The space distribution of the point-wise field variables in the differential formulation, when discretized, gives rise to a discrete distribution of points that can be viewed as the 0-simplices of a simplicial cell-complex. On the contrary, the CM makes use of 3-simplices, or 4-simplices when also time is involved, allowing us to associate global variables with all the dimensions of the 4-cell complex. Using 4-cell complexes instead of 1-cell complex is the topological equivalent of avoiding to apply the Cancelation Rule for limits[14] and is the main reason why the CM is able to take into account the length scales in computational Physics – until the third dimension – while the differential formulation does not.

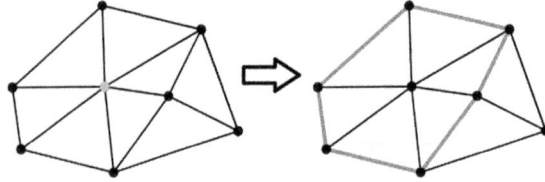

Figure 3.5. A simplex (left) and its link (right).

3.3   Faces and Cofaces

In higher-dimensional geometry, the convex hull[98] of any nonempty subset of the points that define a k-dimensional polytope, or k-polytope, is called a **face** of the k-polytope. Moreover, a polytope A is a **coface** of a polytope B if B is a face of A.

The set of faces of a polytope includes the polytope itself and the empty set, which for consistency may be defined as having dimension -1.

The $(k-1)$-faces of a k-dimensional polytope are called the **facets of the k-polytope**. For example:

- The facets of a line segment are its 0-faces, or vertices.
- The facets of a polygon are its 1-faces, or edges.
- The facets of a polyhedron, or plane tiling, are its 2-faces.
- The facets of a 4-polytope, or 3-honeycomb, are its 3-faces.
- The facets of a 5-polytope, or 4-honeycomb, are its 4-faces.

The term "facet" is also used in reference to simplicial complexes. In this second case, the **facet of a simplicial complex** is any simplex in the complex that is not a face of any larger simplex.

A $(k-2)$-face of a k-dimensional polytope is called a **ridge**. A $(k-3)$-face of a k-dimensional polytope is called a **peak**. The same definitions of ridge and peak apply to polytopes that are also simplices.

When the polytopes are simplices, the faces are simplices themselves. In particular, the convex hull of a subset of size $m+1$ (of the $k+1$ defining points) is an m-simplex, called an m-**face** of the k-simplex.

Excluding from the list of the m-faces the empty set, the total number of faces of a k-simplex, $F(k)$, is always a power of two minus one:

$$F(k) = 2^{k+1} - 1. \tag{3.3.1}$$

In this case, the number of m-faces of a k-simplex is a figurate number.[100] In particular:

---

[100] A **figurate number** is the number of stacking spheres forming some regular geometrical figure on the plane (**polygonal numbers**, or m-gonal numbers), or in the space (**polyhedral numbers**, or m-hedral numbers). Thus, a figurate number represents a regular, discrete geometric pattern. The mathematical study of figurate numbers is said to have originated with Pythagoras of Samos (c. 570 BC–c. 495 BC), possibly based on Babylonian or Egyptian precursors. The modern study of figurate numbers goes back to Pierre de Fermat (17 August 1601 or 1607–12 January

(continues next page)

- The number of 0-faces (vertices) of the k-simplex is the $(k+1)$-th linear number, $P_1(k+1)$.[101]
- The number of 1-faces (edges) of the k-simplex is the k-th triangle number, $P_2(k)$.[102]

(*continues from previous page*)

1665). Later, it became a significant topic for Leonhard Euler (15 April 1707–18 September 1783), who gave an explicit formula for all triangular numbers that are also perfect squares.

[101] A **linear number** is a figurate number that represents a line segment. It counts the aligned spheres in a row of spheres.

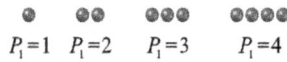

$$P_1=1 \quad P_1=2 \quad P_1=3 \quad P_1=4$$

The first four linear numbers.

The n-th linear number is the sum of the first n elements of the sequence of units:

$$P_0 = 1, 1, 1, 1, 1, 1, 1, 1, 1, 1, 1, 1, 1, 1, 1, 1, 1, \ldots$$

Note that the sequence P_0 is also the sequence of the elements in the left diagonal of Pascal's triangle,[71] the diagonal 0 (Fig. 2.13).

Any positive integer, n, can be viewed as the n-th linear number. Therefore, the sequence of the linear numbers is equal to the sequence of the positive integers:

$$P_1 = 1, 2, 3, 4, 5, 6, 7, 8, 9, 10, 11, 12, 13, 14, 15, 16, 17, 18, \ldots$$

The n-th linear number, $P_1(n)$, is given by the binomial coefficient[70] $\binom{n}{1}$:

$$P_1(n) = \sum_{k=1}^{n} P_0(k) = n = \frac{n}{1} = \binom{n}{1};$$

which represents the number of distinct objects that can be selected from a set of n objects. Being binomial coefficients, the linear numbers are contained in Pascal's triangle. They are collected along the first diagonal from left (entry 1, second position in each row), arranged in order from top to bottom (Fig. 2.13). Due to the symmetry of the elements in Pascal's triangle, the linear numbers also form the first diagonal from right.

[102] A **triangle number**, or **triangular number**, is a figurate number that represents an equilateral triangle. It counts the juxtaposed spheres that are arranged on a plane in the shape of an equilateral triangle. The n-th triangle number is the number of spheres composing a triangle with n spheres on a side and is equal to the sum of the first n linear numbers.[101]

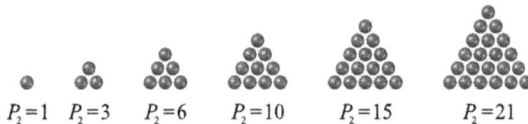

$$P_2=1 \quad P_2=3 \quad P_2=6 \quad P_2=10 \quad P_2=15 \quad P_2=21$$

The first six triangular numbers.

The sequence of the first few triangular numbers is

$$P_2 = 1, 3, 6, 10, 15, 21, 28, 36, 45, 55, 66, 78, 91, 105, 120, \ldots$$

The n-th triangle number, $P_2(n)$, is given by the following explicit formulas:

$$P_2(n) = \sum_{k=1}^{n} P_1(k) = \sum_{k=1}^{n} k = \frac{n(n+1)}{2} = \binom{n+1}{2};$$

(*continues next page*)

- The number of 2-faces (surfaces) of the k-simplex is the $(k-1)$-th tetrahedron number, $P_3(k-1)$.[103]
- The number of 3-faces (volumes) of the k-simplex is the $(k-2)$-th pentatope number, $P_4(k-2)$.[104]
- The number of m-faces of the k-simplex is the $(k-m+1)$-th $(m+1)$-simplex number, $P_{m+1}(k-m+1)$.[105]

(continues from previous page)

where the binomial coefficient[70] $\begin{pmatrix} n+1 \\ 2 \end{pmatrix}$ represents the number of distinct pairs that can be selected from $n+1$ objects,

and it is read aloud as "n plus one choose two". Being binomial coefficients, the triangle numbers are contained in Pascal's triangle.[71] They are in third position (entry two, second diagonal from left), arranged in order from top to bottom (Fig. 2.13). A second set of ordered triangle numbers forms the second diagonal from right.

[103] A **tetrahedron number**, or **tetrahedral number**, or **triangular pyramidal number**, is a figurate number that represents a tetrahedron. It counts the stacking spheres that form a tetrahedron. The n-th tetrahedral number is the sum of the first n triangular numbers.[102]

The sequence of the first few tetrahedral numbers is

$$P_3 = 1, 4, 10, 20, 35, 56, 84, 120, 165, 220, 286, 364, 455, 560, 680, 816, 969, \ldots$$

The n-th tetrahedral number, $P_3(n)$, is represented either by the third rising factorial of n divided by the factorial of three, or by the distinct triples (3-tuples) that can be selected from $n+2$ objects:

$$P_3(n) = \sum_{k=1}^{n} P_2(k) = \frac{n(n+1)(n+2)}{6} = \frac{n^{\overline{3}}}{3!} = \begin{pmatrix} n+2 \\ 3 \end{pmatrix}.$$

Being binomial coefficients,[70] the tetrahedral numbers are contained in Pascal's triangle.[71] They are in fourth position (entry three, third diagonal from left), arranged in order from top to bottom (Fig. 2.13). A second set of ordered tetrahedral numbers forms the third diagonal from right.

[104] A **pentatope number**, or **pentatopic number**, or **pentachoron number**, or **4-simplex number**, or **4-topic number**, belongs in the class of figurate numbers. It counts the number of intersections created when the corners of a polygon of size n are all connected to one another. For example, a triangle has zero intersections, a square has 1 intersection, a pentagon has 5 intersections, a hexagon has 15 intersections, and a heptagon has 35 intersections. The n-th pentatope number is the sum of the first n tetrahedral numbers[103] and equals the number of quads (4-tuples) that can be selected from $n+3$ objects:

$$P_4(n) = \sum_{k=1}^{n} P_3(k) = \frac{n(n+1)(n+2)(n+3)}{24} = \frac{n^{\overline{4}}}{4!} = \begin{pmatrix} n+3 \\ 4 \end{pmatrix}.$$

The sequence of the first few pentatope numbers is

$$P_4 = 1, 5, 15, 35, 70, 126, 210, 330, 495, 715, 1001, 1365, \ldots$$

Pentatope numbers are arranged in order along the fourth diagonal (Fig. 2.13), either from left or right, of Pascal's triangle.[71]

[105] The n-th $(m+1)$-**simplex number**, or n-th $(m+1)$-**topic number**, $P_{m+1}(n)$, is the sum of the first n m-simplex numbers and is given by the following explicit formulas:

$$P_{m+1}(n) = \sum_{k=1}^{n} P_m(k) = \frac{n(n+1)(n+2)\ldots(n+m)}{(m+1)!} = \frac{n^{\overline{m+1}}}{(m+1)!} = \begin{pmatrix} n+m \\ m+1 \end{pmatrix}.$$

Since all the figurate numbers are binomial coefficients,[70] the number of m-faces of the k-simplex is also given by the binomial coefficient $P_{m+1}(k-m+1)$:

$$P_{m+1}(k-m+1) = \binom{k+1}{m+1}. \qquad (3.3.2)$$

Consequently, the numbers of the m-faces of a k-simplex may be found in row $k+1$ and diagonal $m+1$, either from left or right, of Pascal's triangle[71] (Fig. 2.13).

In conclusion, the numbers of the m-faces of a k-simplex are ordered along the $(k+1)$-th row of Pascal's triangle, deprived of the 0-entry (belonging to the left diagonal, diagonal 0). They are arranged in Table 3.1, for $0 \le k \le 10$ and $0 \le m \le 10$.

The total number of faces of a k-simplex is the sum of the elements of the $(k+1)$-th row in Pascal's triangle, minus 1 (the first element of the row). It is also given by the sum of the elements of the $(k+1)$-th row in Table 3.1.

If we introduce the local systems of coordinate with origins in the 0-cells of the k-simplex, the triangular array of Pascal's triangle takes on a further meaning. By way of example, consider the 3-simplex in Fig. 3.6, where a local reference system is given by the three edges outgoing from the same vertex. These three axes define a system of affine coordinates (Section 4.2).

Whichever be the 0-cell where the origin has been fixed, there are always three edges that are parallel to the axes of the local reference system, the ξ, η, and ζ axes. There are also three surfaces that are parallel to the coordinate planes, the ξ/η, η/ζ, and ζ/ξ planes. Finally, there is one volume included inside the local axes and one vertex on the origin of the reference system. In other words, only one of the four vertices is on the origin, only three of the six edges are along the coordinate axes, only three of the four surfaces are on the coordinate planes and the volume of the tetrahedron is also the volume that is included inside coordinate axes.

Now, both the ordered sequence of the m-faces of the tetrahedron (4, 6, 4, 1) and the ordered sequence of those m-faces that obey to conditions of belonging, parallelism, or inclusion (1, 3, 3, 1) are sequences of binomial coefficients. The first sequence may be found in the $(k+1)$-th row of Pascal's triangle, deprived of the first entry, while the second sequence is given by the k-th row of Pascal's triangle (see Fig. 3.7 for $k=3$).

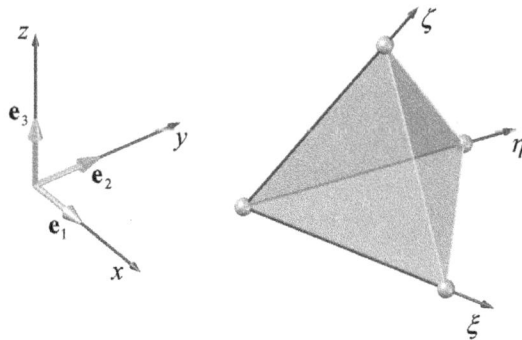

Figure 3.6. Local reference system of the affine coordinates for a 3-symplex (tetrahedron).

Table 3.1. m-faces of a k-simplex for $0 \le k \le 10$ and $0 \le m \le 10$: in the $(k+1)$-th row, the sum of the elements is equal to $2^{k+1} - 1$.

	0-faces (vertices)	1-faces (edges)	2-faces (surfaces)	3-faces (volumes)	4-faces	5-faces	6-faces	7-faces	8-faces	9-faces	10-faces	Sum
0-simplex (point)	1											1
1-simplex (line)	2	1										3
2-simplex (triangle)	3	3	1									7
3-simplex (tetrahedron)	4	6	4	1								15
4-simplex	5	10	10	5	1							31
5-simplex	6	15	20	15	6	1						63
6-simplex	7	21	35	35	21	7	1					127
7-simplex	8	28	56	70	56	28	8	1				255
8-simplex	9	36	84	126	126	84	36	9	1			511
9-simplex	10	45	120	210	252	210	120	45	10	1		1023
10-simplex	11	55	165	330	462	462	330	165	55	11	1	2047

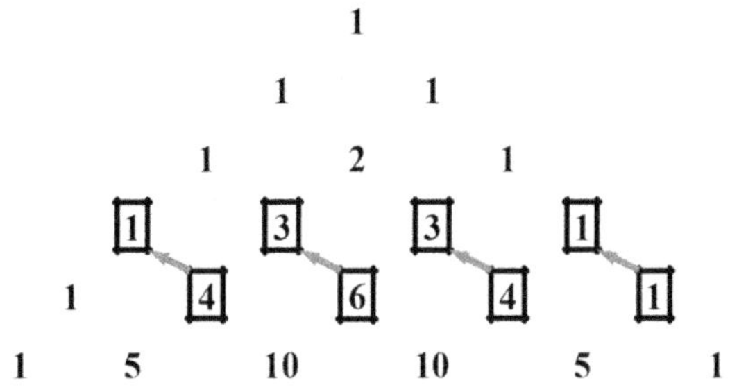

Figure 3.7. How to move from the element providing the total number of m-faces of the 3-simplex to the element providing the number of those m-faces of the 3-simplex that obey to conditions of belonging, parallelism, or inclusion.

By extending the former analysis to the k-simplices, with k not necessarily equal to three, we can define the following general rule:

Given a reference system with the origin in one vertex of the k-simplex and the axes along those edges of the k simplex that are connected with the chosen vertex, for finding how many m-faces of the k-simplex obey to conditions of belonging or inclusion in the reference axes, start from the element in row $k+1$ and diagonal $m+1$ of Pascal's triangle (which provides the total number of m-faces of the k-simplex) and move diagonally, on the number above and to the left.

As can be easily checked in Table 3.1, this means that the number of those m-faces of the k-simplex that obey to conditions of belonging or inclusion is equal to the total number of $(m-1)$-faces of the $(k-1)$-simplex.

In terms of binomial coefficients, the number of those m-faces of the k-simplex that obey to conditions of belonging or inclusion is given by the binomial coefficient $P_m(k-m+1)$:

$$P_m(k-m+1) = \binom{k}{m}. \tag{3.3.3}$$

This number may be found in row k and diagonal m, either from left or right, of Pascal's triangle (Fig. 3.8).

By considering also the empty set as a face of the k-simplex, the numbers of the m-faces of a k-simplex are still ordered along the $(k+1)$-th row of Pascal's triangle, included the 0-entry. In this case, the total number of faces of a k-simplex, $F^O(k)$, is the sum of the elements of the $(k+1)$-th row in Pascal's triangle:

$$F^O(k) = 2^{k+1}. \tag{3.3.4}$$

In doing so, the numbers of the m-faces of a k-simplex equal the numbers of basis $(m+1)$-blades needed in a $(k+1)$-dimensional space to represent arbitrary $(m+1)$-blades (Eq. 2.2.28). This is a consequence of the relationship shown in Fig. 3.7, because the coordinate system of the $(k+1)$-dimensional space can be fixed on one vertex of a $(k+1)$-blade.

Since the m-th diagonal of Pascal's triangle provides the m-th figurate number,[100] moving upwards along right diagonals, as illustrated in Fig. 3.7, modifies the figurate numbers associated with the m-faces of the k-simplex. In particular, the procedure illustrated in Fig. 3.7 substitutes the

$$\binom{0}{0}$$

$$\binom{1}{0} \qquad \binom{1}{1}$$

$$\binom{2}{0} \qquad \binom{2}{1} \qquad \binom{2}{2}$$

$$\binom{3}{0} \qquad \binom{3}{1} \qquad \binom{3}{2} \qquad \binom{3}{3}$$

$$\binom{4}{0} \qquad \binom{4}{1} \qquad \binom{4}{2} \qquad \binom{4}{3} \qquad \binom{4}{4}$$

$$\binom{5}{0} \qquad \binom{5}{1} \qquad \binom{5}{2} \qquad \binom{5}{3} \qquad \binom{5}{4} \qquad \binom{5}{5}$$

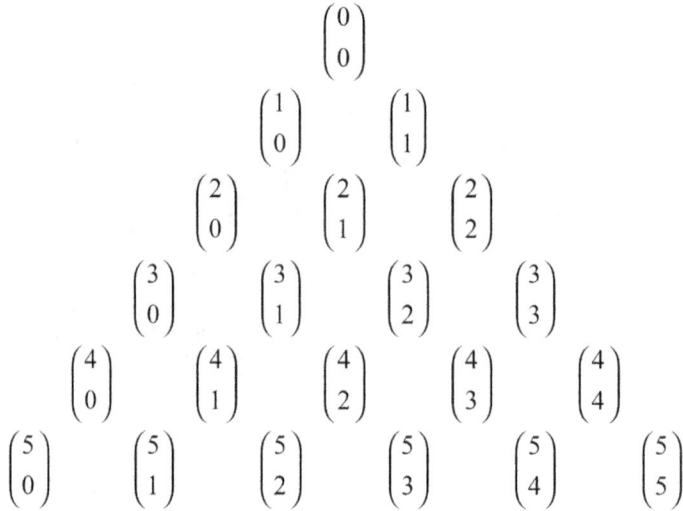

Figure 3.8. The first six rows of Pascal's triangle as Binomial Coefficients.

element in position n along the m-th left diagonal with the element in position n along the $(m-1)$-th left diagonal. Consequently, as far as the conditions of belonging or inclusion are concerned,

- the number of 0-faces (vertices) of the k-simplex is always 1 and is equal to $P_0(k+1)$, the $(k+1)$-th element of the sequence P_0;[101]
- the number of 1-faces (edges) of the k-simplex is the k-th linear number, $P_1(k)$;[101]
- the number of 2-faces (surfaces) of the k-simplex is the $(k-1)$-th triangle number, $P_2(k-1)$;[102]
- the number of 3-faces (volumes) of the k-simplex is the $(k-2)$-th tetrahedron number, $P_3(k-2)$;[103]
- the number of m-faces of the k-simplex is the $(k-m+1)$-th m-simplex number, $P_m(k-m+1)$.

Thus, all the $(k+1)$-simplex numbers are substituted by k-simplex numbers, while the element of the sequence does not change in position.

Let $P_m(n)$ be an element of the k-th row of Pascal's triangle and let $P_{k-m}(n-k+2m)$ be its symmetric element in the triangular pattern. Due to the symmetry of Pascal's triangle, moving upward along the right diagonal that passes through $P_m(n)$ provides the same result as moving upward along the left diagonal that passes through $P_{k-m}(n-k+2m)$ (Fig. 3.9). This suggests us that a second way for finding $P_m(k-m+1)$ is calculating its symmetric element in the triangular array.

By using figurate numbers, we find that, as far as the conditions of belonging or inclusion are concerned,

- the number of 0-faces (vertices) of the k-simplex is also the 1-st k-simplex number, $P_k(1)$;
- the number of 1-faces (edges) of the k-simplex is also the 2-nd $(k-1)$-simplex number, $P_{k-1}(2)$;
- the number of 2-faces (surfaces) of the k-simplex is also the 3-rd $(k-2)$-simplex number, $P_{k-2}(3)$;

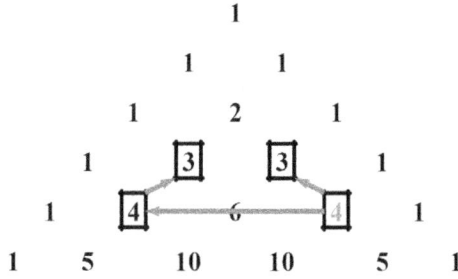

Figure 3.9. Finding $P_m(k-m+1)$ and its equivalent symmetric element, starting from $P_{m+1}(k-m+1)$.

- the number of 3-faces (volumes) of the k-simplex is also the 4-th $(k-3)$-simplex number, $P_{k-3}(4)$;
- the number of m-faces of the k-simplex is also the $(m+1)$-th $(k-m)$-simplex number, $P_{k-m}(m+1)$.

By comparing the ways for finding $P_m(k-m+1)$ and its symmetric element, we can conclude that

- the 1-st k-simplex number is equal to the $(k+1)$-th element of the sequence P_0,[101] that is, is always equal to 1:

$$P_k(1) = P_0(k+1) = 1; \qquad (3.3.5)$$

- the 2-nd $(k-1)$-simplex number is equal to the k-th element of the sequence P_1,[101] that is, to the k-th linear number:

$$P_{k-1}(2) = P_1(k); \qquad (3.3.6)$$

- the 3-rd $(k-2)$-simplex number is equal to the $(k-1)$-th element of the sequence P_2,[102] that is, to the $(k-1)$-th triangle number:

$$P_{k-2}(3) = P_2(k-1); \qquad (3.3.7)$$

- the 4-th $(k-3)$-simplex number is equal to the $(k-2)$-th element of the sequence P_3,[103] that is, to the $(k-2)$-th tetrahedron number:

$$P_{k-3}(4) = P_3(k-2); \qquad (3.3.8)$$

- the $(m+1)$-th $(k-m)$-simplex number is equal to the $(k-m+1)$-th m-simplex number:

$$P_{k-m}(m+1) = P_m(k-m+1). \qquad (3.3.9)$$

Given two disjoint k-simplices of the same simplicial k-complex, the m-faces of the first k-simplex are not parallel to the m-faces of the second k-simplex, in general. The condition of parallelism between m-faces is satisfied only in coordinate cell complexes, which are obtained from coordinate systems. In this special case, all the local 1-cells are parallel to the coordinate axes, and all the local 2-cells are parallel to the coordinate planes of the global coordinate system.

Due to the parallelism between local and global axes and planes, we can extend the findings on the single k-simplex to the whole coordinate cell k-complex. This allows us to speak of **families of m-faces**.

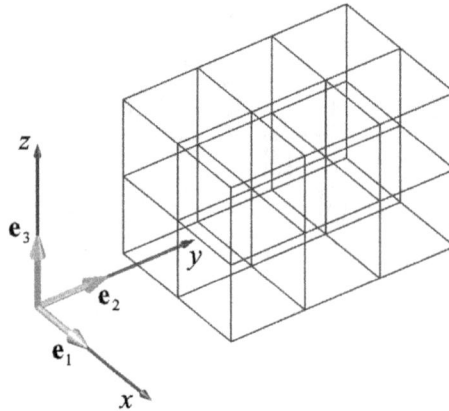

Figure 3.10. Cartesian cell complex in three dimensions.

By way of example, let us consider a Cartesian cell complex in three-dimensional space, a Cartesian 3-cell complex (Fig. 3.10). If we fix a local reference system for every 3-cell of the 3-cell complex, the number of m-faces that satisfy the conditions of belonging or inclusion for each 3-cell is still provided by the third row of Pascal's triangle in Fig. 3.7. This second time, however, all the 1-faces (edges) are parallel to one of the three global coordinate axes.

The subset of 1-faces that are parallel to the first global coordinate axis, x, defines the first family of 1-faces. Analogously, the subset of 1-faces that are parallel to the second global coordinate axis, y, defines the second family of 1-faces, and the subset of 1-faces that are parallel to the third global coordinate axis, z, defines the third family of 1-faces.

As far as the 2-faces are concerned, each of them turns out to be parallel to one of the three global coordinate planes, x/y, y/z, and z/x, defining three families of 2-faces. Finally, in the 3-cell complex of Fig. 3.10 there is one family of points and one family of volumes.

In conclusion, the elements of the third row in Pascal's triangle indicate, in this case, the number of families of m-faces in the 3-cell complex, not only the number of m-faces satisfying a given condition for a single 3-cell.

This result can easily be extended to n-dimensional spaces and cell complexes obtained from other kinds of coordinate systems. In any case, the n-th row of Pascal's triangle provides the number of families of m-faces in n-dimensional space (Fig. 3.11).

By comparison between Fig. 3.11 and Table 3.1, we can conclude that the number of families of m-faces of a given cell k-complex is equal to the total number of $(m-1)$-faces of the $(k-1)$-simplex.

3.4 Some Notions of the Graph Theory

Graphs[106] are one of the prime objects of study in discrete mathematics. They are mathematical structures used to model pair-wise relations between objects.

---

[106] The paper written by Leonhard Euler (15 April 1707–18 September 1783) on the *Seven Bridges of Königsberg* and published in 1736 is regarded as the first paper in the history of graph theory. This paper, as well as the one written by Alexandre-Théophile Vandermonde (28 February 1735–1 January 1796) on the *knight problem,* carried on with the

(continues next page)

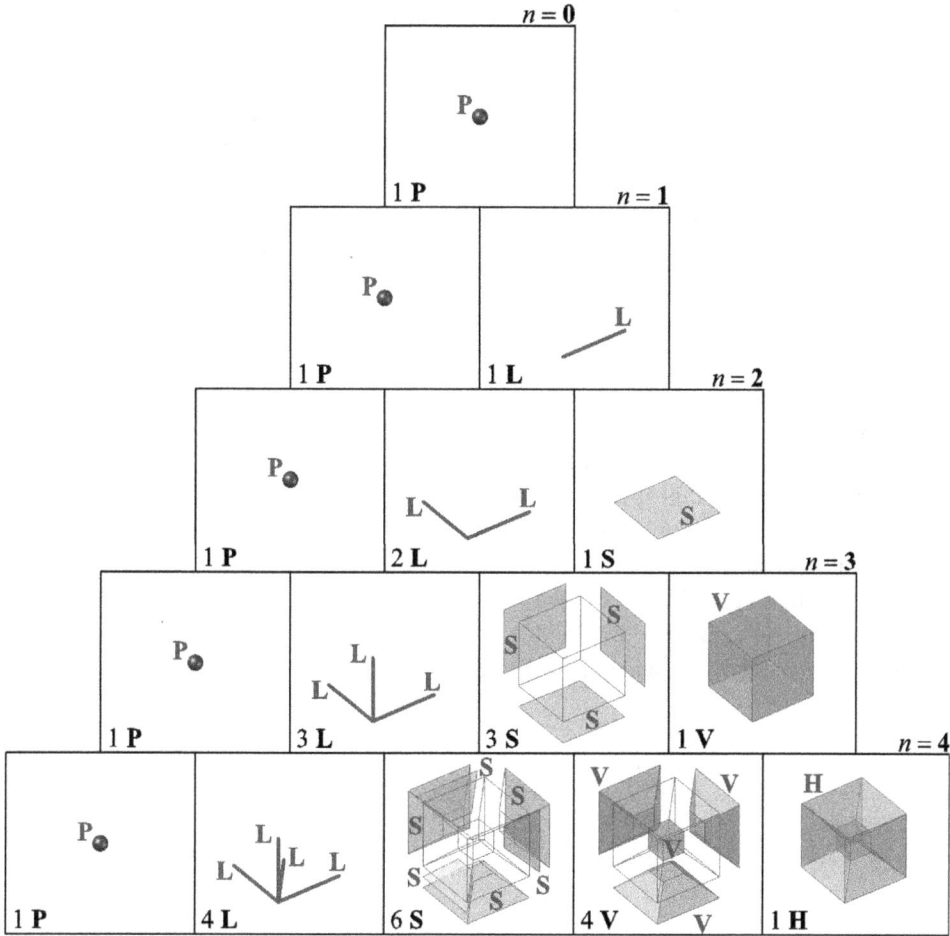

Figure 3.11. Pascal's triangle and the number of families of m-faces in n-dimensional space.

A graph is made up of **vertices**, also called **nodes** or **points**, and segments called **edges**, **lines**, or **arcs**, that connect them. Typically, a graph is depicted in a diagrammatic form as a set of dots for the vertices, joined by lines or curves for the edges. Normally, the vertices of a graph, by their nature as elements of a set, are distinguishable. This kind of graph may be called **vertex-labeled** (Fig. 3.12).

Graphs with labeled edges are called **edge-labeled** graphs. Graphs with labels attached to edges or vertices are more generally designated as **labeled**. Consequently, graphs in which vertices are indistinguishable and edges are indistinguishable are called **unlabeled**.

A **loop** is an edge which starts and ends on the same vertex (Fig. 3.13). Loops may be permitted or not permitted according to the application.

(continues from previous page)

analysis situs initiated by Gottfried Wilhelm von Leibniz (July 1, 1646–November 14, 1716). Euler's formula relating the number of edges, vertices, and faces of a convex polyhedron was studied and generalized by Augustin-Louis Cauchy (21 August 1789–23 May 1857) and Simon Antoine Jean L'Huilier (or L'Huillier) (Geneva, 24 April 1750–Geneva, 28 March 1840) and is at the origin of topology. The term "graph" was introduced by James Joseph Sylvester (3 September 1814–15 March 1897) in a paper published in 1878 in *Nature*, where he draws an analogy between "quantic invariants" and "co-variants" of algebra and molecular diagrams.

A **multigraph** (or **pseudograph**) is a graph which is permitted to have multiple edges (also called "parallel edges"), that is, edges that have the same end nodes, and sometimes loops (Fig. 3.13). Thus two vertices may be connected by more than one edge.

Both a graph and a multigraph may be **undirected** (Figs. 3.12 and 3.13) meaning that edges have no orientation, or their edges may be **directed** from one vertex to another (Fig. 3.14).

An undirected graph is an ordered pair $G = (V, E)$ comprising a set V of vertices or nodes together with a set E of edges or lines, which are 2-element subsets of V (that is, an edge is related with two vertices, and the relation is represented as unordered pair of the vertices with respect to the particular edge). The vertices belonging to an edge are called the **ends**, **end-points**, or **end vertices** of the edge. A vertex may exist in an undirected graph and not belong to an edge.

An undirected graph can be seen as a simplicial complex consisting of 1-simplices (the edges) and 0-simplices (the vertices). As such, cell complexes are generalizations of graphs since they allow for higher-dimensional simplices.

A directed graph, or **digraphs**, is an ordered pair $D = (V, A)$, in which V is a set whose elements are called vertices or nodes, and A is a set of ordered pairs of vertices, called arcs, **directed edges**, or **arrows**. An arc $a = (x, y)$ is considered to be directed from x to y. y is called the **terminal vertex**, or **head**, and x is called the **initial vertex**, or **tail**, of the arc. y is said to be a **direct successor** of x, and x is said to be a **direct predecessor** of y. If a path leads from x to y, then y is said to be a **successor** of x and **reachable** from x, and x is said

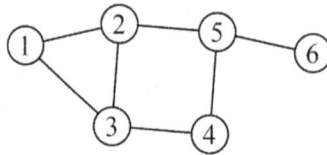

Figure 3.12. A drawing of a labeled graph on 6 vertices and 7 edges.

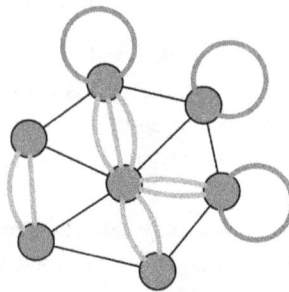

Figure 3.13. A multigraph with multiple edges (light gray) and several loops (dark gray).

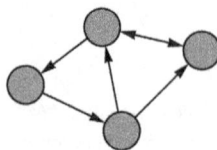

Figure 3.14. A directed graph.

to be a **predecessor** of y. The arc (y,x) is called the **arc** (x,y) **inverted**. A directed graph D is called **symmetric** if, for every arc in D, the corresponding inverted arc also belongs to D. An **oriented graph** is a directed graph in which at most one of (x,y) and (y,x) may be arcs.

An oriented graph can be seen as a simplicial complex consisting of directed 1-simplices (the edges) and 0-simplices (the vertices). As such, oriented cell complexes are generalizations of oriented graphs to higher-dimensional simplices, which allow us to provide also the vertices with an orientation.

An edge and a vertex on that edge are called **incident**. In set theory, the incidence relations[107] are binary relations, while, in graph theory, an incidence relation can either be a binary or a ternary relation. In particular, if the graph is an undirected graph, the incidence relation is a binary relation, while, if the graph is an oriented graph, the incidence relation is a ternary (or triadic) relation.

An **incidence structure** (P,L,I) consists of a set P, whose elements are called the **points**, a disjoint set L, whose elements are called the **lines**, and an incidence relation, I, between them, which is a subset of the Cartesian product $P \times L$. The elements of I are called the **flags**. If we want to take into account also the orientation of elements, such as in oriented graphs, I is a subset of the Cartesian product $P \times L \times O$, where O is the set of the **orientations**. In this second case, the incidence structure is the quadruple (P,L,O,I).

The branch of mathematics that studies incidence structures, mainly in the form of the triples (P,L,I), is called **incidence geometry**.

The matrix that shows all the relationships between two classes of objects in a finite incidence geometry (one with a finite number of points and lines) is called **incidence matrix**.[108] If the first class is X and the second is Y, the matrix has one row for each element of X and one column for each element of Y. The rows of the matrix represent points, while the columns represent lines.

The entries of the incidence matrix are called the **incidence numbers**. An incidence number different from zero in row i and column j means that the point i is incident with the line j. All other incidence numbers are zero. The nonzero incidence numbers are equal to $+1$ if the incidence relation is a binary relation (undirected graphs), $+1$ or -1 if the incidence relation is a ternary relation (oriented graphs). In the latter case, the sign of the incidence number b_{ij} tell us whether the edge x_j leaves or enters vertex v_i (there is not a single convention on how to associate the sign with the two cases).

Topological graph theory is a branch of graph theory. It studies the embedding of graphs in surfaces, spatial embeddings of graphs, and graphs as topological spaces.[92] It also studies immersions of graphs.

Embedding a graph in a surface means that we want to draw the graph on a surface, a sphere for example, without two edges intersecting. In particular, an embedding (also spelled

---

[107] An **incidence relation** describes how subsets meet. In geometry, the relations of incidence are those such as "lies on", between points and lines, lines and planes, or planes and volumes, and "intersects", between pairs of lines, planes, or volumes. Examples of incidence relations include "point **P** lies on line **L**" and "line $\mathbf{L}_1$ intersects line $\mathbf{L}_2$".

[108] Early ideas on the incidence matrices date back to Gustav Robert Kirchhoff (12 March 1824–17 October 1887), to whom we owe the first comprehensive treatment of the electrical network problem, 1847.

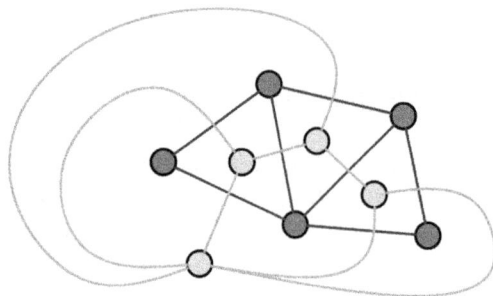

Figure 3.15. The graph in light gray is the dual graph of the graph in dark gray.

imbedding) of a graph G on a surface Σ is a representation of G on Σ in which points of Σ are associated with vertices and simple arcs are associated with edges in such a way that

- the endpoints of the arc associated with an edge e are the points associated with the end vertices of e.
- no arcs include points associated with other vertices.
- two arcs never intersect at a point which is interior to either of the arcs.

A **planar graph** is a graph that can be embedded in the plane, that is, it can be drawn on the plane in such a way that its edges intersect only at their endpoints. Such a drawing is called a **plane graph** or **planar embedding of the graph**.

The **dual graph** of a plane graph G is a graph that has a vertex corresponding to each face of G, and an edge joining two neighboring faces for each edge in G (Fig. 3.15). The term "dual" is used because this property is symmetric, meaning that if H is a dual of G, then G is a dual of H (if G is connected). The dual of a plane graph is a plane multigraph.

Since the dual graph depends on a particular embedding, the dual graph of a planar graph is not unique in the sense that the same planar graph can have non-isomorphic[44] dual graphs.

A **graph algebra** is a way of giving a directed graph an algebraic structure. It was introduced in 1983 (McNulty and Shallon) and has seen many uses in the field of universal algebra since then.[109]

Let $D = (V, A)$ be a directed graph, and 0 an element not in V. The graph algebra associated with D is the set $V \cup \{0\}$ equipped with multiplication defined by the rules:

$$xy = x \text{ if } x, y \in V, (x, y) \in A; \tag{3.4.1}$$

$$xy = 0 \text{ if } x, y \in V \cup \{0\}, (x, y) \notin A. \tag{3.4.2}$$

Algebraic graph theory is a branch of mathematics in which algebraic methods are applied to problems about graphs.

---

[109] Graph algebras have been used, for example, in constructions concerning dualities (Davey et al., 2000), equational theories (Pöschel, 1989), flatness (Delić, 2001), groupoid rings (Lee, 1991), topologies (Lee, 1988), varieties (Oates-Williams, 1984), finite state automata (Kelarev, Miller, and Sokratova, 2005), finite state machines (Kelarev, and Sokratova, 2003), tree languages and tree automata (Kelarev and Sokratova, 2001).

3.5 Boundaries, Coboundaries, and the Incidence Matrices

In three-dimensional geometry, only the $(p-1)$-cells that bound a given p-cell, that is, the facets (Section 3.3) of the p-cell, are called the faces of that p-cell. Since the problems of computational physics pertain to three spatial dimension, at most, in the following we will adopt the terminology of three-dimensional geometry. Consequently, if we consider a cell complex made of p-cells of degree 0, 1, 2 and 3, the faces of a 1-cell are its two end nodes (two 0-cells), the faces of a 2-cell are its edges (1-cells), and the faces of a volume are the surfaces (2-cells) that bound the volume (Fig. 3.16).

The set of faces of a p-cell defines the **boundary** of the p-cell. When the p-cell is a simplex, the boundary of the p-cell is its link (see Section 3.2).

According to the definition of incidence relation (Section 3.4), one can also say that the faces of a p-cell are those $(p-1)$-cells that are incident to the p-cell. In particular, since the four space elements, **P**, **L**, **S** and **V**, are provided with an orientation (Section 2.2) and the cell complexes are, broadly speaking, oriented graphs, the relation of incidence is a ternary relation, which is expressed by the three incidence numbers, 0, +1, and −1.

The incidence numbers can be collected to form the incidence matrices (Section 3.4). For reasons that will become clear later, in the CM we define as incidence matrix a matrix that has one row for each p-cell and one column for each $(p-1)$-cell. Therefore, the incidence matrices of the CM are not simply the generalization of the incidence matrices of the graph theory to higher-dimensional simplices, but their transposes.

The incidence number of a p-cell, e^h_p, with a $(p-1)$-cell, e^k_{p-1}, is the relative integer:[9]

$$q_{hk} \triangleq \left[e^h_p : e^k_{p-1} \right]; \tag{3.5.1}$$

where the first index of q_{hk} refers to the cell of greater dimension. The incidence number is equal to

- 0, if the $(p-1)$-cell is not on the boundary of the p-cell;
- +1, if the $(p-1)$-cell is on the boundary of the p-cell and the orientations of the p-cell and $(p-1)$-cell are compatible[110] (Fig. 3.17);
- −1, if the $(p-1)$-cell is on the boundary of the p-cell and the orientations of the p-cell and $(p-1)$-cell are not compatible (Fig. 3.18).

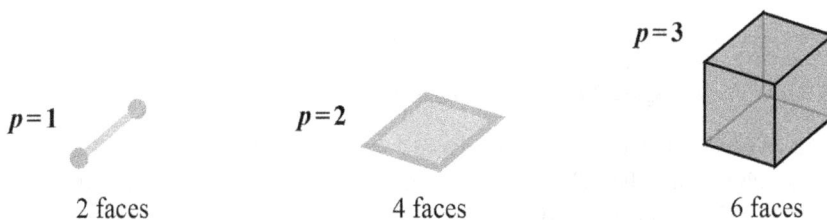

$p=1$ 2 faces $p=2$ 4 faces $p=3$ 6 faces

Figure 3.16. Faces of p-dimensional polytopes, for $1 \le p \le 3$.

---

[110] In cell complexes, the inner orientation of a given p-cell may be fixed independently of the inner orientations of its faces. In the special case when the inner orientation of the p-cell is just the one that would be induced in the p-cell with the inductive procedure described in Section 2.2, the inner orientation of the p-cell is called **compatible** to the inner orientations of its faces.

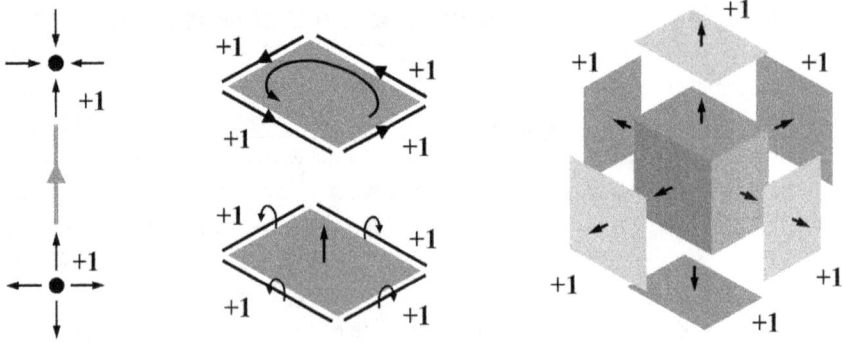

Figure 3.17. Examples of incidence numbers equal to +1.

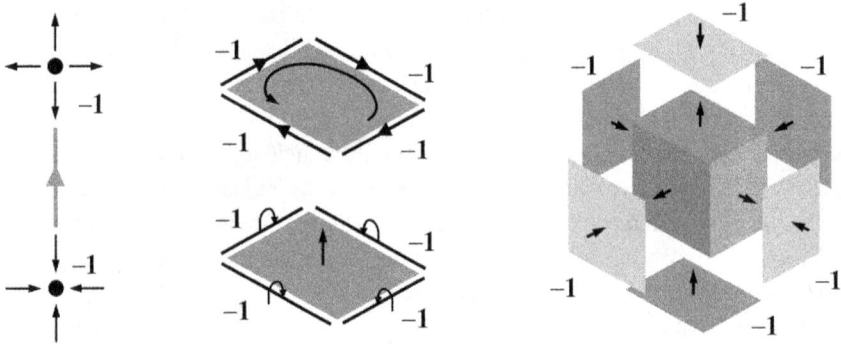

Figure 3.18. Examples of incidence numbers equal to −1.

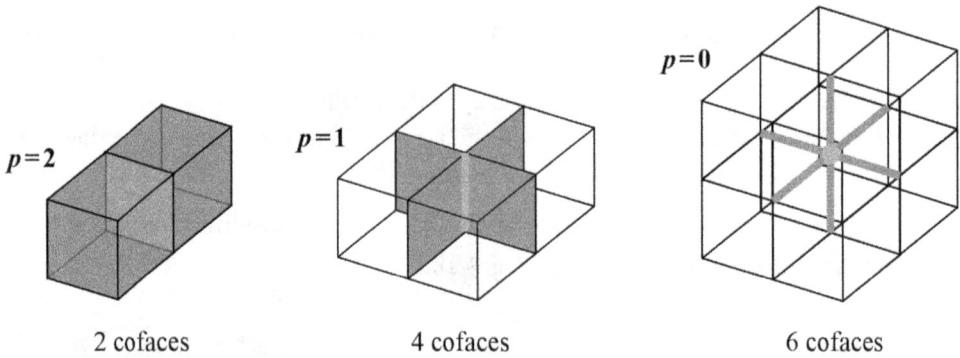

Figure 3.19. Cofaces of a p-cell of degree 0, 1, and 2.

The $(p+1)$-cells that have a given p-cell as a common face are the cofaces of that p-cell (Fig. 3.19). One can also say that the cofaces of a p-cell are those $(p+1)$-cells that are incident to the p-cell. The set of cofaces of a p-cell defines the **coboundary** of the p-cell.

It is worth noting that the notation of Grassmann[111] for vectors may be derived from the definition of incidence numbers and from considering the sinks as the points with positive inner

---

[111] If the vector represents a directed distance or displacement from a point **A** to a point **B**, the point **A** is called the **origin**, **tail**, **base**, or **initial point**, and the point **B** is called the **head**, **tip**, **endpoint**, **terminal point**, or **final**

(*continues next page*)

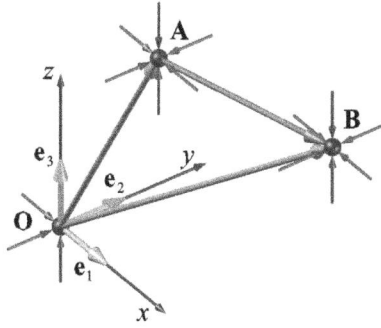

Figure 3.20. The vector $\overrightarrow{AB}$ as the difference between the two radius vectors $\overrightarrow{OB}$ and $\overrightarrow{OA}$.

orientations. Let $\overrightarrow{AB}$ be the vector of initial point **A** and final point **B**. In a coordinate system (x, y, z) of origin **O**, $\overrightarrow{AB}$ can be obtained as the vector sum (Fig. 3.20):

$$\overrightarrow{AB} = \overrightarrow{AO} + \overrightarrow{OB}. \tag{3.5.2}$$

Since the vector $\overrightarrow{AO}$ is the opposite of the radius vector[112] $\overrightarrow{OA}$:

$$\overrightarrow{AO} = -\overrightarrow{OA}; \tag{3.5.3}$$

Eq. (3.5.2) can be re-written as

$$\overrightarrow{AB} = \overrightarrow{OB} - \overrightarrow{OA}; \tag{3.5.4}$$

and the vector $\overrightarrow{AB}$ can be viewed as the difference between the two radius vectors $\overrightarrow{OB}$ and $\overrightarrow{OA}$, that is, the increment of the radius vector from **A** to **B**, or simply the increment from **A** to **B**.

Now, in the theory of oriented graphs any increment over an interval between two elements can be viewed as the sum of the elements, each one multiplied by the incidence number between the element and the interval. In particular, the increment of the radius vector between two points **A** and **B** is the weighted sum of **A** and **B**, where the weights are the incidence numbers between the points and their (oriented) connecting element, the vector $\overrightarrow{AB}$. Therefore, the vector $\overrightarrow{AB}$ can be regarded as either the weighted sum of **A** and **B**, or the weighted sum of **A**, **O**, and **B**.

Since the positive inner orientation of a point is the one of a sink (Section 2.2), the incidence number between **A** and $\overrightarrow{AB}$ is −1, while the incidence number between **B** and $\overrightarrow{AB}$ is +1 (Fig. 3.20). Thus, the weighted sum of **A** and **B** is equal to

$$\overrightarrow{AB} = (+1)\mathbf{B} + (-1)\mathbf{A} = \mathbf{B} - \mathbf{A}; \tag{3.5.5}$$

(*continues from previous page*)

point. Hermann Günther Grassmann (April 15, 1809–September 26, 1877) denotes this vector as the difference **B** − **A** between endpoint and initial point.

[112] A **position** or **position vector**, also known as **location vector** or **radius vector**, is an Euclidean vector[73] which represents the position of a point **P** in space, in relation to an arbitrary reference origin, **O**. It is usually defined as the vector connecting the fixed origin, **O**, with the point **P** and changes sign when the points are oriented as sources, instead than as sinks.

which is the same result provided by the sum of $\mathbf{A}$, $\mathbf{O}$, and $\mathbf{B}$:

$$\overrightarrow{AB} = \overrightarrow{AO} + \overrightarrow{OB} = (+1)\mathbf{O} + (-1)\mathbf{A} + (+1)\mathbf{B} + (-1)\mathbf{O} = \mathbf{B} - \mathbf{A}; \qquad (3.5.6)$$

where, since even the origin $\mathbf{O}$ is a sink, the incidence numbers have been evaluated as follows:

- -1 between $\mathbf{A}$ and $\overrightarrow{AO}$,
- $+1$ between $\mathbf{O}$ and $\overrightarrow{AO}$,
- -1 between $\mathbf{O}$ and $\overrightarrow{OB}$,
- $+1$ between $\mathbf{B}$ and $\overrightarrow{OB}$.

In a three-dimensional space, we can define three incidence matrices

- $\mathbf{G}$: matrix of the incidence numbers between 1-cells and 0-cells,
- $\mathbf{C}$: matrix of the incidence numbers between 2-cells and 1-cells,
- $\mathbf{D}$: matrix of the incidence numbers between 3-cells and 2-cells.

In a two-dimensional space, there exist only the two matrices $\mathbf{G}$ and $\mathbf{C}$. An example of matrices $\mathbf{G}$ and $\mathbf{C}$ for a two-dimensional domain is shown in Fig. 3.21 (where all the points are sinks, as usual).

The comparison between Figs. 3.21 and 2.11 clarifies in which sense the cell complexes and their labeling are equivalent, in the algebraic setting, to the coordinate systems and their continuous maps in $\mathbb{R}^n$, as stated in Section 1.2. In fact, the inner orientation of the cells, the edges, and the points establish, by the incidence matrices, the same type of relationship given by a basis blades in two-dimensional vector spaces. Note, in particular, the analogy between the incidence matrices and Eq. (2.2.24), where the $+1$ and -1 coefficients have the same function

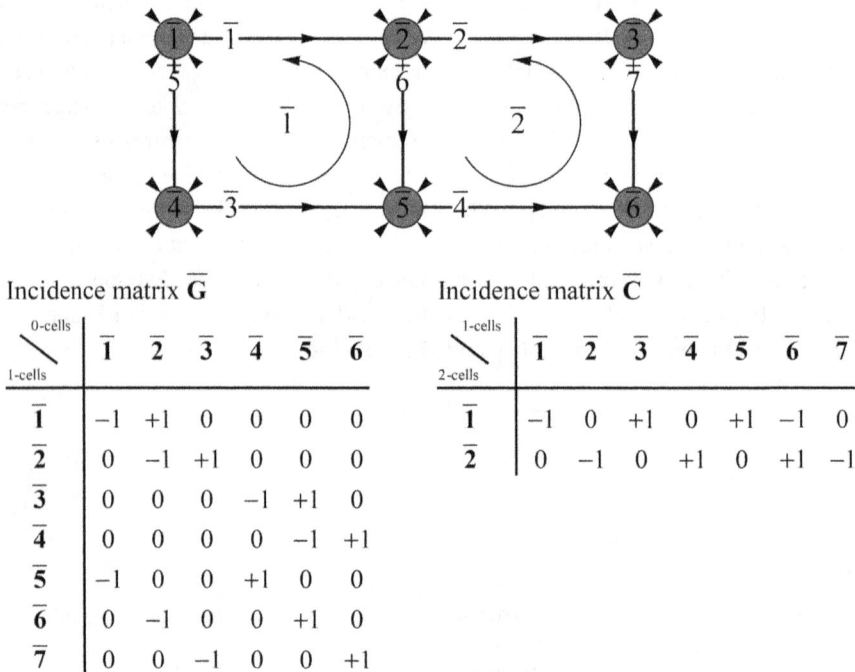

Incidence matrix $\overline{\mathbf{G}}$

1-cells \ 0-cells	$\overline{1}$	$\overline{2}$	$\overline{3}$	$\overline{4}$	$\overline{5}$	$\overline{6}$
$\overline{1}$	-1	$+1$	0	0	0	0
$\overline{2}$	0	-1	$+1$	0	0	0
$\overline{3}$	0	0	0	-1	$+1$	0
$\overline{4}$	0	0	0	0	-1	$+1$
$\overline{5}$	-1	0	0	$+1$	0	0
$\overline{6}$	0	-1	0	0	$+1$	0
$\overline{7}$	0	0	-1	0	0	$+1$

Incidence matrix $\overline{\mathbf{C}}$

2-cells \ 1-cells	$\overline{1}$	$\overline{2}$	$\overline{3}$	$\overline{4}$	$\overline{5}$	$\overline{6}$	$\overline{7}$
$\overline{1}$	-1	0	$+1$	0	$+1$	-1	0
$\overline{2}$	0	-1	0	$+1$	0	$+1$	-1

Figure 3.21. The incidence matrices in a plane cell complex (the symbol " ¯ " over the labels of the p-cells means that the p-cells are endowed with inner orientation).

as the incidence numbers, since their signs depend on the relationship between the order of the vectors in the outer products and the inner orientation of the basis bivector.

Therefore, we can think each cell of a plane cell complex as a two-dimensional space where the points of the cell, with their labeling and inner orientation, play the role of a basis scalar, the edges of the cell, with their labeling and inner orientation, play the role of basis vectors, and the cell itself, with its inner orientation, plays the role of basis bivector, thus generalizing Eq. (2.2.25).

In the special case where the cells of the cell complex are simplices, the number of non-zero incidence numbers in each row of the matrices $\mathbf{G}$, $\mathbf{C}$, and $\mathbf{D}$ (the latter only for three-dimensional domains) is given by the sequence P_1 of the linear numbers,[101] which form the first diagonal of Pascal's triangle.[71] In particular:

- the number of nonzero incidence numbers in each row of $\mathbf{G}$ is equal to $P_1(2) = 2$,
- the number of nonzero incidence numbers in each row of $\mathbf{C}$ is equal to $P_1(3) = 3$,
- the number of nonzero incidence numbers in each row of $\mathbf{D}$ is equal to $P_1(4) = 4$.

3.6 Chains and Cochains Complexes, Boundary and Coboundary Processes

Chain complex and **cochain complex** are algebraic means of representing the relationships between the cycles and boundaries in various dimensions of a topological space.[92]

A chain complex, $(A_\bullet, d_\bullet)$, is formally defined as a sequence of Abelian groups,[113] or modules,[114] ..., $A_{n+2}, A_{n+1}, A_n, A_{n-1}, A_{n-2}, ...$ connected by homomorphisms[115] (called **boundary operators**):

$$d_n : A_n \rightarrow A_{n-1};$$ (3.6.1)

such that the composition of any two consecutive maps is zero for all n:

$$d_n \circ d_{n+1} = 0 \ \forall n.$$ (3.6.2)

They are usually written out as

$$... \xrightarrow{d_{n+2}} A_{n+1} \xrightarrow{d_{n+1}} A_n \xrightarrow{d_n} A_{n-1} \xrightarrow{d_{n-1}} A_{n-2} \xrightarrow{d_{n-2}} ....$$ (3.6.3)

A chain complex is **bounded above** if all degrees above some fixed degree are 0 and is **bounded below** if all degrees below some fixed degree are 0.

---

[113] In abstract algebra,[91] an **Abelian group**, also called a **commutative group**, is a group in which the result of applying the group operation to two group elements does not depend on their order (the axiom of commutativity). Abelian groups generalize the arithmetic of the addition of integers. They are named after the Norwegian mathematician Niels Henrik Abel (5 August 1802–6 April 1829).

[114] In abstract algebra,[91] the concept of a **module** over a ring is a generalization of the notion of vector space over a field, wherein the corresponding scalars are the elements of an arbitrary ring. Modules also generalize the notion of abelian groups, which are modules over the ring of integers.

[115] In abstract algebra,[91] a **homomorphism** (from the ancient Greek language: ὁμός (homos) meaning "same" and μορφή (morphē) meaning "shape") is a structure preserving map between two algebraic structures (such as groups, rings, or vector spaces).

In the special case of a cell complex, a boundary operator is any map from a subset of n p-cells to a subset of m $(p-1)$-cells:

$$d_p : \sum_{i=1}^{n} \mathbf{e}_p^i \rightarrow \sum_{j=1}^{m} \mathbf{e}_{p-1}^j ; \tag{3.6.4}$$

where $\mathbf{e}_p^i$ is the i-th p-cell and $\mathbf{e}_{p-1}^j$ is the j-th $(p-1)$-cell.

When $n=1$ and m equals the number of faces of the only p-cell, the boundary operator is indicated with the symbol ∂:

$$\partial_p : \mathbf{e}_p \rightarrow \sum_{j=1}^{m} \mathbf{e}_{p-1}^j ; \tag{3.6.5}$$

and ∂_p defines the boundary of $\mathbf{e}_p$.

A closed line and a closed surface are p-cells, $\mathbf{e}_p$, without boundary. They thus belong to bounded below chains:

$$\partial_p : \mathbf{e}_p \rightarrow \varnothing. \tag{3.6.6}$$

Let h_i be an integer number associated with the i-th p-cell, $\mathbf{e}_p^i$, then h_i can be viewed as a multiplicity, or a weight of $\mathbf{e}_p^i$. This weight induces a weight on the $(p-1)$-cells of the chains of $\mathbf{e}_p^i$, by a process that is called the **boundary process**. In particular, when m is equal to the number of faces of $\mathbf{e}_p^i$, the weight of $\mathbf{e}_{p-1}^j$, h_j, is obtained by adding the weights of the n_j p-cells on the coboundary of $\mathbf{e}_{p-1}^j$, with the $+1$ or -1 sign according to the mutual incidence numbers:

$$h_j = \sum_{i=1}^{n_j} q_{ij} h_i, \text{ where } n_j = \left| \delta \mathbf{e}_{p-1}^j \right|.^{116} \tag{3.6.7}$$

In other words, each $(p-1)$-cell collects the weights that are spread on the $(p-1)$-cell itself by its cofaces, after having multiplied the weights by the mutual incidence numbers (Fig. 3.22).

The boundary process is defined as the action of the n p-cells, which spread their own weights on their faces, in accordance with the mutual incidence numbers. The pictorial view of the boundary process is provided in Fig. 3.23 for $n=1$ and $p=1,2,3$.

By performing the boundary process twice (Fig. 3.24), we obtain the weight of the k-th $(p-2)$-cell, $\mathbf{e}_{p-2}^k$, which is always equal to zero:

$$h_k = \sum_{j=1}^{n_k} \left(q_{jk} \sum_{i=1}^{n_j} q_{ij} h_i \right) = 0, \text{ where } n_j = \left| \delta \mathbf{e}_{p-1}^j \right|, n_k = \left| \delta \mathbf{e}_{p-2}^k \right|; \tag{3.6.8}$$

where n_k is the number of $(p-1)$-cells on the coboundary of $\mathbf{e}_{p-2}^k$, the k-th $(p-2)$-cell.

---

[116] The symbol δ, defined in Eq. (3.6.16), is used for indicating the coboundary of a cell. Thus, $\delta \mathbf{e}_{p-1}^j$ is the coboundary of the j-th $(p-1)$-cell, $\mathbf{e}_{p-1}^j$. Moreover, $\left| \delta \mathbf{e}_{p-1}^j \right|$ denotes the cardinality of the set $\delta \mathbf{e}_{p-1}^j$, that is, the number of elements of the coboundary of $\mathbf{e}_{p-1}^j$, which is equal to the number of cofaces of $\mathbf{e}_{p-1}^j$. Alternatively, the cardinality of a set A may be denoted by $n(A)$, $\overline{\overline{A}}$, $\mathrm{card}(A)$, or #A.

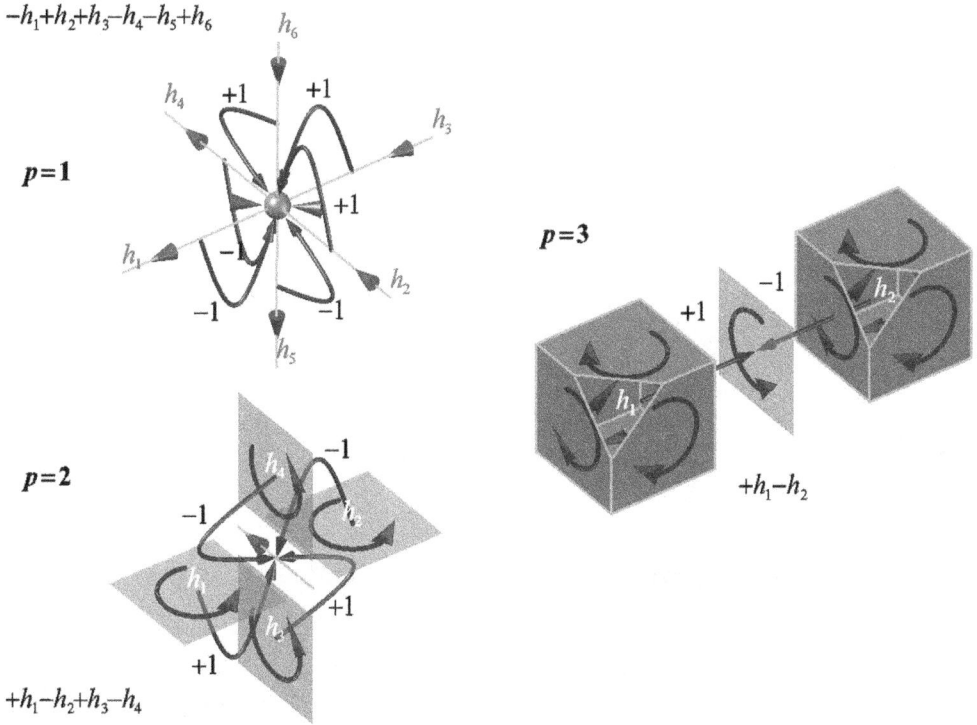

Figure 3.22. How to find the weight of a $(p-1)$-cell starting from the weights of its cofaces.

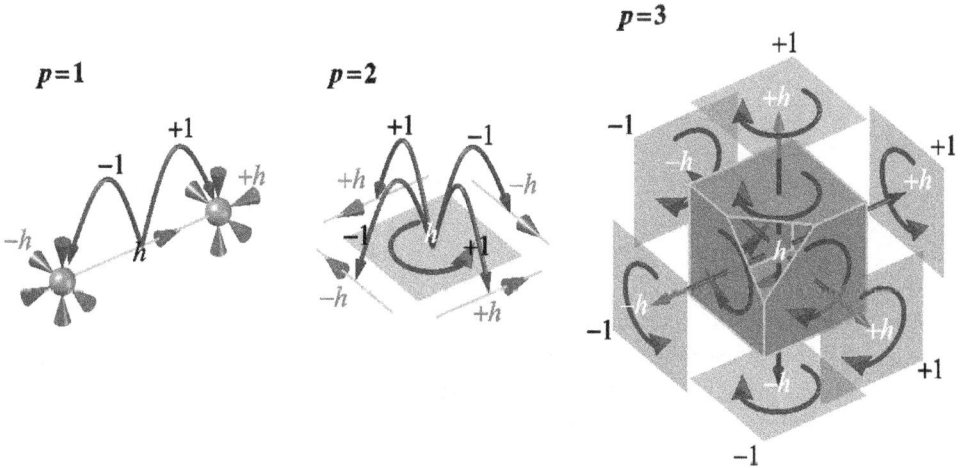

Figure 3.23. Spreading the weight of one p-cell on its faces.

In other words, as for the composition of the consecutive boundary operators in Eq. (3.6.2), even the composition of two consecutive boundary processes is zero for all p:

$$\partial\left(\partial\mathbf{e}_p\right) = \varnothing. \tag{3.6.9}$$

$$p=2$$

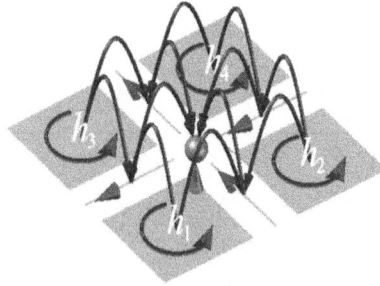

$$-(+h_1-h_3)+(+h_1-h_2)+(+h_2-h_4)-(+h_3-h_4)=0$$

Figure 3.24. How to find the weight of a $(p-2)$-cell starting from the weights of the cofaces of its cofaces: A two-dimensional example.

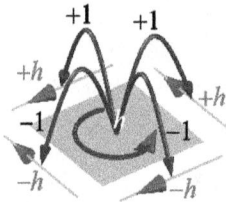

First boundary process   **Second boundary process**

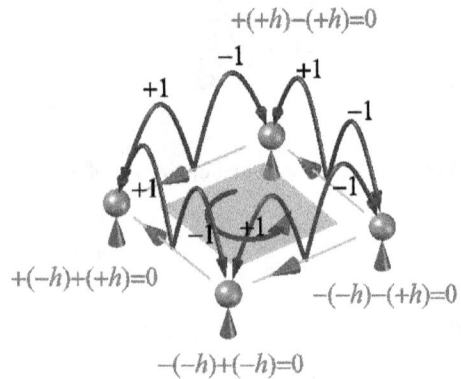

Figure 3.25. Spreading of the weight of one p-cell on the faces of its faces.

The reason for this is that the double action of a p-cell on its faces, the first time, and on the faces of its faces, the second time, generates some weights that, taken in twos, are equal and opposite. This is shown in Fig. 3.25 for $n=1$ and $p=2$.

From Eq. (3.6.8) and the definition of the incidence matrices, **G**, **C**, and **D**, it follows that

$$\mathbf{CG} = \mathbf{0}; \tag{3.6.10}$$

$$\mathbf{DC} = \mathbf{0}; \tag{3.6.11}$$

where **0** is the null matrix.

The only difference in the definitions of chain and cochain complexes is that, in chain complexes, the boundary operators decrease dimension, whereas in cochain complexes they increase dimension. A cochain complex, $(A^\bullet, d^\bullet)$, is formally defined as a sequence of abelian groups, or modules, ..., $A^{n-2}, A^{n-1}, A^n, A^{n+1}, A^{n+2}, ...$ connected by homomorphisms (called **coboundary operators**[117]):

$$d^n : A^n \to A^{n+1}; \tag{3.6.12}$$

such that the composition of any two consecutive maps is zero for all n:

$$d^{n+1} \circ d^n = 0 \ \forall n; \tag{3.6.13}$$

$$... \xrightarrow{d^{n-2}} A^{n-1} \xrightarrow{d^{n-1}} A^n \xrightarrow{d^n} A^{n+1} \xrightarrow{d^{n+1}} A^{n+2} \xrightarrow{d^{n+2}} .... \tag{3.6.14}$$

In the special case of a cell complex, a coboundary operator is any map from a subset of n p-cells to a subset of m $(p+1)$-cell:

$$d^p : \sum_{i=1}^{n} \mathbf{e}_p^i \to \sum_{j=1}^{m} \mathbf{e}_{p+1}^j; \tag{3.6.15}$$

where $\mathbf{e}_p^i$ is the i-th p-cell and $\mathbf{e}_{p+1}^j$ is the j-th $(p+1)$-cell.

When $m = 1$ and n equals the number of cofaces of the $(p+1)$-cell, the coboundary operator is indicated with the symbol δ:

$$\delta^p : \sum_{i=1}^{n} \mathbf{e}_p^i \to \mathbf{e}_{p+1}; \tag{3.6.16}$$

and δ^p defines the coboundary of $\mathbf{e}_{p+1}$.

If $\mathbf{e}_p$ is a closed line or a closed surface, then $\mathbf{e}_p$ belongs to bounded below cochains:

$$\delta^{p-1} : \varnothing \to \mathbf{e}_p. \tag{3.6.17}$$

Let η^i be the weight (an integer number) of the i-th p-cell, $\mathbf{e}_p^i$. This weight induces a weight on the $(p+1)$-cells of the cochains of $\mathbf{e}_p^i$, by a process that is called the **coboundary process**. In particular, when n is equal to the number of cofaces of $\mathbf{e}_{p+1}^j$, the weight of $\mathbf{e}_{p+1}^j$, η^j, is obtained by adding the weights of the n_j p-cells on the boundary of $\mathbf{e}_{p+1}^j$, with the $+1$ or -1 sign according to the mutual incidence numbers:

$$h^j = \sum_{i=1}^{n_j} q_{ji} \eta^i, \text{ where } n_j = \left| \partial \mathbf{e}_{p+1}^j \right|. \tag{3.6.18}$$

Note that the two indices of the incidence numbers have been swapped, because the first index refers to the cell of greater dimension.

Each $(p+1)$-cell collects the weights that are spread on the $(p+1)$-cell itself by its faces, after having multiplied the weights by the mutual incidence numbers (Fig. 3.26).

---

[117] The notion of coboundary operator which raises the degree of a discrete form by one unit corresponds to the notion of exterior differential which raises the degree of an exterior differential form by one unit.

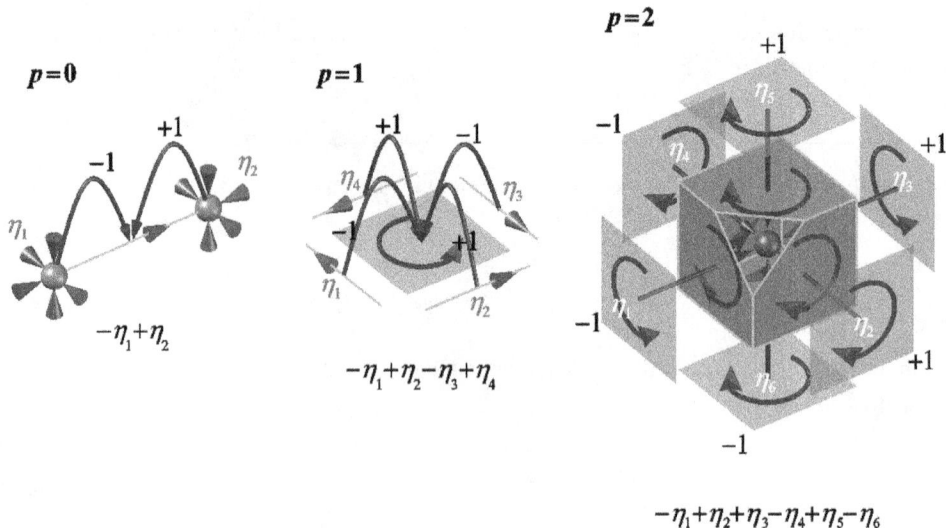

Figure 3.26. How to find the weight of a $(p+1)$-cell starting from the weights of its faces.

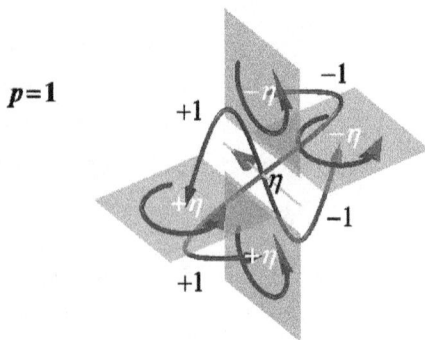

Figure 3.27. The coboundary process.

The coboundary process can be defined as the action of the n p-cells, which spread their own weights on their cofaces, in accordance with the mutual incidence numbers. The pictorial view of the coboundary process is provided in Fig. 3.27 for $n = 1$ and $p = 1$.

By performing the coboundary process twice, we obtain

$$\delta\left(\delta\mathbf{e}_p\right) = \varnothing; \tag{3.6.19}$$

which is the algebraic counterpart of the differential identities:

$$\mathrm{curl}\left(\mathrm{grad}\, f\right) \equiv 0; \tag{3.6.20}$$

$$\mathrm{div}\left(\mathrm{curl}\quad\right) \equiv 0. \tag{3.6.21}$$

3.7 Discrete *p*-forms

A physical variable ϕ associated with one set of p-cells of a cell-complex defines a discrete p-form (or a discrete form of degree p). The potential of a vector field, line integral of a vector, flux and mass content are discrete forms of degree 0, 1, 2, and 3, respectively (Table 3.2).

The discrete p-forms generalize the notion of field functions, because, in a discrete p-form $\Phi[\mathbf{P}]$, $\Phi[\mathbf{L}]$, $\Phi[\mathbf{S}]$, or $\Phi[\mathbf{V}]$, we associate the value of a physical variable with the space elements of degree p, where $p = 0, 1, 2, 3$, while the field functions, $f(\mathbf{P})$, always associate the value of a physical variable with the points of the domain. As a consequence, $\Phi[\mathbf{P}]$, $\Phi[\mathbf{L}]$, $\Phi[\mathbf{S}]$, and $\Phi[\mathbf{V}]$ are set functions (Section 1.3), while $f(\mathbf{P})$ is a point function.

The notion of discrete form is the algebraic version of the exterior differential form, a mathematical formalism that has the great merit of highlighting the geometrical background of physical variables, something ignored by differential calculus, providing a description that is independent of the coordinate system used. Nevertheless, this formalism uses field variables instead of global variables and, for this reason, it must use the notion of derivative.

Let S_p be the set of m p-cells, $\mathbf{e}_p^i$, each one taken with the multiplicity n_i:[118]

$$S_p = \left\{ n_1 \mathbf{e}_p^1, n_2 \mathbf{e}_p^2, ..., n_m \mathbf{e}_p^m \right\};$$ (3.7.1)

and let $\mathscr{G}$ be an additive and commutative group (a scalar, a vector, a matrix, and so forth), then a discrete p-form, c^p, is a linear function on the set S_p with integer-valued coefficients in the group $\mathscr{G}$.

The p-forms are additive:[38]

$$c^p \left(S_p + S_p' \right) = c^p \left(S_p \right) + c^p \left(S_p' \right);$$ (3.7.2)

and homogeneous of degree 1:[39]

$$c^p \left(n S_p \right) = n c^p \left(S_p \right).$$ (3.7.3)

Table 3.2. Examples of discrete p-forms

Variable	Potential of a vector field	Line integral of a vector	Flux	Mass content
Evaluated on	0-cells (points)	1-cells (lines)	2-cells (surfaces)	3-cells (volumes)
Discrete *p*-form	discrete 0-form $\Phi[\mathbf{P}]$	discrete 1-form $\Phi[\mathbf{L}]$	discrete 2-form $\Phi[\mathbf{S}]$	discrete 3-form $\Phi[\mathbf{V}]$

---

[118] The multiplicity, n_i, is an integer. While the p-cells are topological entities, the couples formed by a p-cell, $\mathbf{e}_p^i$, and the multiplicity n_i are algebraic entities. Consequently, while the p-cells cannot be added, the p-cells provided with a multiplicity can be added.

Denoting by g_i the values assumed by the coefficients when the multiplicities of the p-cells are equal to +1, a discrete p-form can be represented by a row vector, whose elements are the coefficients g_i of the discrete form:

$$c^p = \begin{bmatrix} g_1 & g_2 & \cdots & g_m \end{bmatrix}. \tag{3.7.4}$$

The value, g, of the discrete p-form $c^p = \begin{bmatrix} g_1 & g_2 & \cdots & g_m \end{bmatrix}$ on the collection with integer coefficients:

$$\mathbf{c}_p = \sum_{i=1}^{m} n_i \mathbf{e}_p^i; \tag{3.7.5}$$

which can also be described by the column vector

$$\mathbf{c}_p = \begin{bmatrix} n_1 & n_2 & \cdots & n_m \end{bmatrix}^T; \tag{3.7.6}$$

is equal to

$$g = \sum_{i=1}^{m} g_i n_i \equiv \begin{bmatrix} g_1 & g_2 & \cdots & g_m \end{bmatrix} \begin{bmatrix} n_1 \\ n_2 \\ \cdots \\ n_m \end{bmatrix} = \left\langle c^p, \mathbf{c}_p \right\rangle. \tag{3.7.7}$$

Since the p-forms are additive and homogeneous, we can re-write g as

$$g = \left\langle c^p, \mathbf{c}_p \right\rangle = \left\langle c^p, \sum_{i=1}^{m} n_i \mathbf{e}_p^i \right\rangle = \sum_{i=1}^{m} n_i \left\langle c^p, \mathbf{e}_p^i \right\rangle = \sum_{i=1}^{m} n_i g_i. \tag{3.7.8}$$

Eq. (3.7.8) means that a p-form (for example, a line integral, a flux, or a mass content) on one of the three space elements with dimension greater than 0 ($\mathbf{L}$, $\mathbf{S}$, and $\mathbf{V}$) is the sum of the values assumed by the p-form on each of the elementary parts in which the space element can be divided.

In particular, remembering again that the discrete p-forms are homogeneous, from Eq. (3.7.8) it follows the **Oddness Condition** of the p-forms:

$$\left\langle c^p, -\mathbf{e}_p^i \right\rangle = -\left\langle c^p, \mathbf{e}_p^i \right\rangle = -g_i. \tag{3.7.9}$$

The introduction of the discrete p-forms allows us to find the relation between the boundary operator, ∂, and the coboundary operator, δ. Let us evaluate the discrete p-form c^p on the p-dimensional boundary of the $p+1$ collection, $\mathbf{c}_{p+1}$:

$$c^p \left(\partial \mathbf{c}_{p+1} \right) = \left\langle c^p, \partial \mathbf{c}_{p+1} \right\rangle = \left\langle c^p, \sum_{i=1}^{m} n_i \partial \mathbf{e}_{p+1}^i \right\rangle = \sum_{i=1}^{m} n_i \left\langle c^p, \partial \mathbf{e}_{p+1}^i \right\rangle$$

$$= \sum_{i=1}^{m} n_i \left\langle c^p, \sum_{j=1}^{n} q_{ij} \mathbf{e}_p^j \right\rangle = \sum_{i,j} n_i q_{ij} \left\langle c^p, \mathbf{e}_p^j \right\rangle = \sum_{i,j} n_i q_{ij} g_j. \tag{3.7.10}$$

On the other hand, the coboundary of the discrete p-form c^p on the $p+1$ collection, $\mathbf{c}_{p+1}$, provides the same result:

$$\delta c^p\left(\mathbf{c}_{p+1}\right) = \left\langle \delta c^p, \mathbf{c}_{p+1}\right\rangle = \left\langle \delta c^p, \sum_{i=1}^{m} n_i \mathbf{e}^i_{p+1}\right\rangle = \sum_{i=1}^{m} n_i \left\langle \delta c^p, \mathbf{e}^i_{p+1}\right\rangle = \sum_{i,j} n_i q_{ij} g_j. \quad (3.7.11)$$

It follows that

$$\left\langle \delta c^p, \mathbf{c}_{p+1}\right\rangle = \left\langle c^p, \partial \mathbf{c}_{p+1}\right\rangle. \quad (3.7.12)$$

This theorem is the algebraic form of the generalized theorem of Stokes, which includes the theorem of Gauss, the proper theorem of Stokes, and the theorem of Leibnitz:

$$\begin{cases} \int_V \nabla \cdot \mathbf{u} \; \mathrm{d}V = \int_{\partial V} \mathbf{u} \cdot \mathrm{d}S & \text{Gauss} \\ \int_S \left(\nabla \times \mathbf{u}\right) \cdot \mathrm{d}S = \int_{\partial S} \mathbf{v} \cdot \mathrm{d}L & \text{Stokes} \\ \int_V \nabla \phi \cdot \mathrm{d}L = \phi\left(\partial \mathbf{L}\right) = \phi\left(\partial \mathbf{B}\right) - \phi\left(\partial \mathbf{A}\right) & \text{Leibnitz} \end{cases} \quad (3.7.13)$$

For this reason, it is called the **generalized or combinatorial form of Stokes' theorem**. The statement of the theorem is as follows:

The coboundary of the discrete p-form c^p on the $p+1$ collection, $\mathbf{c}_{p+1}$, is equal to the discrete p-form, c^p, on the p-dimensional boundary of the $p+1$ collection, c_{p+1}.

Note that we have derived the theorem of Stokes as a direct consequence of the coboundary process. This means that Stokes' theorem is a purely topological relation, while the various continuity and differentiability conditions usually required in its proof are indeed required by the use of field functions and derivatives. We could say that the most remarkable aspect of the proof of the generalized form of Stokes' theorem lies in highlighting that many differentiability requirements do not belong to physical laws, but are required by the differential apparatus used in their description.

By using Eq. (3.7.12) and Eq. (3.6.9), we can proof that the coboundary of a coboundary vanishes, as stated by Eq. (3.6.19):

$$\delta\left(\delta c^p\right)\left(\mathbf{c}_{p+2}\right) = \left(\delta c^p\right)\left(\partial \mathbf{c}_{p+2}\right) = c^p\left(\partial\partial \mathbf{c}_{p+2}\right) = c^p\left(\varnothing\right) = 0. \quad (3.7.14)$$

In other words, when the coboundary process is performed twice in sequence it gives rise to the null element of the group $\mathcal{G}$.

Analogously, by using Eq. (3.7.12) and Eq. (3.6.19), we can proof Eq. (3.6.9) (the boundary of a boundary vanishes):

$$\partial\left(\partial c^p\right)\left(\mathbf{c}_p\right) = \left(\partial c^p\right)\left(\delta \mathbf{c}_p\right) = c^p\left(\delta\delta \mathbf{c}_p\right) = c^p\left(\varnothing\right) = 0. \quad (3.7.15)$$

3.8   Inner and Outer Orientations of Time Elements

Finding the orientations of the time elements could be viewed in the same way that finding the inner and outer orientations of points and lines in a one-dimensional space. In fact, the time axis defines a one-dimensional cell complex, where the time instants, $\mathbf{I}$, are points (nodes) and the time intervals, $\mathbf{T}$, are the line segments that connect the points (the time instants are the

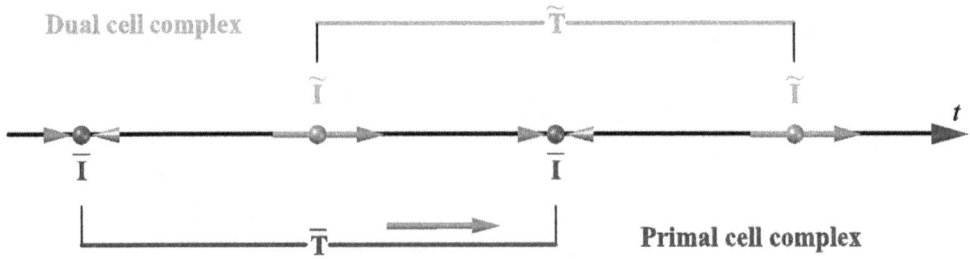

Figure 3.28. Time elements and their duals.

boundary, or the faces, of the time intervals: see Sections 3.3 and 3.5 for details). Moreover, in a one-dimensional space the dual (orthogonal complement) of a point is a line segment and the dual of a line segment is a point. Consequently, the outer orientations could be obtained in accordance with Fig. 2.30, in the special case, where $n = 1$.

As far as the inner orientation is concerned, all the time instants, both those along the positive semi-axis and those along the negative semi-axis, are sinks. Thus, they have an inward inner orientation (Fig. 3.28).

Finally, we can decide that the inner orientation of the time intervals is the same as the orientation of the time axis.

After a more detailed analysis, however, it is clear that building a cell complex in time makes no sense in itself. In fact, in physics time has not importance in itself. It is just a variable, useful for describing how a physical phenomenon evolves. Now, since any physical phenomenon occurs in space, it follows that the time axis must always be related to one or more axes in space. The perception itself of the time is linked to bodies. Therefore, a cell complex in time must be two-dimensional, at least.

When a time axis is added to a three-dimensional cell complex where the cell of maximum dimension has been originated by a trivector $u \wedge v \wedge w$, it gives rise to a four-dimensional space/time cell complex, whose cell of maximum dimension is a **tesseract**.[119] In multilinear algebra,[36] a tesseract, also called a **regular octachoron** or **cubic prism**, is a further element of the (graded) exterior algebra on a vector space (Section 2.1).

The tesseract is the four-dimensional analog of the cube, in the sense that it is to the cube as the cube is to the square. Each edge of a tesseract is of the same length. Just as the surface of the cube consists of six square faces, the hypersurface of the tesseract consists of eight cubical cells (Fig. 3.29).

All in all, a tesseract consists of eight cubes, 24 squares, 32 edges, and 16 vertices. There are four cubes, six squares, and four edges meeting at every vertex.

Since a generalization of the cube to dimensions greater than three is called a "hypercube", or "n-cube", the tesseract is also called the **four-dimensional hypercube**, or **4-cube**.

This structure is not easily imagined, but it is possible to project tesseracts into three- or two-dimensional spaces. One possibility is that of forming the three-dimensional shadows of the tesseract. If the wireframe of a cube is lit from above, the resulting shadow is a square within a square with

---

[119] The tesseract is one of the six convex regular 4-polytopes[120] (four-dimensional analogue of a polyhedron). The simplest possible convex regular 4-polytope is the pentachoron.[99]

Figure 3.29. Unfolding of a tesseract in three-dimensional space.

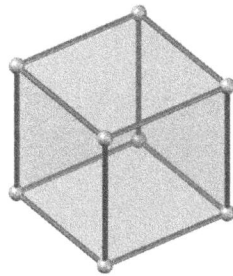

Figure 3.30. Cell-first perspective projection of the tesseract into three dimensions, with hidden surfaces culled.

the corresponding corners connected. Similarly, if the wireframe of a tesseract were lit from "above" (in the fourth direction), its shadow would be that of a three-dimensional cube within another three-dimensional cube (Fig. 3.30), where five faces are obscured by the visible faces. Similarly, seven cells of the tesseract are not seen in Fig. 3.30, because they are obscured by the visible cell.

The tesseract can be constructed in a number of ways:

- As a regular polytope[120] with three cubes folded together around every edge[121] (Fig. 3.31). In this case, it can be represented by the Schläfli symbol[122] {4, 3, 3}.

---

[120] **Regular polytopes** are the generalized analog in any number of dimensions of regular polygons and regular polyhedra. A regular polytope in n dimensions may be defined as having regular facets (Section 3.3), that is, $(n-1)$-faces, and regular vertex figures, which are the figures exposed when the corners are sliced off. These two conditions are sufficient to ensure that all faces are alike and all vertices are alike.

[121] This construction also explains the name "tesseract" given to this geometrical object. In fact, according to the Oxford English Dictionary, the word tesseract was coined and first used in 1888 by Charles Howard Hinton (1853, UK–30 April 1907, Washington D.C., USA), in his book "A New Era of Thought", referring to the four lines (in Greek τέσσερεις ακτίνες: four rays) from each vertex to other vertices.

[122] The **Schläfli symbol**, named after the Swiss geometer and mathematician Ludwig Schläfli (15 January 1814–20 March 1895), is a notation of the form $\{p, q, r, ...\}$ that defines regular polytopes and tessellations. It is a recursive description, starting with a p-sided regular polygon as $\{p\}$. For example, {3} is an equilateral triangle, {4} is a square, and so on. A regular polyhedron which has q regular p-sided polygon faces around each vertex is represented by $\{p, q\}$. For example, the cube has three squares around each vertex and is represented by {4, 3}. A regular four-dimensional polytope, with r $\{p, q\}$ regular polyhedral cells around each edge, is represented by $\{p, q, r\}$, and so on.

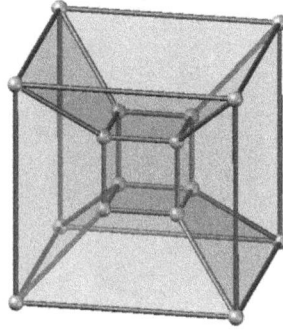

Figure 3.31. The eight cubical cells of the tesseract folded, in threes, around the same edge.

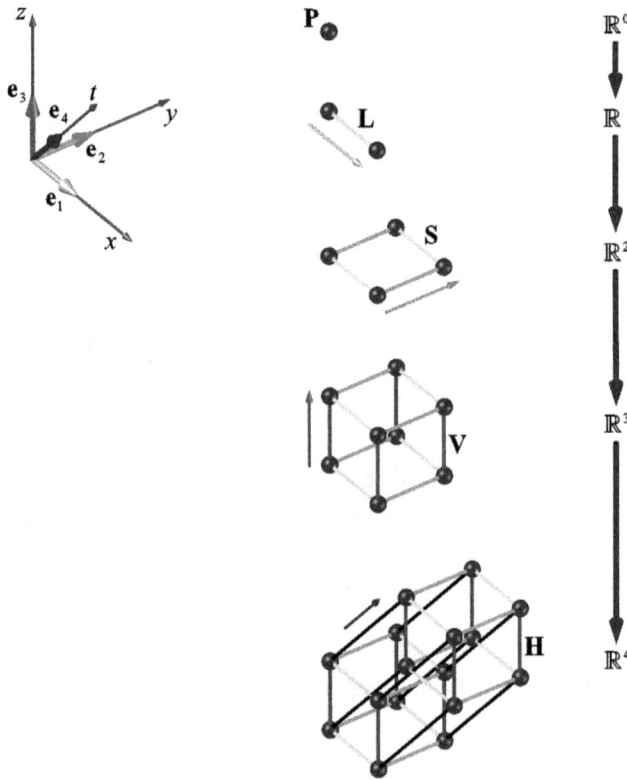

Figure 3.32. Inductive construction of a 4D hyperprism from dimension 0 to dimension 4, by adding one dimension at a time.

- As a 4D hyperprism. In this case, it is made of two parallel cubes (Fig. 3.32). The scheme is similar to the construction of a cube from two squares: juxtapose two copies of the lower-dimensional cube and connect the corresponding vertices.
- As a duoprism.[123] In this case, it is represented by the Cartesian product of two squares.

---

[123] In geometry of four dimensions or higher, a **duoprism** is a polytope resulting from the Cartesian product of two polytopes, each of two dimensions or higher. The Cartesian product of an n-polytope and an m-polytope is an $(n+m)$-polytope, where n and m are 2 (polygon) or higher.

- As an orthotope.[124]
- As the convex hull[98] of the points $(\pm 1, \pm 1, \pm 1, \pm 1)$ in Euclidean 4-space. In this case, it consists of the points:

$$\left\{\left(x_1, x_2, x_3, x_4\right) \in \mathbb{R}^4 : -1 \le x_i \le +1\right\} \tag{3.8.1}$$

and is bounded by eight hyperplanes ($x_i = \pm 1$).
- As bipartite graphs.[125] In this case, the eight cubes of the tesseract, taken in twos, are independent sets of points.
- As the **4-vector** (multivector of degree four), or Clifford number of degree four (Section 2.1.1), $\mathbf{u} \wedge \mathbf{v} \wedge \mathbf{w} \wedge \mathbf{t}$ of the exterior algebra, $\Lambda(V)$, on the vector space, V. This is the construction we will prefer in this book.

Note that the elements of a CM space/time 4-vector (a tesseract) are of different nature, since some p-cells are associated with a variation of the space variables, some other p-cells are associated with a variation of the time variables, and some other p-cells are associated with a variation of both the space and time variables. In particular, the points are associated with a variation of both the space and time variables. Therefore, we can say that there exists just one kind of points. As far as the others p-cells are concerned, on the contrary, we can define two different kinds of cells for each $p = 1, 2, 3$. We will denote

- 1-cells of the kind "space": the 1-cells that connect points associated with the same time instant, that is, the edges of the trivector $\mathbf{u} \wedge \mathbf{v} \wedge \mathbf{w}$ at a given instant (Fig. 3.33);
- 1-cells of the kind "time": the 1-cells that connect points associated with two adjacent time instants, that is, the time intervals (Fig. 3.33);
- 2-cells of the kind "space": the 2-cells that connect edges associated with the same time instant, that is, the faces of the trivector $\mathbf{u} \wedge \mathbf{v} \wedge \mathbf{w}$ at a given instant (Fig. 3.34);
- 2-cells of the kind "space/time": the 2-cells that connect edges associated with two adjacent time instants. The area of one of these faces is given by the product between a time interval and an edge of the trivector $\mathbf{u} \wedge \mathbf{v} \wedge \mathbf{w}$ (Fig. 3.34);
- 3-cells of the kind "space": the 3-cells that connect faces associated with the same time instant, that is, the volume of the trivector $\mathbf{u} \wedge \mathbf{v} \wedge \mathbf{w}$ at a given instant (Fig. 3.35);
- 3-cells of the kind "space/time": the 3-cells that are enclosed within faces associated with two adjacent time instants. The volume of one of these 3-cells is given by the product between a time interval and two edges of the trivector $\mathbf{u} \wedge \mathbf{v} \wedge \mathbf{w}$ (Fig. 3.35).

We can find a formally similar classification in the four-dimensional Minkowski spacetime, where the spacetime interval between two events is either space-like, light-like (null), or time-like.[61]

---

[124] In geometry, an **orthotope**, also called a **hyperrectangle** or a **box**, is the generalization of a rectangle for higher dimensions, formally defined as the Cartesian product of intervals. A special case of an n-orthotope, where all edges are equal length, is the n-hypercube.

[125] In the mathematical field of graph theory (Section 3.4), a **bipartite graph** (or **bigraph**) is a graph whose vertices can be divided into two disjoint sets U and V such that every edge connects a vertex in U to one in V, that is, U and V are each independent sets.

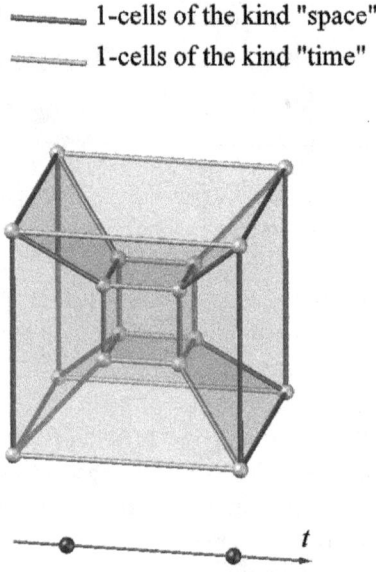

Figure 3.33. Different kinds of 1-cells in a space/time tesseract.

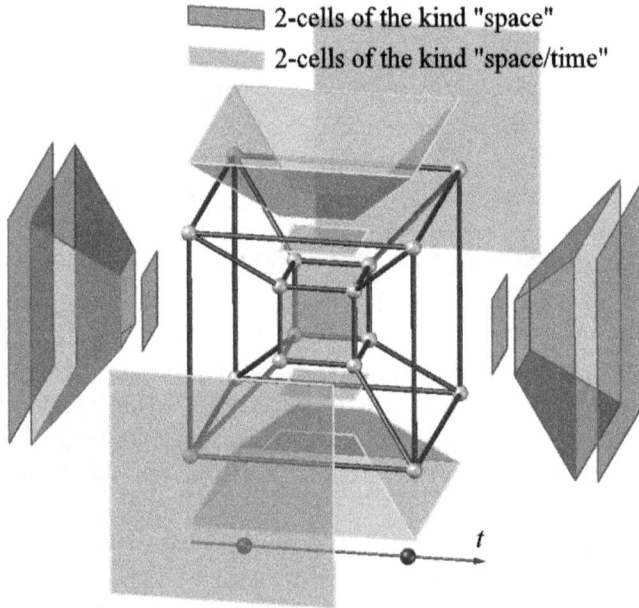

Figure 3.34. Different kinds of 2-cells in a space/time tesseract.

The inner and outer orientations of the 4-vector are shown in Fig. 3.36 for the faces of each of the eight cubes. In order to comply with the natural time sequence, from previous time instants to subsequent time instants, we will adopt a different orientation. In particular, by associating the elements of the left cube with the previous instant and the elements of the right cube with the subsequent instant, the eight edges connecting the left to the right cube turn

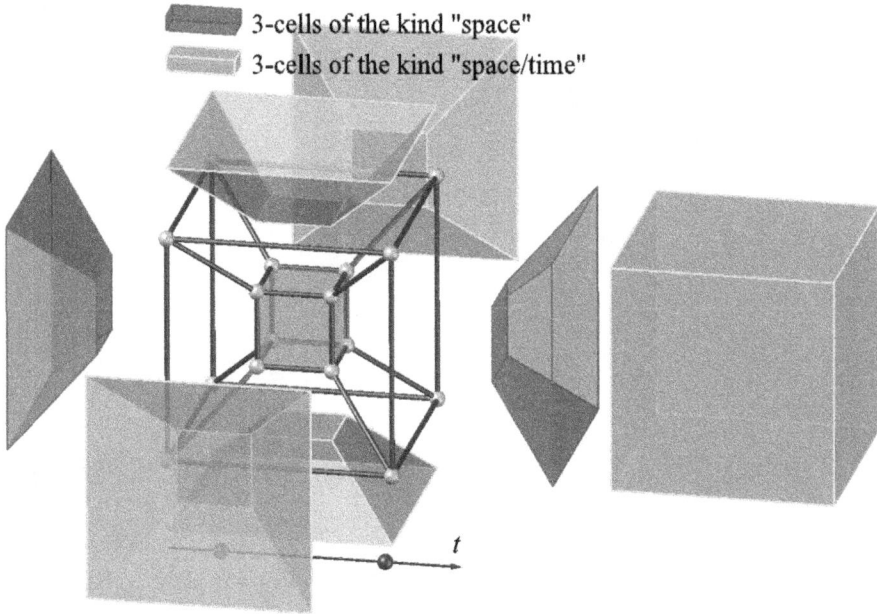

Figure 3.35. Different kinds of 3-cells in a space/time tesseract.

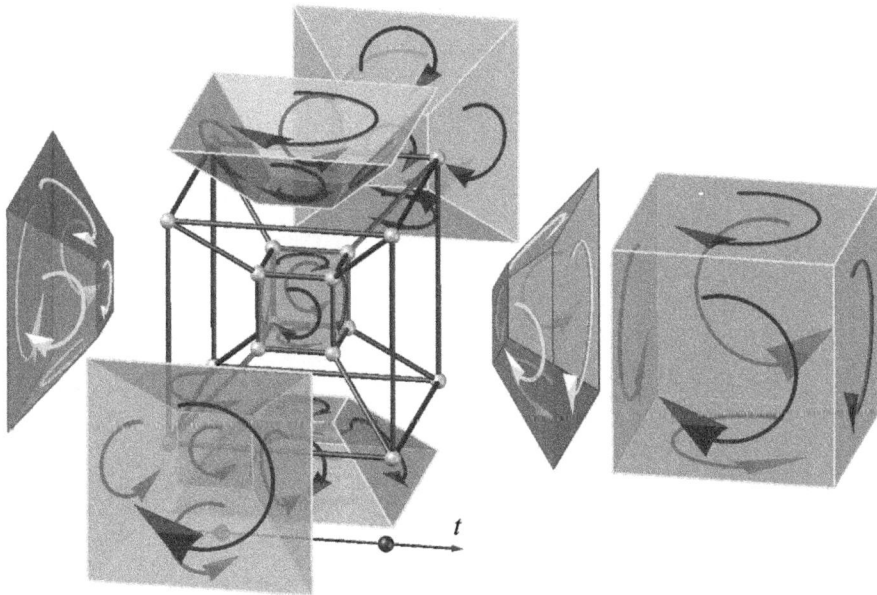

Figure 3.36. Inner orientations on the 2-cells of the 4-vector.

out to have an inner orientation from left to right, that is, the same orientation of the time axis (Fig. 3.37).

When we associate the global space and time variables with the oriented elements of a 4-vector, we obtain the algebraic version of a four-dimensional Minkowski continuum,[61] called

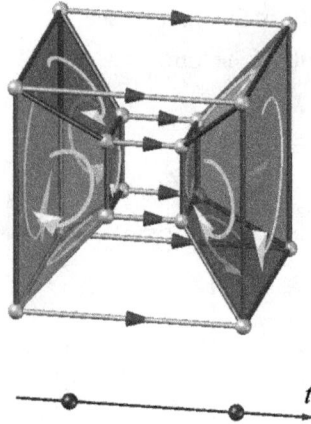

Figure 3.37. The CM tesseract: inner orientations on the 3-cells of the kind space and the 1-cells of the kind time.

spacetime, whose metric treats the time dimension differently from the three spatial dimensions. Spacetime is thus not a Euclidean space.

In the algebraic formulation, the inner orientation of the 1-cells of the kind "space/time" (time intervals) is the same as the orientation of the time axis (from past to future, in the natural time sequence). Moreover, since the same point of a four-dimensional space denotes both a point in space and a point in time (a time instant), it follows that the time instants have an inward inner orientation, that is, they are sinks.

CHAPTER 4

CLASSIFICATION OF THE GLOBAL VARIABLES AND THEIR RELATIONSHIPS

In Section 4.1, global variables are classified according to the role they play in a theory. This second type of classification distinguishes between configuration, source, and energetic variables. It is discussed how the configuration and the source variables have two different orientations, inner rather than outer orientation. This, together with the need to provide a description of vector spaces that is independent of the orientation of the embedding space, require to use two different cell complexes, whose geometrical elements stand by each other in relation of duality. The two cell complexes are called the primal and the dual cell complex. The elements of the first cell complex in space and first cell complex in time are associated with the variables endowed with an inner orientation, while the elements of the second cell complex in space and second cell complex in time are associated with the variables endowed with an outer orientation. As a consequence, the source variables are associated with the elements of the dual complex, and the configuration variables are associated with the elements of the primal complex. The configuration variables with their topological equations, on the one hand, and the source variables with their topological equations, on the other hand, define two vector spaces that are a bialgebra and its dual algebra. The operators of these topological equations are generated by the outer product of the geometric algebra, for the primal vector space, and by the dual product of the dual algebra, for the dual vector space. The topological equations in the primal cell complex are coboundary processes on even exterior discrete p-forms, while the topological equations in the dual cell complex are coboundary processes on odd exterior discrete p-forms. Being expressed by coboundary processes in two different vector spaces, compatibility and equilibrium can be enforced at the same time, with compatibility enforced on the primal cell complex and equilibrium enforced on the dual cell complex. The variables and the equations of the two vector spaces are stored in a classification diagram, made of two columns, one for each vector space. This diagram has the great merit of providing a unifying classification of the physical theories, both in algebraic and differential formulations. The equations and the operators of the classification diagram are analyzed in detail.

The classification diagram is also used, together with the properties of the boundary and coboundary operators, for finding the algebraic form of the virtual work theorem (Section 4.2).

Finally, the incidence matrices are derived for both the primal and dual cell complexes in space and space/time domains, and compared to each other (Sections 4.3 and 4.4).

4.1 Configuration, Source, and Energetic Variables

In Section 1.2, we have classified the physical variables as field and global variables. A further criterion for classifying the physical variables is based on the role they play in a theory. According to this second criterion, all physical variables belong to one of the following three classes:[126]

- **Configuration variables**, describing the field configuration (displacements for solid mechanics, spatial velocity for fluidodynamics, electric potential for electrostatics, temperature for thermal conduction, and so forth). All variables linked to configuration variables by operations of sum, difference, division by a length, division by an area, division by a volume, division by an interval, limit process, time and space derivatives, line integrals, surface integrals, volume integrals, and time integrals are configuration variables if and only if the operations do not contain physical variables.
- **Source variables**, describing the field sources (forces for solid mechanics and fluidodynamics, masses for geodesy, electric charges for electrostatics, electric currents for magnetostatics, heat sources for thermal conduction, and so forth). All variables linked to source variables by operations of sum, difference, division by a length, division by an area, division by a volume, division by an interval, limit process, time and space derivatives, line integrals, surface integrals, volume integrals, and time integrals are source variables if and only if the operations do not contain physical variables.
- **Energetic variables**, resulting from the multiplication of a configuration variable by a source variable (elastic energy density for solid mechanics, kinetic energy for dynamics, electrostatic energy for electrostatics, magnetostatic energy for magnetostatics, heat for thermal conduction, and so forth). All variables linked to energetic variables by operations of sum, difference, division by a length, division by an area, division by a volume, division by an interval, time and space derivatives, line integrals, surface integrals, volume integrals, and time integrals are energetic variables.

The equations used to relate the configuration variables of the same physical theory to each other and the source variables of the same physical theory to each other are known as **topological equations**. They can also be defined as those equations that express a relationship between a variable associated with a space element and a variable associated with the boundary of the same space element. Let $\mathbf{M}$ be a space element and let $\partial\mathbf{M}$ be its boundary, broadly speaking a topological equation:

$$A[\mathbf{M}] = \pm B[\partial\mathbf{M}]; \qquad (4.1.1)$$

is therefore expressed by one of the two following maps:

$$t_1 : \partial\mathbf{M} \to \mathbf{M}; \qquad (4.1.2)$$

$$t_2 : \mathbf{M} \to \partial\mathbf{M}. \qquad (4.1.3)$$

---

[126] The terms "configuration variables" and "source variables" were introduced by Tonti[1] in 1972. They correspond to the "geometric variables" and "force variables", respectively, used by Penfield and Haus (since 1967).

In the algebraic formulation, topological equations share some common features:

- They are valid for whatever shape and extent of the space elements involved. This is the reason why they can be called topological equations.[127]
- They are valid both in large scale and in small scale, that is, they are global equations.
- They do not contain material or system parameters; hence they are valid even across material discontinuities. This is the reason why they are used to find the jump conditions in the interfaces between two media.

The equations that relate configuration to source variables, of the same physical theory, are known as **constitutive equations**,[128] or **material equations**. They are phenomenological equations and specify the behavior of a material, a substance, or a media. Their main properties are

- they are local;
- they contain physical parameters and material constants, with vacuum being considered as a particular medium;
- they contain metric notions.

The constitutive relations can be reversible or irreversible.

The equations providing the configuration of a system, once the sources are assigned, are called the **fundamental equations**, and the related problem is called the **fundamental problem**. The set of fundamental equations defines the **fundamental system** of equations.

The topological equations of a fundamental problem are always maps of the type t_1 in Eq. (4.1.2). They can therefore be described in algebraic topology by using discrete p-forms (Section 3.7) and the coboundary operators.[129]

When the constitutive relations are reversible, it is also possible to find the sources once the configuration of the system is assigned. In this last case, the solving system is called the system of the **dual fundamental problem**, or **dual fundamental system**, and its equations are called the **dual fundamental equations**.

The topological equations of a dual fundamental problem are still maps of the type t_1. Therefore, even the topological equations of the dual fundamental problem can be described in algebraic topology by using discrete p-forms and coboundary operators. Consequently, the coboundary process plays a key role in both the fundamental problems in physics.

In Section 1.2 we have observed that the global physical variables have a natural association with one of the four space elements, **P**, **L**, **S**, and **V**, and one of the two time elements,

---

[127] In the differential formulation, the topological equations lose their topological nature, since they are often mixed with metric notions.

[128] In physics and engineering, a **constitutive equation**, or **constitutive law**, or **constitutive relation** is a relation between two physical quantities (especially kinetic quantities as related to kinematic quantities) that is specific to a material or substance and approximates the response of that material to external stimuli, usually as applied fields or forces. The first constitutive equation was developed by Robert Hooke (28 July [O.S. 18 July] 1635–3 March 1703) and is known as Hooke's law. It deals with the case of linear elastic materials. Walter Noll (born January 7, 1925) advanced the use of constitutive equations, clarifying their classification and the role of invariance requirements, constraints, and definitions of terms like "material", "isotropic", "aeolotropic", and so forth.

[129] In the differential formulation, the topological equations can be described by scalar or vector valued exterior differential forms and by the exterior differential.

I and **T**. Moreover, in Sections 2.2.2 and 3.8 we have seen how the space and time elements are provided with two kinds of orientations, inner and outer, in relation of duality between them. Thus, global physical variables are associated with oriented space and time elements.

A more accurate analysis of this association shows that some global variables are associated with space or time elements provided inner orientations,[130] while some other global variables are associated with space or time elements provided with outer orientations.[131]

The main criterion[132] for discriminating whether a physical variable is associated with a space element endowed with inner or outer orientation is given by the **Oddness principle**,[133] which is based on an experimental evidence for many global variables:

---

[130] The magnetic flux, Φ, and the vortex flux, W, are associated with surfaces endowed with inner orientations. The voltage, E, the line integral of the velocity, Γ, and the work, W, of a force in a force field are associated with lines endowed with inner orientations. The hypothetical magnetic charge, $\overset{\vee}{G}$, is associated with a volume endowed with inner orientation.

[131] The electric flux, Ψ, the mass flow, M^f, the energy flow, E^f, the charge flow, Q^f, the entropy flow, S^f, the momentum flow, $\mathbf{P}^f$, and the heat, Q, are associated with surfaces endowed with outer orientations. The magnetomotive force, F_m, is associated with a line endowed with outer orientation. Almost all the contents, but the hypothetical magnetic charge, $\overset{\vee}{G}$, are associated with volumes endowed with outer orientations, for example, the mass content, M^c, the energy content, E^c, the charge content, Q^c, the entropy content, S^c, the momentum content, $\mathbf{P}^c$, and the angular momentum content, $\overset{\vee}{L^c}$.

[132] Another criterion is given by the name itself of the variable Q:

- Names containing the terms **"produced"**, **"stored"**, **"released"**, **"generated"**, **"dissipated"**, **"source"**, and **"supply"** are typical of variables associated with volumes endowed with outer orientation and time intervals endowed with inner or outer orientation. Consequently, $Q = Q\left[\overline{\mathbf{T}}, \widetilde{\mathbf{V}}\right]$, or $Q = Q\left[\widetilde{\mathbf{T}}, \widetilde{\mathbf{V}}\right]$.
- Names containing the terms **"content"**, **"amount"**, **"volume density"**, and **"specific"** are typical of variables associated with volumes endowed with outer orientation and time instants endowed with inner or outer orientation. Consequently, $Q = Q\left[\overline{\mathbf{I}}, \widetilde{\mathbf{V}}\right]$, or $Q = Q\left[\widetilde{\mathbf{I}}, \widetilde{\mathbf{V}}\right]$.
- Names containing the terms **"absorbed"**, **"emitted"**, **"transmitted"**, **"flow"**, and **"current"** are typical of variables associated with surfaces endowed with outer orientation and time intervals endowed with inner or outer orientation. Consequently, $Q = Q\left[\overline{\mathbf{T}}, \widetilde{\mathbf{S}}\right]$, or $Q = Q\left[\widetilde{\mathbf{T}}, \widetilde{\mathbf{S}}\right]$.
- Names containing the term **"flux"** are typical of variables associated with surfaces endowed with inner or outer orientation and time instants endowed with inner or outer orientation. Consequently, $Q = Q\left[\overline{\mathbf{I}}, \widetilde{\mathbf{S}}\right]$, $Q = Q\left[\widetilde{\mathbf{I}}, \widetilde{\mathbf{S}}\right]$, $Q = Q\left[\overline{\mathbf{I}}, \overline{\mathbf{S}}\right]$, or $Q = Q\left[\widetilde{\mathbf{I}}, \overline{\mathbf{S}}\right]$.
- Names containing the terms **"potential difference"**, **"voltage"**, **"strength"**, **"thermodynamic force"**, are typical of variables associated with lines endowed with inner or outer orientation and time intervals endowed with inner or outer orientation. Consequently, $Q = Q\left[\overline{\mathbf{T}}, \widetilde{\mathbf{L}}\right]$, $Q = Q\left[\widetilde{\mathbf{T}}, \widetilde{\mathbf{L}}\right]$, $Q = Q\left[\overline{\mathbf{T}}, \overline{\mathbf{L}}\right]$, or $Q = Q\left[\widetilde{\mathbf{T}}, \overline{\mathbf{L}}\right]$.
- Names containing the term **"circulation"** are typical of variables associated with lines endowed with inner or outer orientation and time instants endowed with inner or outer orientation. Consequently, $Q = Q\left[\overline{\mathbf{I}}, \widetilde{\mathbf{L}}\right]$, $Q = Q\left[\widetilde{\mathbf{I}}, \widetilde{\mathbf{L}}\right]$, $Q = Q\left[\overline{\mathbf{I}}, \overline{\mathbf{L}}\right]$, or $Q = Q\left[\widetilde{\mathbf{I}}, \overline{\mathbf{L}}\right]$.
- Names containing the term **"potential"** are typical of variables associated with points endowed with inner or outer orientation and with time intervals endowed with inner or outer orientation. Consequently, $Q = Q\left[\overline{\mathbf{T}}, \widetilde{\mathbf{P}}\right]$, $Q = Q\left[\widetilde{\mathbf{T}}, \widetilde{\mathbf{P}}\right]$, $Q = Q\left[\overline{\mathbf{T}}, \overline{\mathbf{P}}\right]$, or $Q = Q\left[\widetilde{\mathbf{T}}, \overline{\mathbf{P}}\right]$.
- Names containing the term **"function"** are typical of variables associated with points endowed with inner or outer orientation and time instants endowed with inner or outer orientation. Consequently, $Q = Q\left[\overline{\mathbf{I}}, \widetilde{\mathbf{P}}\right]$, $Q = Q\left[\widetilde{\mathbf{I}}, \widetilde{\mathbf{P}}\right]$, $Q = Q\left[\overline{\mathbf{I}}, \overline{\mathbf{P}}\right]$, or $Q = Q\left[\widetilde{\mathbf{I}}, \overline{\mathbf{P}}\right]$.

[133] The Oddness principle is implicit in the requirement of additivity for physical variables. It is also required for the validity of Green's and Stokes' theorems. The Oddness principle expresses the independence on the orientation of the space and time elements of the form of the physical laws. In this sense, it may be considered the analogous, for the algebraic formulation, of the principle according to which the physical laws must be independent of the coordinate system chosen. In effect, the need of introducing this principle is a direct consequence of having substituted the discrete

(continues next page)

A global physical variable associated with an oriented space or time element changes in sign when the orientation of the element is inverted.

In particular, if the sign of a space variable, Q, changes when we reverse the inner orientation of the related space element, then the variable Q is associated with an element endowed with inner orientation. Conversely, if the sign does not change, then the space variable Q is associated with an element endowed with outer orientation.

Analogously, if the sign of a time variable, Q, changes when we perform a reversal of the motion, then the variable Q is associated or with time intervals endowed with inner orientation, hence $Q\left[\overline{T}\right]$, or with time instants endowed with outer orientation, hence $Q\left[\tilde{I}\right]$, depending on which time element it is associated with. In contrast, if the sign does not change, then the variable Q is associated with time instants endowed with inner orientation, hence $Q\left[\overline{I}\right]$, or with time intervals endowed with outer orientation, hence $Q\left[\tilde{T}\right]$, depending on which time element it is associated with.

If we consider all the combinations between oriented space and oriented time elements related to physical variables, we see that there are 32 possible couples of space/time elements in physics. These 32 couples can, in turn, be divided into two groups (Fig. 4.1), each consisting of 16 elements. In the first group, there are the couples of time and space elements that are endowed with the same kind of orientation, either inner or outer, while, in the second group, there are the couples of time and space elements that are endowed with opposite kinds of orientation, one inner and the other outer.

Belonging to one group rather than another group is not accidental for the physical variables. In fact, the physical variables of the first group are those of the mechanical theories,[134] while the physical variables of the second group are those of the field theories.[135]

As far as the field theories are concerned, it was found that the configuration variables of the fundamental problem of any field theory are associated with space elements endowed with an inner orientation, $\overline{P}, \overline{L}, \overline{S}$, and $\overline{V}$, while the source variables are associated with space elements endowed with an outer orientation, $\tilde{P}, \tilde{L}, \tilde{S}$, and $\tilde{V}$.

This becomes a key point in computational physics, when we relate it with the discussion on the inner and outer orientations of a vector space and its dual vector space (Sections 2.2.2.1 and 2.2.2.2). In fact, we have seen that, by providing the elements of a vector space with an inner orientation, the elements of the dual vector space turn out to be automatically provided with an outer orientation,[136] as a consequence of the Riesz representation theorem

---

(continues from previous page)

set of points and their discrete mapping, established by the coordinate systems of the differential setting, with the cell complexes and their labeling.

[134] Examples of mechanical theories include particle mechanics, analytical mechanics, mechanics of fluids, and fluid dynamics.

[135] A field theory is a physical theory that describes how one or more physical fields interact with matter. Field theory usually refers to a construction of the dynamics of a field, that is, a specification of how a field changes with time or with respect to other independent physical variables on which the field depends. Examples of field theories include electromagnetism, gravitation, thermal conduction, diffusion, and irreversible thermodynamics.

[136] It is always true also the vice-versa: by providing the elements of a dual vector space with an inner orientation, the elements of the related vector space turn out to be automatically provided with an outer orientation.

	$\bar{I}$	$\bar{T}$		$\tilde{I}$	$\tilde{T}$		$\tilde{I}$	$\tilde{T}$		$\bar{I}$	$\bar{T}$
$\bar{P}$	$[\bar{I},\bar{P}]$	$[\bar{T},\bar{P}]$	$\tilde{P}$	$[\tilde{I},\tilde{P}]$	$[\tilde{T},\tilde{P}]$	$\bar{P}$	$[\tilde{I},\bar{P}]$	$[\tilde{T},\bar{P}]$	$\tilde{P}$	$[\bar{I},\tilde{P}]$	$[\bar{T},\tilde{P}]$
$\bar{L}$	$[\bar{I},\bar{L}]$	$[\bar{T},\bar{L}]$	$\tilde{L}$	$[\tilde{I},\tilde{L}]$	$[\tilde{T},\tilde{L}]$	$\bar{L}$	$[\tilde{I},\bar{L}]$	$[\tilde{T},\bar{L}]$	$\tilde{L}$	$[\bar{I},\tilde{L}]$	$[\bar{T},\tilde{L}]$
$\bar{S}$	$[\bar{I},\bar{S}]$	$[\bar{T},\bar{S}]$	$\tilde{S}$	$[\tilde{I},\tilde{S}]$	$[\tilde{T},\tilde{S}]$	$\bar{S}$	$[\tilde{I},\bar{S}]$	$[\tilde{T},\bar{S}]$	$\tilde{S}$	$[\bar{I},\tilde{S}]$	$[\bar{T},\tilde{S}]$
$\bar{V}$	$[\bar{I},\bar{V}]$	$[\bar{T},\bar{V}]$	$\tilde{V}$	$[\tilde{I},\tilde{V}]$	$[\tilde{T},\tilde{V}]$	$\bar{V}$	$[\tilde{I},\bar{V}]$	$[\tilde{T},\bar{V}]$	$\tilde{V}$	$[\bar{I},\tilde{V}]$	$[\bar{T},\tilde{V}]$

first group   second group

Figure 4.1. Classification of the space and time elements related to the physical variables.

(Section 2.1.2). Now, due to the geometrical interpretation of the elements of the vector spaces, given by the geometric algebra, we can associate the elements of the two vector spaces with the geometrical elements of two cell complexes, where the elements of the second cell complex are the orthogonal complements of the corresponding elements in the first cell complex. Due to this association, by providing the elements of the first cell complex with an inner (or an outer) orientation, we induce an outer (or an inner) orientation on the second cell complex. This suggests us two considerations:

- Due to the inner rather than outer orientations of the configuration and source variables, the space of the configuration variables may be viewed as a real (or complex) inner product Hilbert space,[45] H, and the space of the source variables may be viewed as its dual space, consisting of all continuous linear functionals from H into the field $\mathbb{R}$ (or the field $\mathbb{C}$). For example, a force (which is a source variable) is a covector on the space of the configuration variables because the force acts on the displacement vector (which is a configuration variable) by originating the real scalar that represents the work of the force.
- Since the source variables requires an outer orientation, a proper description of a given physical phenomenon requires to use two cell complexes in relation of duality,[137] not just one, as usually was done in computational physics before the introduction of the CM. In fact, it is true that the inner orientation of the elements of a vector space also induces an outer orientation on the elements of the same vector space and this may allow us to think that a single cell complex would be sufficient. Nevertheless, remember that the association between the two orientations of the same cell complex is not automatic (Section 2.2.2.2). There are always two possible criteria for establishing the correspondence between the two orientations, which depend on the orientation of the embedding space. Conversely, the relationship between inner (or outer) orientation of a cell complex and outer (or inner) orientation of its dual cell complex is derived from the Riesz representation theorem and does not depend on the orientation of the embedding space. Therefore, choosing to use two cell complexes, the one the dual of the other, instead

---

[137] The association of physical variables to elements of a cell complex and its dual was introduced by Okada and Onodera (1951) and Branin (1966). In particular, Branin had the idea of extending the duality that exists in graph theory, between a graph and its dual (Section 3.4), to the cell complexes, and used a cell complex in space and its dual for describing an electromagnetic field.

of one single cell complex, is motivated by the need to provide a description of vector spaces that is independent of the orientation of the embedding space.

We will call the first complexes in space and time the **primal cell complexes**, or **primal complexes**, and the second complexes in space and time the **dual cell complexes**, or **dual complexes**.

Remembering that the cell complexes are generalizations of the oriented graphs, all the properties of the dual graphs naturally extend to the dual cell complexes. In particular, we have seen that the dual graphs depend on a particular embedding (Section 3.4). Since even the orthogonal complements (that is, the isomorphic dual vectors) and the outer orientation depend on the embedding (Section 2.2.2.2), we will associate the outer orientation with the dual cell complex and will retain the inner orientation for the primal cell complex (Fig. 4.2). This is why we will not take into consideration the possibility of providing the primal cell complex with an outer orientation and the dual cell complex with an inner orientation.

In doing so, the elements of the first cell complex in space and first cell complex in time can be associated with those variables that require an inner orientation of the cell complex, while the elements of the second cell complex in space and second cell complex in time can be associated with those variables that require an outer orientation of the cell complex.

From the relationship between global variables and orientations, it follows that source variables are always associated with the elements of the dual complex only, and configuration variables are always associated with the elements of the primal complex only (some examples for different physical theories are collected in Fig. 4.3).

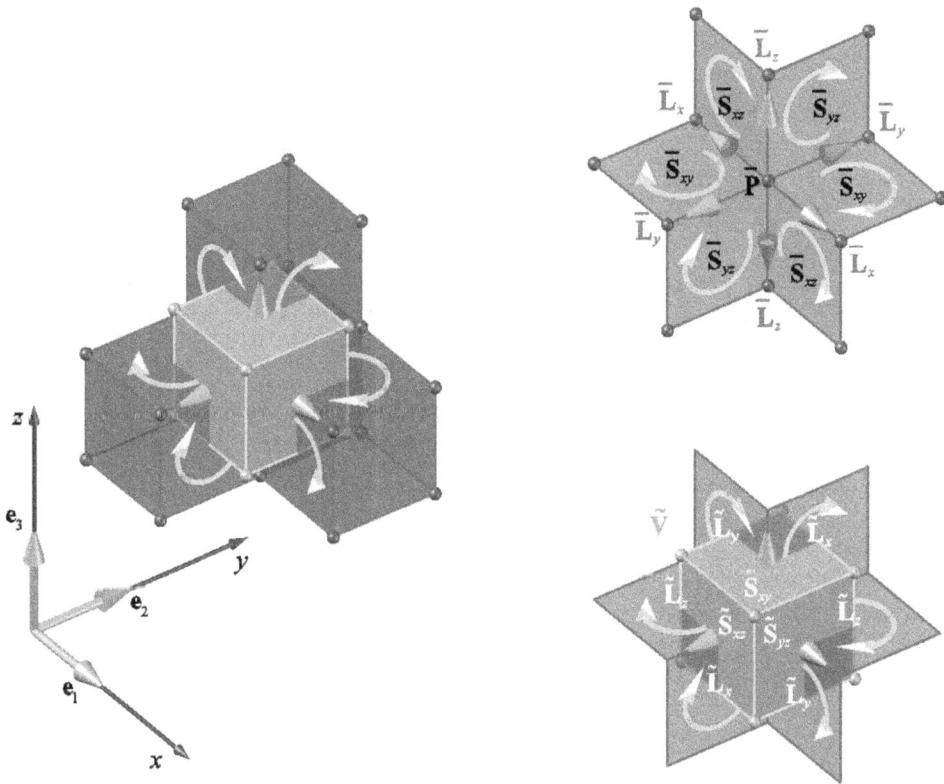

Figure 4.2. Relation of duality in three-dimensional space, between inner orientations of the primal cells and outer orientations of the dual cells.

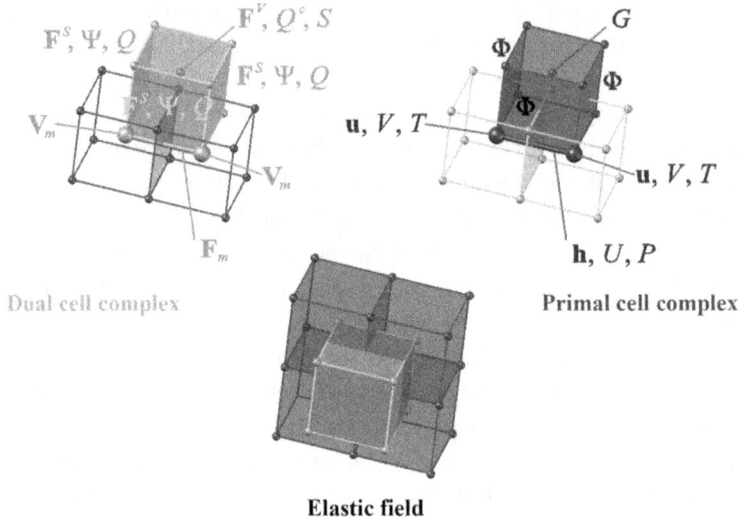

Elastic field

The **volume force** $\mathbf{F}^V$ is referred to Volumes $\tilde{\mathrm{V}}$ of the dual cell complex	The **displacement** $\mathbf{u}$ is referred to Points $\overline{\mathrm{P}}$ of the primal cell complex
The **surface force** $\mathbf{F}^S$ is referred to Surfaces $\tilde{\mathrm{S}}$ of the dual cell complex	The **relative displacement** $\mathbf{h}$ is referred to Lines $\overline{\mathrm{L}}$ of the primal cell complex

Electric field

The **electric charge content** Q^c is referred to Volumes $\tilde{\mathrm{V}}$ of the dual cell complex	The **potential** V is referred to Points $\overline{\mathrm{P}}$ of the primal cell complex
The **electric flux** Ψ is referred to Surfaces $\tilde{\mathrm{S}}$ of the dual cell complex	The **tension** U is referred to Lines $\overline{\mathrm{L}}$ of the primal cell complex

Magnetic field

The **magnetic tension** F_m is referred to Lines $\tilde{\mathrm{L}}$ of the dual cell complex	The **magnetic flux** Φ is referred to Surfaces $\overline{\mathrm{S}}$ of the primal cell complex
The **magnetic potential** V_m is referred to Points $\tilde{\mathrm{P}}$ of the dual cell complex	The **magnetic charge** G is referred to Volumes $\overline{\mathrm{V}}$ of the primal cell complex

Thermal field

The **heat production** S is referred to Volumes $\tilde{\mathrm{V}}$ of the dual cell complex	The **temperature** T is referred to Points $\overline{\mathrm{P}}$ of the primal cell complex
The **heat** Q is referred to Surfaces $\tilde{\mathrm{S}}$ of the dual cell complex	The **temperature difference** P is referred to Lines $\overline{\mathrm{L}}$ of the primal cell complex

Figure 4.3. Association between global variables and space elements of the primal and dual cell complexes, in different physical theories.

The most natural way for building the two cell complexes is starting from a primal cell complex made of simplices and providing this first cell complex with an arbitrary inner orientation. The set of the dual elements can then be chosen as any arbitrary set of staggered elements whose outer orientations provide the (known) inner orientations of the primal p-cells. In this sense, we can say that the outer orientations of the dual p-cells are induced by the inner orientations of the primal p-cells (Fig. 4.2).

We have spoken of "any" arbitrary set of staggered elements because, as already discussed in Section 2.2.2.2, since the dual elements are equipped with the strong topology, there is not a unique way for defining the dual elements. In particular, they may also overlap. When they do not overlap, each p-space element of the dual cell complex can be put in dual correspondence with one $(n-p)$-space element of the primal cell complex, staggered with respect to the former one, where n is the dimension of the space. In particular, in three-dimensional space

- each node of the dual complex is contained in one volume of the primal complex,
- each edge of the dual complex intersects a face of the primal complex,
- each face of the dual complex is intersected by one edge of the primal complex,
- each volume of the dual complex contains one node of the primal complex.

By associating the configuration variables with the primal p-cells, the set of topological equations between global configuration variables defines a geometric algebra (Section 2.2) on the space of global configuration variables, provided with a geometric product. The operators of these topological equations are generated by the outer product of the geometric algebra, which is equal to the exterior product of the enclosed exterior algebra. The dual algebra of the enclosed exterior algebra is the space of global source variables, associated with the dual p-cells, and is provided with a dual product that is compatible with the exterior product of the exterior algebra. The topological equations between global source variables arise from the adjoint operators of the primal operators. Finally, the pairing between the exterior algebra and its dual gives rise to the energetic variables, by the interior product. Since the reversible constitutive relations may be written in terms of energetic variables, because energy is the potential of the reversible constitutive relations, the reversible constitutive relations realize the pairing between the exterior algebra and its dual.

In algebraic topology, the topological equations on the primal cell complex are coboundary processes on even exterior discrete p-forms, while the topological equations on the dual cell complex are coboundary processes on odd exterior discrete p-forms.

When we deduce the field variables from the corresponding global variables, the exterior discrete forms become exterior differential forms. In particular, while the configuration variables can be described by exterior differential forms of even kind, the source variables can be described by differential forms of odd kind (that is, twisted differential forms). Moreover, the elements of the group $\mathcal{G}$ are vectors instead of scalars, and the corresponding differential form is a vector valued differential form.

The association between the physical variables, with their topological equations, and two vector spaces, which are a bialgebra and its dual algebra, suggests us to store the global variables in a classification diagram made of two columns,[138] that is, the column of the primal vector space,

---

[138] The classification diagram of physical quantities and equations is known in the literature as **Tonti diagram**, named after Enzo Tonti (born October 30, 1935).[1]

(*continues next page*)

Source variables
Dual vector space

Configuration variables
Primal vector space

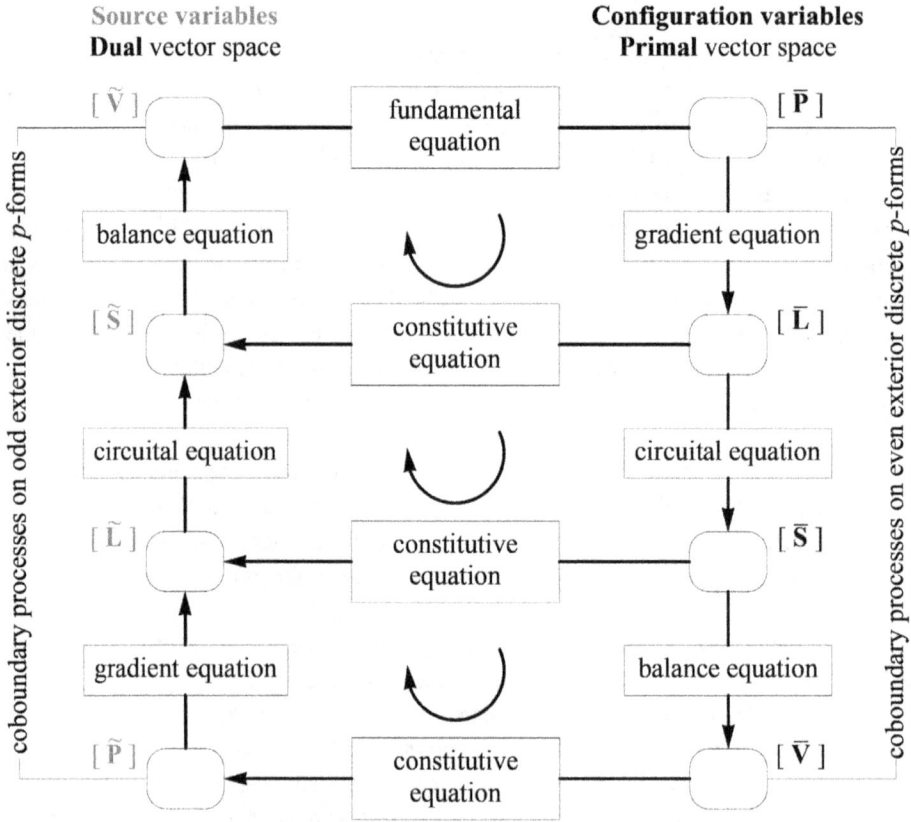

Figure 4.4. Classification diagram of the physical variables in the fundamental problem.

composed of the configuration variables with their topological equations, and the column of the dual vector space, composed of the source variables with their topological equations (Fig. 4.4).

The configuration variables are arranged from top to bottom in their column, in order of increasing multiplicity of the associated space element, thus realizing a downward cochain. Conversely, the source variables are arranged from bottom to top in their column, in order of increasing multiplicity of the associated space element, thus realizing an upward cochain. With this choice, each primal p-cell is at the same level of its dual $(n - p)$-cell.

(*continues from previous page*)

The idea of associating the physical variables with algebraic tools originally dates back to Gabriel Kron (1901–1968), a Hungarian American electrical engineer of General Electric who promoted the use of methods of linear algebra, multilinear algebra, and differential geometry in the field. Though he published widely, his methods were slow to be assimilated. The topic was later developed by Paul Roth, a mathematician who realized the role of algebraic topology in network analysis. In 1955, Roth introduced a diagram made of chains and cochains for studying electrical circuits. He considered only static and stationary fields. Therefore, Roth's diagrams do not include time. In 1966, Branin completed the diagrams of Roth by adding the time derivatives, thus coupling the diagram of electrostatics with the one of magnetostatics. Moreover, since Branin also introduced in electromagnetism the use of two cell complexes,[137] the one the dual of the other, he abandoned the chain and cochain sequences of Roth in favour of two cochain sequences. In 1981, Deschamps published a paper dealing with exterior differential forms in electromagnetism, organized in sequences of cochains.

The solid lines indicate the constitutive relations between (primal) configuration variables and (dual) source variables. They also represent the pairing between configuration and source variables.

Note that the structure of the classification diagram is the same both for the global and the field variables of every physical theory of the macrocosm. The importance of this diagram stands just in its ability of providing a concise description of physical variables, without distinguishing between the physical theories.

As observed in Tonti (2013), even the variables and the equations of relativistic quantum mechanics for particles with integer spins can be arranged in a diagram, which is formally similar to the classification diagram. This leads us to assume that even the operators used in quantum mechanics for describing the microcosm can be associated with space and time elements. As a proof of how this is actually possible, Tonti remarks that both Bohm[139] and Schönberg[140] used algebraic topology for treating quantum mechanics. Viewing the algebraic formulation of the CM as a geometric algebra allows us, now, to provide an explanation of why the tables for quantum mechanics, shown in Tonti (2013), are formally similar to the classification diagrams provided by Tonti for the macrocosm. In fact, being a geometric algebra and, as such, a quantization of the exterior algebra (Section 2.2.1), the algebraic formulation of the CM, as the geometric algebra, can be successfully employed for providing compact and intuitive descriptions even in quantum mechanics.

The strict relationship between geometric algebras, on one hand, and quantum physics and quantum field theory, on the other hand, was highlighted by Schönberg himself, in a series of publications of 1957/1958. Schönberg pointed out that those algebras can be described in terms of extensions of the commutative and the anti-commutative Grassmann algebras (Section 2.1),

---

[139] **David Joseph Bohm** (20 December 1917–27 October 1992) was an American theoretical physicist who contributed innovative and unorthodox ideas to quantum theory, philosophy of mind, and neuropsychology. He is widely considered to be one of the most significant theoretical physicists of the 20th century. As a post-graduate at Berkeley, he developed a theory of plasmas, discovering the electron phenomenon known now as Bohm-diffusion. His first book, *Quantum Theory* published in 1951, was well received by Einstein, among others. However, Bohm became dissatisfied with the orthodox interpretation of quantum theory, which he had written about in that book. Bohm's aim was not to set out a deterministic, mechanical viewpoint, but rather to show that it was possible to attribute properties to an underlying reality, in contrast to the conventional approach. He began to develop his own interpretation (De Broglie–Bohm theory), the predictions of which agree perfectly with the nondeterministic quantum theory. He initially referred to his approach as a hidden variable theory, but later referred to it as *ontological theory*, reflecting his view that a stochastic process that would underlie the phenomena described by his theory may be found. Bohm and his colleague Basil J. Hiley (born 1935) later stated that they found their own choice of terms of an "interpretation in terms of hidden variables" to be too restrictive, in particular as their variables, position and momentum, "are not actually hidden". Bohm's work and the EPR argument[188] became the major factor motivating John Bell's inequality,[191] which rules out local hidden variable theories. Due to his youthful Communist affiliations, Bohm was targeted during the McCarthy era, leading him to leave the United States. He pursued his scientific career in several countries, becoming a Brazilian, and later, a British citizen.

[140] **Mário Schenberg** (var. *Mário Schönberg, Mario Schonberg, Mário Schoenberg;* July 2, 1914–November 10, 1990), was a Jewish Brazilian electrical engineer, physicist, art critic and writer. In the University of São Paulo, Schönberg had interacted closely with David Bohm[139] during the final years of Bohm's exile in Brazil, and in 1954 Schönberg demonstrated a link among the quantized motion of the Madelung fluid and the trajectories of the de Broglie–Bohm theory. In a paper published in 1958, Schönberg suggested to add a new idempotent to the Heisenberg algebra, and this suggestion was taken up and expanded upon in the 1980s by Basil J. Hiley (born 1935) and his co-workers in their work on algebraic formulations of quantum mechanics. Schönberg's ideas have also been cited in connection with algebraic approaches to describe relativistic phase space. His work has been cited, together with that of Marcel Riesz (16 November 1886–4 September 1969), the younger brother of the mathematician Frigyes Riesz,[46] for its importance to Clifford algebras and mathematical physics in the proceedings of a workshop held in France in 1989, which had been dedicated to these two mathematicians.

which have the same structure as the *boson algebra* and the *fermion algebra* of creation and annihilation operators.[80] These algebras, in turn, are related to the symplectic algebra[141] and Clifford algebra, respectively. Also Hiley[142] worked on the algebraic descriptions of quantum physics in terms of underlying symplectic and orthogonal Clifford algebras.

When the physical phenomenon evolves in time, we have so many classification diagrams of the type shown in Fig. 4.4 as the time instants are. Since it is not possible to draw a classification diagram for each time instant, we simply double the diagram in Fig. 4.4 and shift it to the rear (Fig. 4.5).

The choice of two mutually dual cell complexes also allows us to improve the description of global variables in computational physics. In fact, in the spirit of geometric algebra (Section 2.2), where the oriented space elements are p-vectors generated by the exterior product, the attitude vectors of the p-cells are given by the inner orientation of their dual elements, the $(n-p)$-cells. This means that two mutually dual cell complexes allows us to describe all the attributes of the p-vectors, that is, attitude vector, orientation, and magnitude. Conversely, by using just one cell complex, we cannot describe the attitude vector, but only the (unoriented attitude).

In conclusion, by associating the global variables with the elements of two mutually dual cell complexes, the consequence is twofold:

- The set of the configuration variables, together with their topological equations, is a particular case of bialgebra. This leads us to enforce compatibility and equilibrium at the same time, with compatibility enforced on the primal cell complex and equilibrium enforced on the dual cell complex.

---

[141] A **symplectic vector space** is a vector space V (over a field, for example the real numbers $\mathbb{R}$) equipped with a bilinear form:

$$\omega : V \times V \to \mathbb{R} \, ;$$

that is:

- Skew-symmetric:
$$\omega(u,v) = -\omega(v,u) \quad \text{for all } u, v \in V \, .$$

- Totally isotropic:
$$\omega(v,v) = 0 \quad \text{for all } v \in V \, .$$

- Nondegenerate: if
$$\omega(u,v) = 0 \quad \text{for all } v \in V \, ;$$

then
$$u = 0 \, .$$

The bilinear form ω is said to be a **symplectic form**.

[142] **Basil J. Hiley** (born 1935), was a long-time co-worker of David Bohm.[139] Hiley co-authored the book *The Undivided Universe* with David Bohm, which is considered the main reference for Bohm's interpretation of quantum theory. Hiley and his co-worker Fabio A. M. Frescura expanded on the notion of an *implicate order* by building on the work of Fritz Eduard Josef Maria Sauter (1906–1983) and Marcel Riesz,[140] who had identified spinors with minimal left ideals of an algebra. The identification of *algebraic spinors* with minimal left ideals, which can be seen as a generalization of the ordinary spinor was to become central to the Birkbeck group's work on algebraic approaches to quantum mechanics and quantum field theory. Frescura and Hiley considered algebras that had been developed in the 19th century by the mathematicians Hermann Günther Grassmann (April 15, 1809–September 26, 1877), Sir William Rowan Hamilton (midnight, 3–4 August 1805–2 September 1865), and William Kingdon Clifford (4 May 1845–3 March 1879).

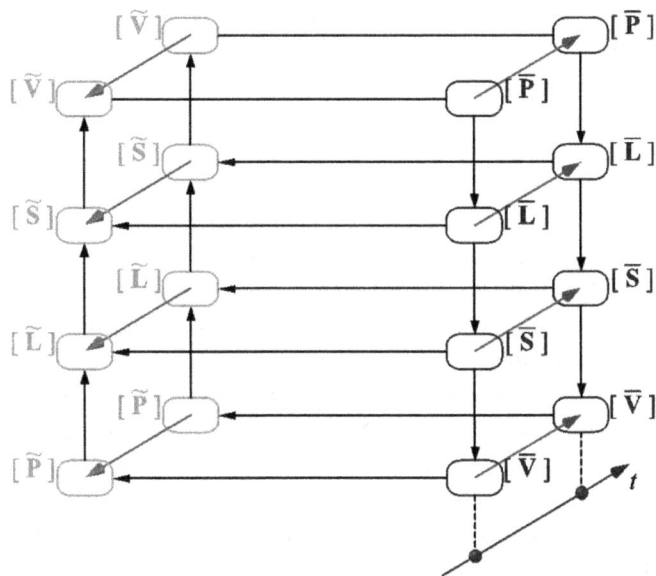

Figure 4.5. Space–time classification diagram of the physical variables.

- The description of both the configuration and the source variables is improved, by allowing us to automatically take into account the attitude vectors of the p-vectors, which is impossible when the outer orientation of cell complexes is ignored.

By overturning the point of view, that is, by assuming these two conclusions as our starting point, and not as the consequence, we can find in these properties, in particular the first one, the reason why the configuration variables are associated with space elements endowed with a kind of orientation and the source variables are associated with space elements endowed with the other kind of orientation. In effect, the fact that the equilibrium operators in the fundamental problem of a given physical theory are adjoint operators of the compatibility operators does not depend on the used computational tool. It does not even depend on computation. It is a general property of the fundamental problem and, consequently, we can take it as our starting point.

In particular, by assuming for the orientation of volumes their positive orientation, the inward orientation, the relationships between equilibrium operators on source variables, grad*, div*, and curl*, and compatibility operators on configuration variables, grad, div, and curl, are

$$\text{div*} = \text{grad}^{\text{T}}; \tag{4.1.4}$$

$$\text{curl*} = \text{curl}^{\text{T}}; \tag{4.1.5}$$

$$\text{grad*} = \text{div}^{\text{T}}; \tag{4.1.6}$$

while by assuming for the orientation of volumes their negative orientation, the outward orientation (as usual), Eq. (4.1.4) is changed in

$$\text{div*} = -\text{grad}^{\text{T}}. \tag{4.1.7}$$

Due to the relationship between a basis of a given vector space and its dual basis (Section 2.1.3), the adjoints in Eqs. (4.1.4 – 4.1.6) indicate that it is always possible to choose the orientation of volumes in the way that the set of configuration variables, with their topological equations, is a bialgebra. Being elements of a space vector, the configuration variables are provided with inner orientations and their covectors—which, in this case, are the source variables—are provided with outer orientations.

Finally, the possibility of formulating a dual fundamental problem when the constitutive laws are reversible suggests us that, in this second case, the role of bialgebra is played by the source variables, together with the dual exterior product (leading to the topological equations between source variables). Thus, the source variables are now provided with inner orientations, while the configuration variables of the dual exterior algebra are provided with outer orientations. In this second case, we will denote the source variables as the dual configuration variables and the configuration variables as the dual source variables. The classification diagram for the dual fundamental problem is shown in Fig. 4.6.

Consequently, for the computational solution of the dual fundamental problems, we have to associate the source variables (dual configuration variables) with the elements of the primal cell complex and the configuration variables (dual source variables) with the elements of the dual cell complex. This is always possible as, in a relation of mutual duality, defining which one of the two vector spaces is the exterior algebra and which one is the dual exterior algebra is just a convention.

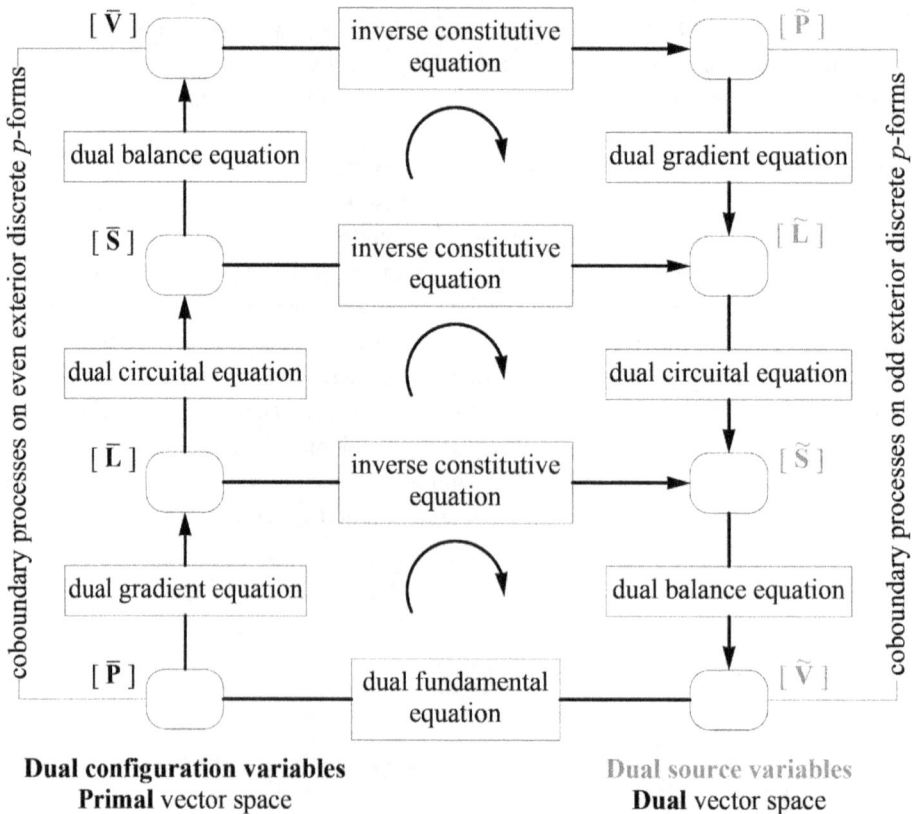

Figure 4.6. Classification diagram of the physical variables in the dual fundamental problem.

4.2 The Mathematical Structure of the Classification Diagram

Each rounded box of the classification diagram contains a finite-dimensional vector sub-space, and each rectangular box contains an algebraic operator.

Being elements of an exterior algebra and a dual exterior algebra (Sections 2.1.1, 2.1.2), all the properties of dual spaces naturally extend to the vector spaces of the two columns of the classification diagram and their operators. In particular, two vector spaces of the same level are isomorphic, that is, it is possible to establish a one-to-one map between the elements of the two vector spaces, which preserves the composition laws of the two spaces.

Two elements $\tilde{a}$ (source variables) and $\overline{b}$ (configuration variables) of the two columns are put in duality by a bilinear form[43] between the two spaces, $\langle \tilde{a}, \overline{b} \rangle$, and the pairing (Section 2.1.1) between the exterior algebra and its dual is given by the scalar product, $\tilde{a} \cdot \overline{b}$, between the elements of the two vector spaces. In particular, if $\overline{b}$ is a configuration variable associated with the zero-dimensional primal space elements, the primal points, the bilinear form $\langle \tilde{a}, \overline{b} \rangle$ has the geometric interpretation shown in Fig. 2.2a, where the planes represent the equipotential surfaces of the source variables and the vector v represents the configuration variables of the same level.

The bilinear form also represents the pairing of a functional in the dual space (column of the source variables) and an element of the primal space (column of the configuration variables).

Given a basis $B = \{e_1, ..., e_n\}$ in the primal space, B and its dual basis, $B^* = \{e^1, ..., e^n\}$, form a biorthogonal system.

An element $\tilde{a}$, of the dual complex, and an element $\overline{b}$, of the primal complex, for which:

$$\langle \tilde{a}, \overline{b} \rangle = 0; \tag{4.2.1}$$

are called orthogonal elements. Only the null vector of one space is orthogonal to all vectors of the other space.

All the bilinear forms between primal and dual complexes in space have the dimension of an energy, while those dealing with space–time, that is, with relativity, have the dimension of an action (energy × time),[143] or a power (work/time).[144]

---

[143] The action is the product of two global variables that lie at the same level. Thus, while the scalar product of a force to a displacement has the dimensions of an energy, the scalar product of the impulse (time global variable of the force) to a displacement has the dimension of an action. Hence, while the field functions that lie on the same level of a diagram are conjugated with respect to energy, the corresponding global variables are conjugated with respect to the action.

[144] The power is the rate of work. It is defined as

$$P(t) \triangleq \mathbf{F} \cdot \mathbf{v}.$$

Since the velocity changes sign under reversal of the motion, while not the force:

$$\mathscr{R}\mathbf{v}(t) = -\mathbf{v}(t);$$

$$\mathscr{R}\mathbf{F}(t) = \mathbf{F}(t);$$

where we have denoted by $\mathscr{R}$ the operation of reversal of motion, the power absorbed by a system becomes power released by the system when the motion reverses:

$$\mathscr{R}P(t) = -P(t).$$

(continues next page)

The configuration and source variables of the fundamental problem are expressed in covariant and contravariant bases (Section 2.1.3), respectively. Let:

$$\tilde{\mathbf{a}}^k = a_h \mathbf{e}_k^h \tag{4.2.2}$$

be a source variable in controvariant basis, and

$$\overline{\mathbf{b}}_k = b^h \mathbf{e}_h^k, \tag{4.2.3}$$

a configuration variable in covariant basis, with $\tilde{\mathbf{a}}^k$ at the same level of $\overline{\mathbf{b}}_k$ in the classification diagram of the physical variables; then the scalar product between the two variables is an invariant:

$$\left\langle \tilde{\mathbf{a}}^k, \overline{\mathbf{b}}_k \right\rangle = \left\langle \overline{\mathbf{b}}_k, \tilde{\mathbf{a}}^k \right\rangle = a_h b^h. \tag{4.2.4}$$

In effect, being the product between two conjugate variables,[145] a configuration variable and a source variable, the product $\left\langle \tilde{\mathbf{a}}^k, \overline{\mathbf{b}}_k \right\rangle$ is an energetic variable and, as such, cannot depend on the coordinate system. The consequence is that the global variables that belong to the same level, k, of the classification diagram have the same tensorial nature, besides the opposed tensorial variance. This is true also for those global variables whose product is an action or a power, since even action and power are invariant for a change of basis.

The constitutive equations, when they are linear, are formed by symmetric operators.

Moreover, in the special case of a bilinear form that gives a work, from the generalized form of Stokes' theorem in Eq. (3.7.12) follows the algebraic form of one of the most important identities of physics, which we will call the **generalized form of the virtual work theorem**.

In order to prove this identity, let us evaluate the boundary of the discrete $(p+3)$-form on the scalar product between the $p+2$ dual collection $\tilde{\mathbf{c}}_{p+2}^a$ and the p primal collection $\overline{\mathbf{c}}_p^b$, where the two collections, a and b, are not related by any constitutive relation, in general, and $p = 0$:

$$\left\langle \partial c^{p+3}, \left\langle \tilde{\mathbf{c}}_{p+2}^a, \overline{\mathbf{c}}_p^b \right\rangle \right\rangle = \left\langle c^{p+3}, \delta \left\langle \tilde{\mathbf{c}}_{p+2}^a, \overline{\mathbf{c}}_p^b \right\rangle \right\rangle = \left\langle c^{p+3}, \left\langle \delta \tilde{\mathbf{c}}_{p+2}^a, \overline{\mathbf{c}}_p^b \right\rangle \right\rangle + \left\langle c^{p+3}, \left\langle \tilde{\mathbf{c}}_{p+2}^a, \delta \overline{\mathbf{c}}_p^b \right\rangle \right\rangle$$

$$= \left\langle c^{p+3}, \left\langle \tilde{\mathbf{c}}_{p+3}^a, \overline{\mathbf{c}}_p^b \right\rangle \right\rangle + \left\langle c^{p+3}, \left\langle \tilde{\mathbf{c}}_{p+2}^a, \overline{\mathbf{c}}_{p+1}^b \right\rangle \right\rangle. \tag{4.2.5}$$

Since a and b are not related by material parameters, the discrete $(p+2)$-form, ∂c^{p+3}, and the discrete $(p+3)$-form, c^{p+3}, are virtual works. By equating the first and last terms, we find

$$\left\langle \partial c^{p+3}, \left\langle \tilde{\mathbf{c}}_{p+2}^a, \overline{\mathbf{c}}_p^b \right\rangle \right\rangle = \left\langle c^{p+3}, \left\langle \tilde{\mathbf{c}}_{p+3}^a, \overline{\mathbf{c}}_p^b \right\rangle \right\rangle + \left\langle c^{p+3}, \left\langle \tilde{\mathbf{c}}_{p+2}^a, \overline{\mathbf{c}}_{p+1}^b \right\rangle \right\rangle.^{[146]} \tag{4.2.6}$$

(*continues from previous page*)

This means that the power is time odd under reversal of motion. Consequently, also the work is time odd.

[145] In physics and engineering, the term "conjugate" or "canonically conjugate" refers to those pairs of physical variables whose product gives an energy (thermodynamics), an action (analytical mechanics and quantum mechanics) or a power (system theory and network theory).

[146] As far as the sign of the work is concerned, it is worth to recall that Eq. (4.2.6) has been derived in the assumption that the outward orientation of volumes is the inward orientation.

In continuum mechanics, a slightly different form of this theorem is known as the **virtual work theorem**, or **fundamental identity of solid mechanics:**[147]

External virtual work is equal to internal virtual work when equilibrated forces and stresses undergo unrelated but consistent displacements and strains.

In symbols, the theorem of virtual work states that:

$$\int_S \mathbf{T}^T \hat{\mathbf{u}} \, dS + \int_V \mathbf{f}^T \hat{\mathbf{u}} \, dV = \int_V \boldsymbol{\sigma}^T \hat{\boldsymbol{\varepsilon}} \, dV \; ; \qquad (4.2.7)$$

where as usual in continuum mechanics, the volumes are outward oriented and

- the system of the external surface forces, $\mathbf{T}$, the external body forces, $\mathbf{f}$, and the internal stresses of the stress tensor $\boldsymbol{\sigma}$, is a system in equilibrium;
- the system of the continuous displacements, $\hat{\mathbf{u}}$, and the strains of the strain tensor $\hat{\boldsymbol{\varepsilon}}$ is a consistent system;
- the symbol "^" emphasizes that the two systems are unrelated.

The left-hand side of Eq. (4.2.7) is the total external virtual work, which is done by $\mathbf{T}$ and $\mathbf{f}$. It is therefore called the external virtual work, while the right-hand side is called the internal virtual work.

The theorem of virtual work includes the theorem of virtual work for rigid bodies as a special case where the internal virtual work is zero.

If we adopt the positive outer orientation, that is, the inward orientation, as the outer orientation of volumes, the second term on the left-hand side becomes negative. Consequently, after rearrangement of the terms, we can re-write the virtual work theorem in terms of equivalence between the work that is done on the volume and the work that is done through the surface, as follows directly from Eq. (4.2.6):

$$\int_S \mathbf{T}^T \hat{\mathbf{u}} \, dS = \int_V \mathbf{f}^T \hat{\mathbf{u}} \, dV + \int_V \boldsymbol{\sigma}^T \hat{\boldsymbol{\varepsilon}} \, dV . \qquad (4.2.8)$$

A further example of how the outer orientation of volumes affects the signs of the work can be found in thermodynamics. In fact, according to the first law of thermodynamics for a closed

---

[147] The dynamic analogue to the principle of virtual work is the D'Alembert's principle, named after the French mathematician, mechanician, physicist, philosopher, and music theorist **Jean-Baptiste le Rond d'Alembert** (16 November 1717–29 October 1783):

$$\sum_i \left(\mathbf{F}_i - m_i \mathbf{a}_i \right) \cdot \delta \mathbf{r}_i = 0 ;$$

where

- i is an integer used to indicate (via subscript) a variable corresponding to a particular particle in the system,
- $\mathbf{F}_i$ is the total applied force (excluding constraint forces) on the i-th particle,
- m_i is the mass of the i-th particle,
- $\mathbf{a}_i$ is the acceleration of the i-th particle,
- $m_i \mathbf{a}_i$ together as product represents the time derivative of the momentum of the i-th particle,
- $\delta \mathbf{r}_i$ is the virtual displacement of the i-th particle, consistent with the constraints.

system, any net increase in the internal energy U must be fully accounted for, in terms of heat δQ entering the system and the work δW done by the system:

$$dU = \delta Q - \delta W .^{148} \qquad (4.2.9)$$

The minus sign in front of δW indicates that a positive amount of work done by the system leads to energy being lost from the system. This sign convention entails that a nonzero quantity of isochoric work always has a negative sign, because of the second law of thermodynamics. An alternate sign convention is to consider the work performed on the system by its surroundings as positive. This leads to a change in sign of the work, so that

$$dU = \delta Q + \delta W . \qquad (4.2.10)$$

The two operators—one in the left and the other in the right column—that lie on the same level of the classification diagram are one the adjoint of the other. More precisely, the two operators are expressed by matrices that are one the transpose of the other, with the same or the opposite sign depending on whether the outer orientation of the dual cells of the upper level is or is not the outer orientation induced by the inner orientation of the primal cells of the upper level.

It is worth noting that the operator "curl" and its adjoint operator lie on the same level of the classification diagram. As a consequence, in differential formulation, which uses one single function space, rather than two separate function spaces for the configuration and the source variables, the operator "curl" is adjoint of itself, that is, is a self-adjoint operator.[55] In algebraic formulation, on the contrary, the algebraic version of the operator "curl" operates in one function space while its adjoint operator operates in the dual function space. Therefore, there cannot exist self-adjoint operators in algebraic formulation, and the algebraic operators and their adjoint operators can never coincide.

The source variable of the higher level is the source of the fundamental problem, while the configuration variable of the higher level represents the potential of the fundamental problem.

In the neighborhood of a point, each regular[149] potential can always be represented by an affine field.[150] In particular, in a three-dimensional Cartesian coordinate system an affine scalar field, assumed continuous and with continuous derivatives, has the scalar affine equation:[151]

---

[148] The letter d indicates an exact differential, expressing that internal energy U is a property of the state of the system, that is, it depends only on the original state and the final state, and not upon the path taken. In contrast, the Greek deltas (δ's) in this equation reflect the fact that the heat transfer and the work transfer are not properties of the final state of the system. Given only the initial state and the final state of the system, one can only say what the total change in internal energy was, not how much of the energy went out as heat, and how much as work. This can be summarized by saying that heat and work are not state functions of the system.

[149] A field that is regular in a region is continuous and with continuous variations.

[150] An affine field is the sum of a uniform field (that is, invariant under translation) and a linear field. It is obtained from the development of a function in a Taylor series, ignoring the terms higher than the first order.

[151] Eq. (4.2.11) is originated by translating the origin of the reference frame on the point of coordinates $\left(x_0, y_0, z_0\right)$. The scalar form of the affine scalar field in the reference frame (O, x, y, z) is

$$\phi(x, y, z) = a' + h'_x\left(x - x_0\right) + h'_y\left(y - y_0\right) + h'_z\left(z - z_0\right) ;$$

(continues next page)

$$\phi(x,y,z) = a + h_x x + h_y y + h_z z; \tag{4.2.11}$$

where

$$a = \phi(0,0,0); \tag{4.2.12}$$

(*continues from previous page*)

where

$$a' = \phi(x_0, y_0, z_0);$$

$$h'_x = \frac{\partial \phi(x,y,z)}{\partial x}\bigg|_{(x_0,y_0,z_0)};$$

$$h'_y = \frac{\partial \phi(x,y,z)}{\partial y}\bigg|_{(x_0,y_0,z_0)};$$

$$h'_z = \frac{\partial \phi(x,y,z)}{\partial z}\bigg|_{(x_0,y_0,z_0)}.$$

The equation

$$\phi(x,y,z) = \phi(x_0,y_0,z_0) + \frac{\partial \phi(x,y,z)}{\partial x}\bigg|_{(x_0,y_0,z_0)}(x-x_0) + \frac{\partial \phi(x,y,z)}{\partial y}\bigg|_{(x_0,y_0,z_0)}(y-y_0)$$
$$+ \frac{\partial \phi(x,y,z)}{\partial z}\bigg|_{(x_0,y_0,z_0)}(z-z_0);$$

is the equation of the tangent plane to the hyper-surface of the function $\phi(x,y,z)$, for $x = x_0$, $y = y_0$, $z = z_0$.

Eq. (4.2.11) can then be obtained by the change of coordinates:

$$\overline{x} = x - x_0;$$

$$\overline{y} = y - y_0;$$

$$\overline{z} = z - z_0;$$

which provides

$$\phi(x,y,z) = a + h_x \overline{x} + h_y \overline{y} + h_z \overline{z}.$$

In a two-dimensional reference frame (O, x, y), the scalar form of the affine scalar field reduces to

$$\phi(x,y) = \phi(x_0,y_0) + \frac{\partial \phi(x,y)}{\partial x}\bigg|_{(x_0,y_0)}(x-x_0) + \frac{\partial \phi(x,y)}{\partial y}\bigg|_{(x_0,y_0)}(y-y_0),$$

which is the equation of the tangent plane to the surface of the function $\phi(x,y)$, for $x = x_0$, $y = y_0$.

$$h_x = \left.\frac{\partial \phi(x,y,z)}{\partial x}\right|_{(0,0,0)} ; \tag{4.2.13}$$

$$h_y = \left.\frac{\partial \phi(x,y,z)}{\partial y}\right|_{(0,0,0)} ; \tag{4.2.14}$$

$$h_z = \left.\frac{\partial \phi(x,y,z)}{\partial z}\right|_{(0,0,0)} . \tag{4.2.15}$$

In the algebraic formulation, the derivatives are substituted by the divided differences.[10] This means that the affine scalar field retains the form in Eq. (4.2.11), where the coefficients h_x, h_y, and h_z are now provided by the increments Δ_x, Δ_y, and Δ_z, along the directions x, y, and z, respectively,

$$h_x = \left.\frac{\Delta_x \phi(x,y,z)}{\Delta x}\right|_{(0,0,0)} ; \tag{4.2.16}$$

$$h_y = \left.\frac{\Delta_y \phi(x,y,z)}{\Delta y}\right|_{(0,0,0)} ; \tag{4.2.17}$$

$$h_z = \left.\frac{\Delta_z \phi(x,y,z)}{\Delta z}\right|_{(0,0,0)} . \tag{4.2.18}$$

Eq. (4.2.11) has a linear behavior. From the geometrical viewpoint, it represent the tangent plane to the hyper-surface of the function $\phi(x,y,z)$, for $x = y = z = 0$.

The vector form of Eq. (4.2.11) is

$$\phi = a + \mathbf{h} \cdot \mathbf{r}. \tag{4.2.19}$$

An affine vector field in a three-dimensional Cartesian coordinate system has the scalar form:

$$\begin{cases} v_x = a_x + h_{xx}x + h_{xy}y + h_{xz}z \\ v_y = a_y + h_{yx}x + h_{yy}y + h_{yz}z \\ v_z = a_z + h_{zx}x + h_{zy}y + h_{zz}z \end{cases} \tag{4.2.20}$$

characterized by the 12 unknown a_x, a_y, a_z, h_{xx}, h_{xy}, h_{xz}, h_{yx}, h_{yy}, h_{yz}, h_{zx}, h_{zy}, and h_{zz}, and the vector form:

$$\mathbf{v} = \mathbf{a} + \mathbf{H}\mathbf{r}; \tag{4.2.21}$$

where $\mathbf{H}$ is called the gradient of the vector $\mathbf{v}$. From the geometrical viewpoint, even the three scalar equations (4.2.20) are planes.

The equipotential surfaces, that is, the surfaces along which the scalar field or the components of the vector field have a constant value, are parallel planes.

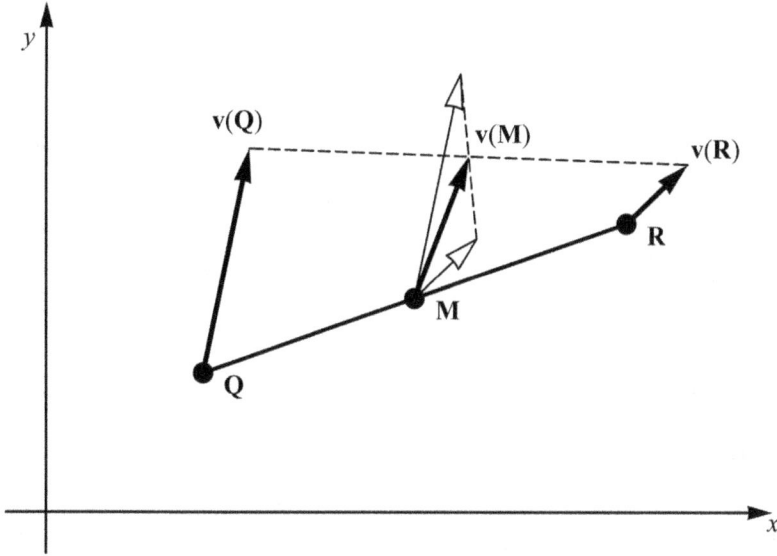

Figure 4.7. How to achieve the vector in the middle point of a segment, once the two end vectors are known.

From the linear behavior of the affine equations, many useful properties of the affine fields may be derived. Among them, there is an interesting property of the scalar affine fields:

The value of an affine function in the middle point of a segment is the average of the values of the function at the two extremes.

An analogue property can be proven also for the vector affine fields (Fig. 4.7):

The value of an affine vector valued function in the middle point, **M**, *of a segment is the mean of the values of the affine vector valued function at the two extremes,* **Q** *and* **R**.

$$v(M) = \frac{1}{2}[v(Q) + v(R)]. \qquad (4.2.22)$$

Among the other important theorems that can be proven for the vector affine fields, we recall the following four theorems:

The line integral of an affine vector field along a straight line segment of unit normal **t** *is equal to the scalar product of the vector, evaluated in the middle point,* **M**, *of the segment, for the vector* $(R - Q)$, *which describes the oriented segment:*

$$\int_Q^R v \cdot t \, dL = v(M) \cdot (R - Q). \qquad (4.2.23)$$

The circulation, Γ, *of an affine vector field along the boundary* ∂S *of a plane surface is proportional to the area,* A, *of the surface:*

$$\int_\partial v \cdot t \, dL \propto A. \qquad (4.2.24)$$

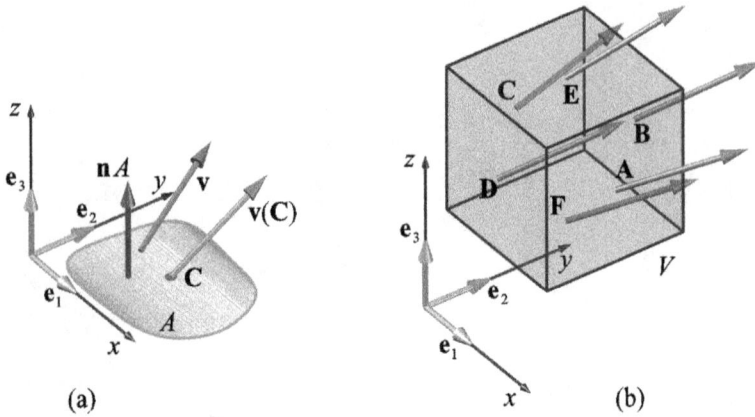

Figure 4.8. The flux across a plane surface (a) and across the boundary of a cube (b).

The flux, Φ, of an affine vector field across a plane surface of area A and unit normal
n *is equal to the scalar product of the vector evaluated in the centroid[152] of the plane*
*surface for the vector A**n**, which describes the surface (Fig. 4.8a):*

$$\Phi \triangleq \int_S \mathbf{v}_P \cdot \mathbf{n} \; dS = \mathbf{v}_C \cdot \mathbf{n}A. \tag{4.2.25}$$

---

[152] The centroid of a plane surface is the point of coordinates:

$$x_C \triangleq \frac{\int_S x dS}{\int_S dS} ;$$

$$y_C \triangleq \frac{\int_S y dS}{\int_S dS} .$$

The centroid of a set of plane surfaces of areas A_i and centroids (x_i, y_i) is the point of coordinates:

$$x_C \triangleq \frac{\sum_i x_i A_i}{\sum_i A_i} ;$$

$$y_C \triangleq \frac{\sum_i y_i A_i}{\sum_i A_i} .$$

If the material of the plane surface is homogeneous, the centroid coincides with the center of mass, which is the point of coordinates:

$$x_G \triangleq \frac{\int_S \rho x dS}{\int_S \rho dS} ;$$

$$y_G \triangleq \frac{\int_S \rho y dS}{\int_S \rho dS} ,$$

where ρ is the mass density.

The flux, Φ, of an affine vector field on a closed surface is proportional to the volume, V, enclosed by the surface (Fig. 4.8b):

$$\Phi \triangleq \int_S \mathbf{v} \cdot \mathbf{n} \ dS \propto V. \tag{4.2.26}$$

4.3 The Incidence Matrices of the Two Cell Complexes in Space Domain

In the three-dimensional space the dual correspondence is established between primal (dual) cells of dimension p and dual (primal) cells of dimension $3 - p$.

After having labeled the primal p-cells, it is thus natural to assign to each $3 - p$ of the dual cell complex the same label of the corresponding primal p-cell. Using this criterion, when the outer orientation of the dual complex is induced by the inner orientation of the primal one, the incidence number between a p-cell and a $(p-1)$-cell of the primal cell complex is equal to the incidence number between the corresponding dual cells.

The criterion of dual labeling is shown in Fig. 4.9 for a two-dimensional space. The dual cell complex in Fig. 4.9 is built on the primal cell complex of Fig. 3.21. Since the outer orientation of the dual cells in Fig. 4.9 is equal to the inner orientation of the corresponding primal element, each dual cell is inward oriented (because the primal nodes are sinks) and each dual side is crossed in the sense indicated by its primal side. Finally, the dual nodes in Fig. 4.9 are sources.

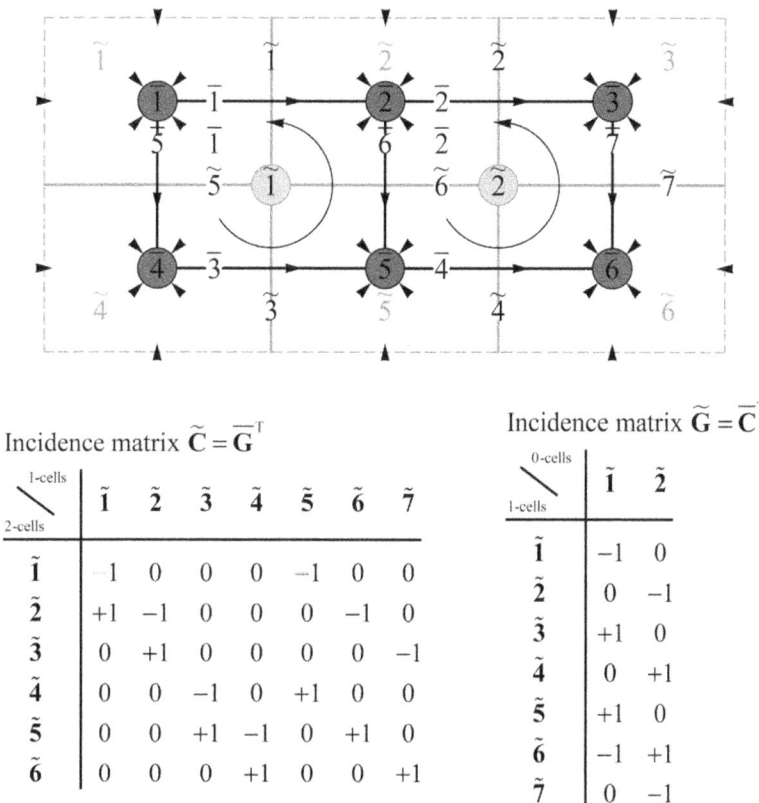

Incidence matrix $\widetilde{\mathbf{C}} = \overline{\mathbf{G}}^{\mathrm{T}}$

2-cells \ 1-cells	$\tilde{1}$	$\tilde{2}$	$\tilde{3}$	$\tilde{4}$	$\tilde{5}$	$\tilde{6}$	$\tilde{7}$
$\tilde{1}$	-1	0	0	0	-1	0	0
$\tilde{2}$	$+1$	-1	0	0	0	-1	0
$\tilde{3}$	0	$+1$	0	0	0	0	-1
$\tilde{4}$	0	0	-1	0	$+1$	0	0
$\tilde{5}$	0	0	$+1$	-1	0	$+1$	0
$\tilde{6}$	0	0	0	$+1$	0	0	$+1$

Incidence matrix $\widetilde{\mathbf{G}} = \overline{\mathbf{C}}^{\mathrm{T}}$

1-cells \ 0-cells	$\tilde{1}$	$\tilde{2}$
$\tilde{1}$	-1	0
$\tilde{2}$	0	-1
$\tilde{3}$	$+1$	0
$\tilde{4}$	0	$+1$
$\tilde{5}$	$+1$	0
$\tilde{6}$	-1	$+1$
$\tilde{7}$	0	-1

Figure 4.9. The incidence matrices in a plane dual cell complex.

In the most general case of a three-dimensional space:

$$\overline{\mathbf{G}} \triangleq \left[\, \overline{g}_{ih} \,\right]; \tag{4.3.1}$$

where

$$\overline{g}_{ih} \triangleq \left[\, \overline{\mathbf{l}}^i : \overline{\mathbf{p}}^h \,\right]. \tag{4.3.2}$$

On the other hand:

$$\widetilde{\mathbf{D}} \triangleq \left[\, \tilde{d}_{hi} \,\right]; \tag{4.3.3}$$

where, since, for historical reasons,[83] the volumes have a negative outer orientation (outward orientation), the outer orientation of a volume is the opposite of the orientation that would be induced by the primal point inside it:

$$\tilde{d}_{hi} \triangleq \left[\, \tilde{\mathbf{v}}^h : \tilde{\mathbf{s}}^i \,\right] = -\left[\, \overline{\mathbf{l}}^i : \overline{\mathbf{p}}^h \,\right]; \tag{4.3.4}$$

and, consequently:

$$\widetilde{\mathbf{D}} = -\overline{\mathbf{G}}^{\mathrm{T}}. \tag{4.3.5}$$

Analogously, we find

$$\overline{\mathbf{C}} \triangleq \left[\, \overline{c}_{ji} \,\right], \text{ where } \overline{c}_{ji} \triangleq \left[\, \overline{\mathbf{s}}^j : \overline{\mathbf{l}}^i \,\right]; \tag{4.3.6}$$

$$\widetilde{\mathbf{C}} \triangleq \left[\, \tilde{c}_{ij} \,\right], \text{ where } \tilde{c}_{ij} \triangleq \left[\, \tilde{\mathbf{s}}^i : \tilde{\mathbf{l}}^j \,\right]; \tag{4.3.7}$$

$$\overline{\mathbf{D}} \triangleq \left[\, \overline{d}_{kj} \,\right], \text{ where } \overline{d}_{kj} \triangleq \left[\, \overline{\mathbf{v}}^k : \overline{\mathbf{s}}^j \,\right]; \tag{4.3.8}$$

$$\widetilde{\mathbf{G}} \triangleq \left[\, \tilde{g}_{jk} \,\right], \text{ where } \tilde{g}_{jk} \triangleq \left[\, \tilde{\mathbf{l}}^j : \tilde{\mathbf{p}}^k \,\right] = \left[\, \overline{\mathbf{v}}^k : \overline{\mathbf{s}}^j \,\right]; \tag{4.3.9}$$

and, consequently

$$\widetilde{\mathbf{C}} = \overline{\mathbf{C}}^{\mathrm{T}}; \tag{4.3.10}$$

$$\widetilde{\mathbf{G}} = \overline{\mathbf{D}}^{\mathrm{T}}. \tag{4.3.11}$$

For consistency between the two-dimensional and three-dimensional cases, we can think both the primal and the dual plane domains as provided with a unit thickness. Consequently:

- The primal and dual cell complexes will result staggered in thickness (along the direction of the observer).
- The oriented 0-cells become oriented 1-cells. In particular, the primal points become edges that enter in the plane, while the dual nodes become edges that come out from the plane.
- The oriented 1-cells become oriented 2-cells.
- The oriented 2-cells become oriented 3-cells.
- The matrices $\overline{\mathbf{C}}$ and $\overline{\mathbf{G}}$ remain unaltered, provided that we do not account for the new 2-cells and 1-cells.
- The matrix $\widetilde{\mathbf{C}}$ in Fig. 4.9 becomes the matrix of the incidence numbers between dual 3-cells and dual 2-cells (provided that we do not account for the new 3-cells and 2-cells). If, in accordance with the three-dimensional case, we change the outer orientation of the

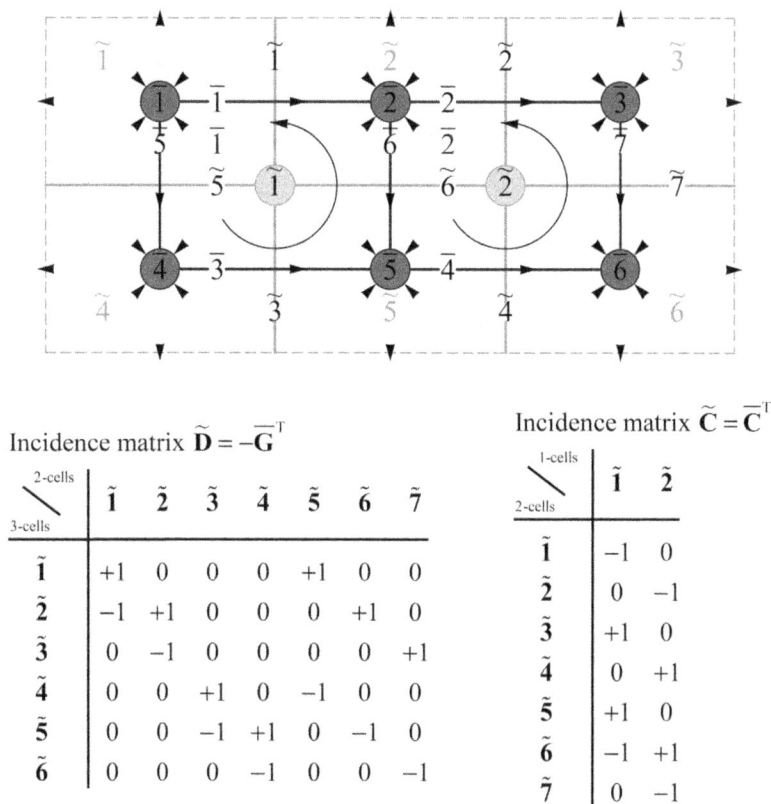

Incidence matrix $\widetilde{\mathbf{D}} = -\overline{\mathbf{G}}^{\mathrm{T}}$

2-cells 3-cells	$\tilde{1}$	$\tilde{2}$	$\tilde{3}$	$\tilde{4}$	$\tilde{5}$	$\tilde{6}$	$\tilde{7}$
$\tilde{1}$	+1	0	0	0	+1	0	0
$\tilde{2}$	−1	+1	0	0	0	+1	0
$\tilde{3}$	0	−1	0	0	0	0	+1
$\tilde{4}$	0	0	+1	0	−1	0	0
$\tilde{5}$	0	0	−1	+1	0	−1	0
$\tilde{6}$	0	0	0	−1	0	0	−1

Incidence matrix $\widetilde{\mathbf{C}} = \overline{\mathbf{C}}^{\mathrm{T}}$

2-cells	1-cells $\tilde{1}$	$\tilde{2}$
$\tilde{1}$	−1	0
$\tilde{2}$	0	−1
$\tilde{3}$	+1	0
$\tilde{4}$	0	+1
$\tilde{5}$	+1	0
$\tilde{6}$	−1	+1
$\tilde{7}$	0	−1

Figure 4.10. The modified incidence matrices in a plane dual cell complex with unit thickness.

3-cells from inward to outward, we obtain a matrix $\widetilde{\mathbf{D}}$ that satisfy the relation $\widetilde{\mathbf{D}} = -\overline{\mathbf{G}}^{\mathrm{T}}$ (Fig. 4.10).

- The matrix $\widetilde{\mathbf{G}}$ in Fig. 4.9 becomes the matrix $\widetilde{\mathbf{C}}$ of the incidence numbers between dual 2-cells and dual 1-cells (Fig. 4.10), and satisfies the relation $\widetilde{\mathbf{C}} = \overline{\mathbf{C}}^{\mathrm{T}}$.
- The matrices $\overline{\mathbf{D}}$ and $\widetilde{\mathbf{G}}$ remain undetermined, but they are not needed in two-dimensional computation.

4.4 Primal and Dual Cell Complexes in Space/Time Domain and Their Incidence Matrices

When we add a time axis to a three-dimensional primal cell complex in space, we obtain a primal four-dimensional cell complex. In particular, if the 3-cell of the primal space cell complex is a cube, the 4-cell of the primal space/time cell complex is a tesseract (Section 3.8).

According to the convention of considering the time instant as sinks and the time intervals as oriented from the preceding to the following time instant, the theory of oriented graphs applied to the time axis allows us to describe the relationship between primal time instant and primal time intervals in terms of incidence numbers and incidence matrices. For the three time intervals and four time instants of the one-dimensional example in Fig. 4.11 (obtained by

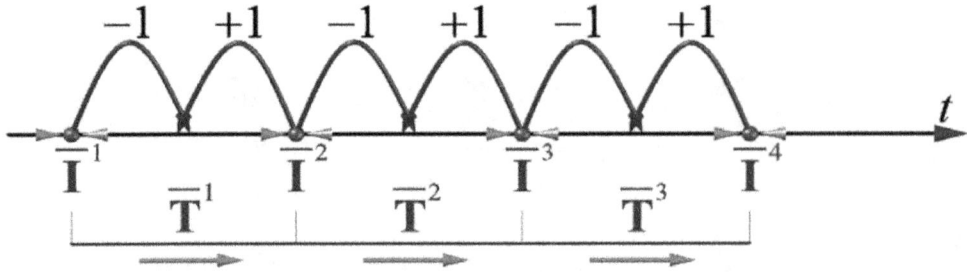

Figure 4.11. Inner orientations of the primal time elements for the case of four time instants.

projection from the four-dimensional space to the one-dimensional space of the time axis), we find an incidence matrix of the same shape as the incidence matrix **G** in Fig. 3.21:

	$\bar{1}$	$\bar{2}$	$\bar{3}$	$\bar{4}$
$\bar{1}$	-1	$+1$	0	0
$\bar{2}$	0	-1	$+1$	0
$\bar{3}$	0	0	-1	$+1$

(4.4.1)

0-cells (top), 1-cells (left)

which suggests a similarity between the time incidence matrix and the matrix $\mathbf{G} = \overline{\mathbf{G}}$.

The velocity is one example of variable associated with the primal time intervals, rather than with the primal space points. The velocity is computed as the increment of the radius vector in the unit time, where the radius vector is associated with primal time instants and primal space points. Therefore, in the special case of unit time intervals and with reference to the velocity along the x axis, we can find the three velocities $\bar{v}_{x1}$, $\bar{v}_{x2}$, and $\bar{v}_{x3}$, associated with the three primal time intervals, $\overline{T}^1$, $\overline{T}^2$, and $\overline{T}^3$, by

1. spreading the scalar variables of the 0-form $\overline{\Phi}^0 = \left[\bar{x}_1, \bar{x}_2, \bar{x}_3, \bar{x}_4\right]$ on the primal time intervals, $\overline{T}^1$, $\overline{T}^2$, and $\overline{T}^3$, according to the mutual incidence numbers;
2. adding the variables that have been spread on the same primal time interval.

Spreading the scalar variables according to the mutual incidence numbers is equivalent to performing the matrix product between the incidence matrix in Eq. (4.4.1) and the vector of the scalar variables, $\overline{\Phi}^0 = \left[\bar{x}_1 \quad \bar{x}_2 \quad \bar{x}_3 \quad \bar{x}_4\right]^T$:

$$\begin{bmatrix} \bar{v}_{x1} \\ \bar{v}_{x2} \\ \bar{v}_{x3} \end{bmatrix} = \begin{bmatrix} -1 & +1 & 0 & 0 \\ 0 & -1 & +1 & 0 \\ 0 & 0 & -1 & +1 \end{bmatrix} \begin{bmatrix} \bar{x}_1 \\ \bar{x}_2 \\ \bar{x}_3 \\ \bar{x}_4 \end{bmatrix}.$$

(4.4.2)

In the more general case where the time intervals are not unit time intervals but still have the same duration, the incidence matrix must be left-multiplied by the inverse of the time interval. Let Δt be the time interval, we find:

$$
\begin{bmatrix} \overline{v}_{x1} \\ \overline{v}_{x2} \\ \overline{v}_{x3} \end{bmatrix} = \frac{1}{\Delta t} \begin{bmatrix} -1 & +1 & 0 & 0 \\ 0 & -1 & +1 & 0 \\ 0 & 0 & -1 & +1 \end{bmatrix} \begin{bmatrix} \overline{x}_1 \\ \overline{x}_2 \\ \overline{x}_3 \\ \overline{x}_4 \end{bmatrix}.
\tag{4.4.3}
$$

Consequently, the average velocities along the x axis for the three primal time intervals, $\overline{T}^1, \overline{T}^2$, and $\overline{T}^3$, are

$$
\overline{v}_{x1} = \frac{\overline{x}_2 - \overline{x}_1}{\Delta t};
\tag{4.4.4}
$$

$$
\overline{v}_{x2} = \frac{\overline{x}_3 - \overline{x}_2}{\Delta t};
\tag{4.4.5}
$$

$$
\overline{v}_{x3} = \frac{\overline{x}_4 - \overline{x}_3}{\Delta t}.
\tag{4.4.6}
$$

Analogous formulas follow for the velocities along the y and z axes.

This process generates a 1-form, $\overline{\Gamma}^1 = \left[\overline{v}_{x1}, \overline{v}_{x2}, \overline{v}_{x3} \right]$, which is a 1-form on the primal cell complex because the velocities change sign under reversal of motion (see the criterion adopted for distinguishing between variables of the primal and dual cell complexes (Section 4.1), based on the oddness principle[133]).

Due to the duality between primal time intervals and dual time instants, that is, the pairing between primal a and dual basis in time, the process also generates a 0-form on the dual time instants: $\widetilde{\Phi}^0 = \left[\tilde{v}_{x1}, \tilde{v}_{x2}, \tilde{v}_{x3}, \tilde{v}_{x4} \right]$.

For building the dual cell complex in space/time, let us observe that the increment in time of a configuration variable generates a configuration variable, as in the former case, where the increment of the radius vector has generated the velocity vector. Analogously, the increment in time of a source variable generates a source variable. This means that primal and dual cell complexes in time are reciprocally translated along the time axis. This also means that the same cell complex in space/time has two dual cell complexes, one evaluated in space and one valuated in time (Fig. 4.12). We will call these two dual cell complexes the space-dual cell complex and the time-dual cell complex.

By defining the dual time instants as the central points of the primal time intervals (assumed of the same duration), the time-dual cell complex of the primal cell complex in space/time is obtained by shifting the primal cell complex in space along the time axis, for half the duration of the time intervals (Fig. 4.12). In fact, primal and dual time axes are superposed.

Analogously, the time-dual cell complex of the space-dual cell complex in space/time is obtained by shifting the space-dual cell complex along the time axis, for half the duration of the time intervals. We will call this cell complex the space/time-dual cell complex.

The dual time intervals connect the dual time instants. They have the same duration of the primal time intervals.

If we choose to provide the dual time elements with the outer orientation that would be induced on them by the inner orientation of the primal time elements, the positive inner orientation of the primal time instants, $\overline{I}$, would induce the positive outer orientation on the dual

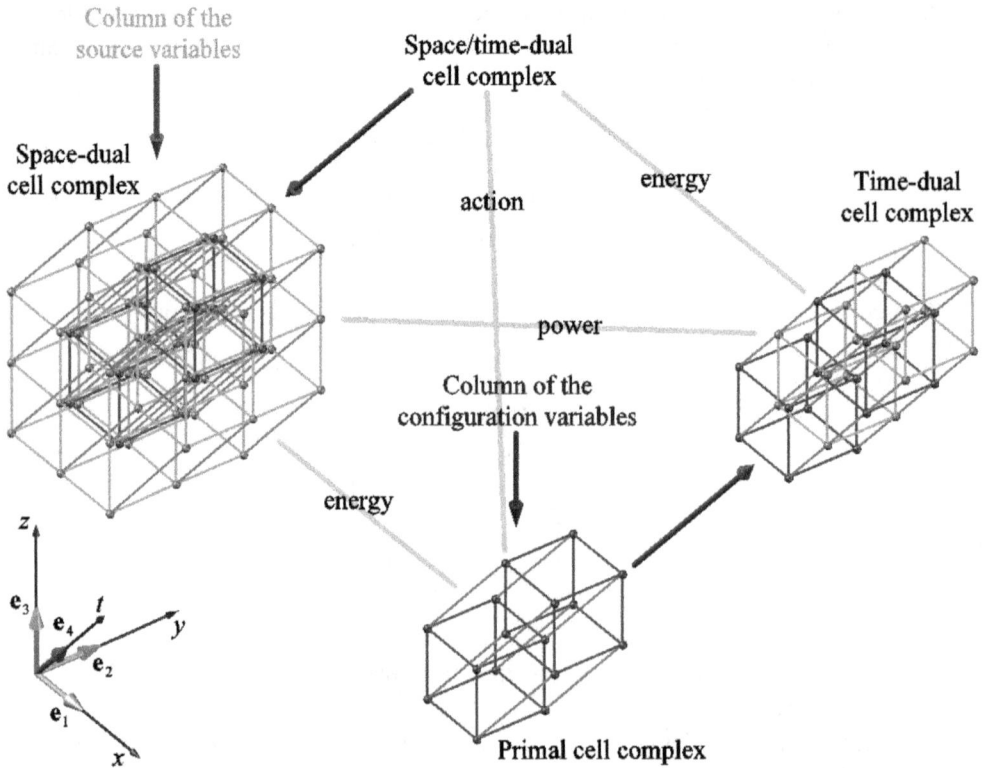

Figure 4.12. Primal and dual cell complexes in space/time.

time intervals, $\widetilde{T}$, that is, the inward orientation. Moreover, the positive inner orientation of the primal time intervals, $\overline{T}$, would induce the positive outer orientation on the dual time instants, $\widetilde{I}$, that is, the same orientation of the time axis. Nevertheless, this would contradict the fact that the positive increments of the dual time variables must be taken in the same direction as the positive increments of the primal time variables, since primal and dual axes have the same orientation. Consequently, we must change the outer orientation or of the dual time instants, or of the dual time intervals. From this moment forth, the positive outer orientation of the dual time instants will be taken equal to the orientation of the time axis and the dual time intervals will be taken outward directed.

As we can verify for the one-dimensional example in Fig. 4.13, this choice of outer orientation for the dual elements generates, for the dual time elements, $\widetilde{I}$ and $\widetilde{T}$, the same incidence matrix that puts in relationship the primal time elements, $\overline{I}$ and $\overline{T}$:

0-cells 1-cells	$\widetilde{1}$	$\widetilde{2}$	$\widetilde{3}$	$\widetilde{4}$
$\widetilde{1}$	-1	$+1$	0	0
$\widetilde{2}$	0	-1	$+1$	0
$\widetilde{3}$	0	0	-1	$+1$

(4.4.7)

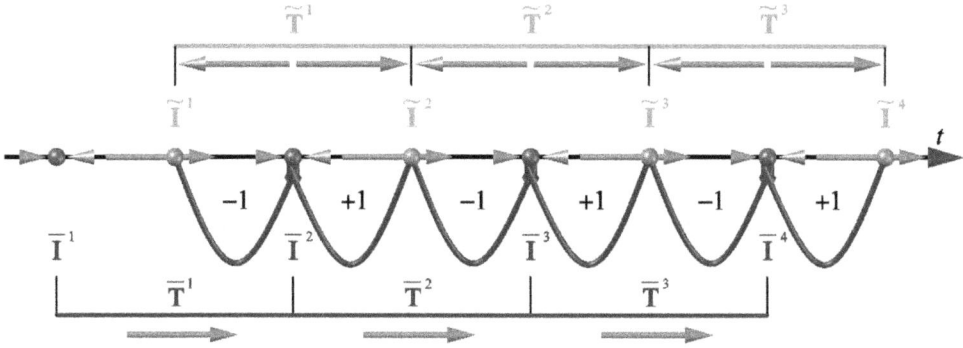

Figure 4.13. Orientations of the primal and dual time elements for the case of four instants.

Note that, as in the four-dimensional Minkowski continuum,[61] in the algebraic formulation the time dimension is treated differently from the three spatial dimensions for two reasons:

- we have oriented the tesseract differently from the related 4-vector (Section 3.8);
- we have taken, not the outer orientation induced by the primal time instants, but its opposite as the positive outer orientation for the dual time intervals.

The acceleration is one example of variable associated with the dual time intervals, rather than with the primal space points. The acceleration is computed as the increment of the velocity vector in the unit time, where the velocity is associated with the dual time instants and the primal space points. In the special case of unit time intervals and with reference to the acceleration along the x axis, we can find the three accelerations $\tilde{a}_{x1}$, $\tilde{a}_{x2}$, and $\tilde{a}_{x3}$, associated with the three dual time intervals, $\widetilde{T}^1$, $\widetilde{T}^2$, and $\widetilde{T}^3$, by

- spreading the scalar variables of the 0-form $\widetilde{\varPhi}^0 = \left[\tilde{v}_{x1}, \tilde{v}_{x2}, \tilde{v}_{x3}, \tilde{v}_{x4}\right]$ on the dual time intervals, $\widetilde{T}^1$, $\widetilde{T}^2$, and $\widetilde{T}^3$, according to the mutual incidence numbers;
- adding the variables that have been spread on the same dual time interval.

Spreading the scalar variables according to the mutual incidence numbers is equivalent to performing the matrix product between the incidence matrix in Eq. (4.4.7) and the vector of the scalar variables, $\widetilde{\varPhi}^0 = \left[\tilde{v}_{x1} \quad \tilde{v}_{x2} \quad \tilde{v}_{x3} \quad \tilde{v}_{x4}\right]^T$:

$$\begin{bmatrix} \tilde{a}_{x1} \\ \tilde{a}_{x2} \\ \tilde{a}_{x3} \end{bmatrix} = \begin{bmatrix} -1 & +1 & 0 & 0 \\ 0 & -1 & +1 & 0 \\ 0 & 0 & -1 & +1 \end{bmatrix} \begin{bmatrix} \tilde{v}_{x1} \\ \tilde{v}_{x2} \\ \tilde{v}_{x3} \\ \tilde{v}_{x4} \end{bmatrix}. \tag{4.4.8}$$

In the more general case, where the time intervals are not unit time intervals but still have the same duration, Δt, the incidence matrix must be left-multiplied by $1/\Delta t$:

$$\begin{bmatrix} \tilde{a}_{x1} \\ \tilde{a}_{x2} \\ \tilde{a}_{x3} \end{bmatrix} = \frac{1}{\Delta t} \begin{bmatrix} -1 & +1 & 0 & 0 \\ 0 & -1 & +1 & 0 \\ 0 & 0 & -1 & +1 \end{bmatrix} \begin{bmatrix} \tilde{v}_{x1} \\ \tilde{v}_{x2} \\ \tilde{v}_{x3} \\ \tilde{v}_{x4} \end{bmatrix}. \tag{4.4.9}$$

The average accelerations on the three dual time intervals, $\tilde{T}^1$, $\tilde{T}^2$, and $\tilde{T}^3$, are

$$\tilde{a}_{x1} = \frac{\tilde{v}_{x2} - \tilde{v}_{x1}}{\Delta t}; \tag{4.4.10}$$

$$\tilde{a}_{x2} = \frac{\tilde{v}_{x3} - \tilde{v}_{x2}}{\Delta t}; \tag{4.4.11}$$

$$\tilde{a}_{x3} = \frac{\tilde{v}_{x4} - \tilde{v}_{x3}}{\Delta t}. \tag{4.4.12}$$

Due to the duality between dual time intervals and primal time instants, this process generates both a 1-form for the dual time intervals and a second 0-form for the primal time instants. In particular, the 1-form $\tilde{\Gamma}^1 = \left[\tilde{a}_{x1}, \tilde{a}_{x2}, \tilde{a}_{x3} \right]$ is a 1-form on the dual cell complex because the accelerations do not change sign under reversal of motion. The reason for this is that the acceleration is symmetric with respect to the radius vector:

$$\tilde{a}_{x2} = \frac{\bar{x}_3 - 2\bar{x}_2 + \bar{x}_1}{\Delta t^2}. \tag{4.4.13}$$

The belonging of the acceleration to the dual complex in time allows us to establish a relationship between the space-dual and the time-dual cell complexes. In fact, since Newton's second law,

$$\mathbf{F} = m\mathbf{a}, \tag{4.4.14}$$

relates an element of the time-dual cell complex, the acceleration, to an element of the space-dual cell complex, the force (source), we can say that the two dual cell complexes in space and time are related by Newton's second law (Fig. 4.14). The time-dual and space-dual cell complexes are also related by some dissipative laws.

In order to distinguish between the impressed sources (assigned by the problem and living in the space-dual cell complex) and the sources that are induced by the configuration variables, due to Newton's second law or dissipative laws, these latter sources will be denoted as the **induced sources**.

In conclusion, velocity and acceleration are space/time variables associated with the points and the edges of the space/time tesseract. The possible space/time combinations between the four space elements, **P**, **L**, **S**, and **V**, and the two time elements, **I** and **T**, living in this projection are in number of 8. Considering that both the space and the time elements can be of primal or dual kind, we find

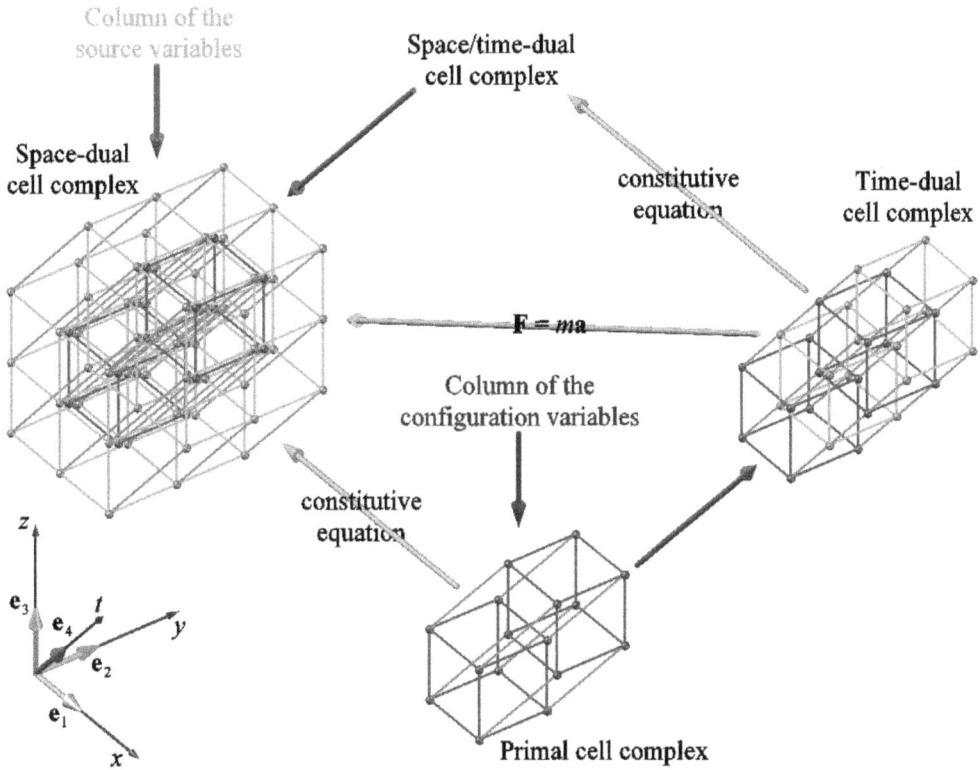

Figure 4.14. Relationship between dual cell complexes in space and time.

a) 8 couples of space/time primal elements,
b) 8 couples of space/time dual elements,
c) 8 + 8 couples of space/time primal and dual elements.

These are exactly the same 32 combination of Fig. 4.1, with the couples of kind (a) and (b) that define the first group of variables in Fig. 4.1 and the couples of kind (c) that define the second group.

CHAPTER 5

THE STRUCTURE OF THE GOVERNING
EQUATIONS IN THE CELL METHOD

In Section 5.1, we discuss the role played by the coboundary process performed on discrete p-forms of degree 2, 1, and 0, both in space and space/time domains. The three coboundary operators, δ^D, δ^C, and δ^G, respectively, are the discrete versions of the differential operators "div", "curl", and "grad". We show how to obtain the matrix form of the three coboundary operators as the sum of expanded local matrices, by using an assembling procedure that is derived from the assembling procedure of the stiffness matrix. In space/time-domain, the coboundary process extends to all the geometrical elements of the CM tesseract. As a consequence, a given coboundary process on the discrete p-form generates discrete $(p+1)$-forms on all the $(p+1)$-dimensional elements of the tesseract, that is, on both the $(p+1)$-dimensional elements of the kind "space" and the $(p+1)$-dimensional elements of the kind "time". The process for the generation of the $(p+1)$-forms of the kind "space" is the algebraic counterpart of finding the divergence, the curl, or the gradient, while the process for the generation of the $(p+1)$-forms of the kind "time" is the algebraic version of finding the time derivative of a function. Therefore, the same coboundary process performed on a discrete p-form in space/time gives rise to two operators at the same time, one in space and one in time. In the special case of a discrete 0-form in space/time, the coboundary process is the algebraic counterpart of the spacetime gradient in spacetime algebra (Section 5.1.2).

In Section 5.2, the topological equations are combined with the constitutive equation, in order to derive the general form of the fundamental problem.

The similarities between the classification diagrams and the fundamental equations of different physical theories are discussed in Section 5.3. They are the main reason for the analogies between physical theories, explaining why a relation or a statement is invariant in different physical theories under the exchanges of the elements involved in them.

In Section 5.4, we clarify when a physical theory has a reversible constitutive relation, making it possible to formulate a dual fundamental equation.

The chapter ends with some considerations on how to choose primal and dual cell complexes for the numerical modeling with the CM (Section 5.5).

5.1 The Role of the Coboundary Process in the Algebraic Formulation

In a differential formulation, the topological equations of the fundamental problem (Section 4.1) are expressed by the first order differential operators such as the divergence, the curl, and the gradient.

As we have discussed in Section 4.1, the coboundary process is analogous, in an algebraic setting, to the exterior differentiation on exterior differential forms, which is used for deriving the topological equations of any physical theory. In particular, balance, circuital equations, and equations forming differences can be expressed by the coboundary process performed on discrete p-forms of degree 2, 1, and 0, respectively.

In fact, let $A_d[\mathbf{V}]$ be a physical variable associated with a volume, which can be endowed with inner or outer orientation, and let $B_d[\partial\mathbf{V}]$ be a physical variable associated with the boundary of the same volume. Since the balance equation:

$$A_d[\mathbf{V}] = \pm B_d[\partial\mathbf{V}], \qquad (5.1.1)$$

is expressed by a map from the boundary of a space element to the space element itself (as the map t_1 shown in Eq. (4.1.2)):

$$\delta^{\mathrm{D}} : \partial\mathbf{V} \to \mathbf{V}, \qquad (5.1.2)$$

where δ^{D} is the coboundary operator defined in Eq. (3.6.16), then the balance equation (Eq. 5.1.1) is a topological equation defined on discrete p-forms of degree 2.

For the same reason, also the circuital equations:

$$A_c[\mathbf{S}] = \pm B_c[\partial\mathbf{S}], \qquad (5.1.3)$$

where $A_c[\mathbf{S}]$ is a physical variable associated with a surface, which can be endowed with inner or outer orientation, and $B_c[\partial\mathbf{S}]$ is a physical variable associated with its boundary, and space differences:

$$A_g[\mathbf{L}] = \pm B_g[\partial\mathbf{L}], \qquad (5.1.4)$$

where $A_g[\mathbf{L}]$ is a physical variable associated with a line, which can be endowed with inner or outer orientation, and $B_g[\partial\mathbf{L}]$ is a physical variable associated with its boundary, are topological equations and can be expressed by using coboundary operators. Their maps, δ^{C} and δ^{G}, respectively, are of the kind t_1, shown in Eq. (4.1.2):

$$\delta^{\mathrm{C}} : \partial\mathbf{S} \to \mathbf{S}; \qquad (5.1.5)$$

$$\delta^{\mathrm{G}} : \partial\mathbf{L} \to \mathbf{L}. \qquad (5.1.6)$$

Once all the p-cells have been opportunely labeled, the three operators δ^{D}, δ^{C}, and δ^{G} specialize in three matrices that can be obtained from the three incidence matrices, $\mathbf{D}$, $\mathbf{C}$, and $\mathbf{G}$, respectively.

In conclusion, δ^{D}, δ^{C}, and δ^{G} are the discrete versions of the differential operators "div", "curl", and "grad", respectively. This justifies the choice of the bold capital letters "$\mathbf{D}$", "$\mathbf{C}$", and "$\mathbf{G}$" for denoting the discrete operators and the corresponding incidence matrices.

It is worth noting that, while δ^{D}, δ^{C}, and δ^{G} are algebraic tensors, thus independent of the labeling,[42] the incidence numbers–which are the entries of the three matrices $\mathbf{D}$, $\mathbf{C}$, and $\mathbf{G}$–depend on the particular choice of labeling. This corresponds to the condition for which the operators "div", "curl", and "grad" are expressed by tensors, which do not depend on the coordinate basis, while the matrices that represent them in a coordinate system depend on the basis vectors. This further proves how the labeling of cells is the algebraic equivalent of mapping the points in $\mathbb{R}^3$ by means of the coordinate systems of the differential setting.

The positions of the coboundary operators δ^D, δ^C, and δ^G in the classification diagram of space elements are shown in Fig. 5.1.

Each topological equation is a coboundary process on a discrete p-form, c^p, which generates a discrete $(p+1)$-form, c^{p+1}. This is performed in the following two steps:

1. For each p-cell, e^i_p, we evaluate the value ϕ^i_p, assumed by the physical variable ϕ on the i-th p-cell, and spread this value on each coface e^j_{p+1} (of degree $p+1$) of the i-th p-cell, after having multiplied it by the mutual incidence number, q_{ji}, and the inverse of the magnitude, $\left\| e^j_{p+1} \right\|$, of e^j_{p+1}.

2. For each $(p+1)$-cell, e^j_{p+1}, we sum the values $\phi^{1,2,\ldots,n_j}_p$ coming from all the faces (of degree p) of its boundary.

Symbolically:

$$\phi^j_{p+1} = \sum_{i=1}^{n_j} \left(\frac{q_{ji}}{\left\| e^j_{p+1} \right\|} \phi^i_p \right), \text{ where } n_j = \left| \partial e^j_{p+1} \right|; \tag{5.1.7}$$

$$c^{p+1} = \delta c^p. \tag{5.1.8}$$

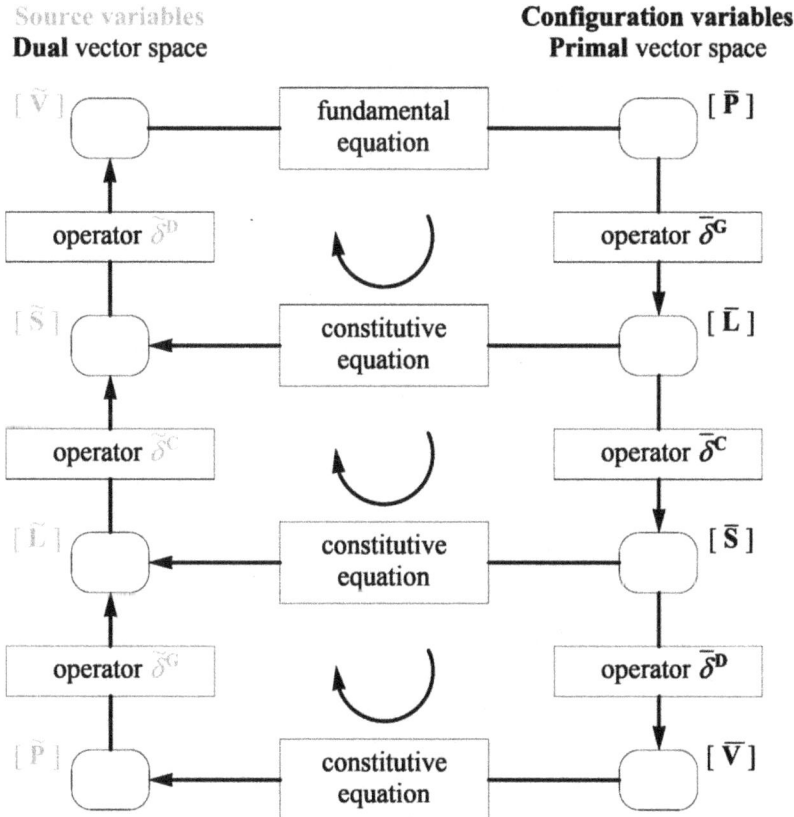

Figure 5.1. Coboundary operators in three-dimensional space for a primal and a dual cell complex.

By way of example, let us consider the primal and dual cell complexes of the plane domain in Fig. 5.2, where the primal nodes, sides, and areas have been labeled according to an arbitrary criterion. The labeling of the dual cell complex is induced by the labeling of the primal cell complex, by providing each dual element with the same label of the corresponding element in the primal cell complex.

For the sake of simplicity, let us assume that each primal side has unit length:

$$\overline{l}_i = 1, \; i = 1, 2, ..., 12. \tag{5.1.9}$$

Let Φ be a scalar variable defined on the nine primal nodes. This assignment generates a discrete 0-form on the nine primal nodes.

The process of spreading the values $\left[\varphi_1, \varphi_2, ..., \varphi_9 \right]$ of the discrete 0-form Φ^0,

$$\Phi^0 = \begin{bmatrix} \varphi_1 & \varphi_2 & \varphi_3 & \varphi_4 & \varphi_5 & \varphi_6 & \varphi_7 & \varphi_8 & \varphi_9 \end{bmatrix}^T, \tag{5.1.10}$$

from the nine primal nodes to the 12 primal sides, according to the mutual incidence numbers, is shown in Fig. 5.3.

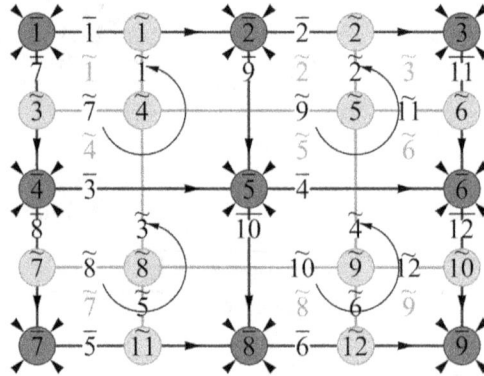

Figure 5.2. An example of plane primal complex and its dual complex.

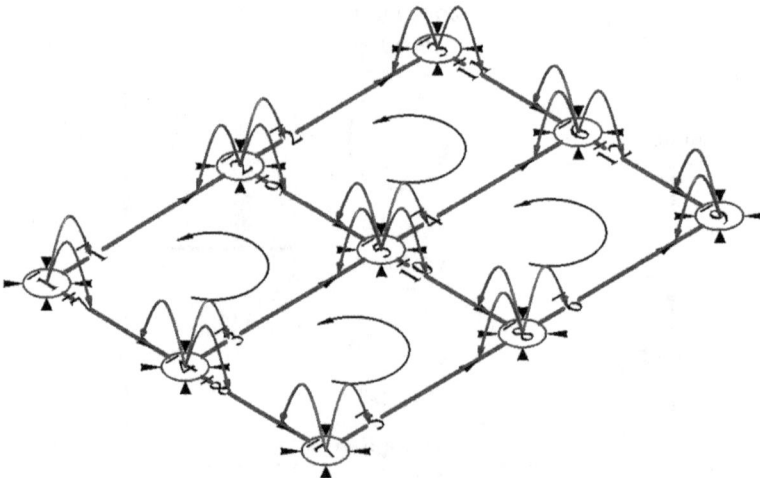

Figure 5.3. Coboundary process performed on the primal cell complex.

By denoting U the global variable that defines the discrete 1-form on the primal sides, the 12 values $\left[u_1, u_2, ..., u_{12} \right]$ of the 1-form U^1:

$$\mathbf{U}^1 = \begin{bmatrix} u_1 & u_2 & u_3 & u_4 & u_5 & u_6 & u_7 & u_8 & u_9 & u_{10} & u_{11} & u_{12} \end{bmatrix}^{\mathrm{T}}; \qquad (5.1.11)$$

are obtained by adding the scalar variables that are spread on the sides by the end nodes, together with the mutual incidence numbers. Note that, since each primal side has unit length, we do not need to normalize the elements of the 0-form by the lengths of the primal sides, and the coboundary process,

$$U^1 = \delta \Phi^0, \qquad (5.1.12)$$

is described by the matrix product $\overline{\mathbf{G}}\Phi^0$, between $\overline{\mathbf{G}}$, the matrix of the incidence numbers between primal 1-cells and primal 0-cells, and the vector Φ^0 of the 0-form Φ^0:

$$\mathbf{U}^1 = \overline{\mathbf{G}}\Phi^0; \qquad (5.1.13)$$

$$\begin{bmatrix} u_1 \\ u_2 \\ u_3 \\ u_4 \\ u_5 \\ u_6 \\ u_7 \\ u_8 \\ u_9 \\ u_{10} \\ u_{11} \\ u_{12} \end{bmatrix} = \begin{bmatrix} -1 & +1 & 0 & 0 & 0 & 0 & 0 & 0 & 0 \\ 0 & -1 & +1 & 0 & 0 & 0 & 0 & 0 & 0 \\ 0 & 0 & 0 & -1 & +1 & 0 & 0 & 0 & 0 \\ 0 & 0 & 0 & 0 & -1 & +1 & 0 & 0 & 0 \\ 0 & 0 & 0 & 0 & 0 & 0 & -1 & +1 & 0 \\ 0 & 0 & 0 & 0 & 0 & 0 & 0 & -1 & +1 \\ -1 & 0 & 0 & +1 & 0 & 0 & 0 & 0 & 0 \\ 0 & 0 & 0 & -1 & 0 & 0 & +1 & 0 & 0 \\ 0 & -1 & 0 & 0 & +1 & 0 & 0 & 0 & 0 \\ 0 & 0 & 0 & 0 & -1 & 0 & 0 & +1 & 0 \\ 0 & 0 & -1 & 0 & 0 & +1 & 0 & 0 & 0 \\ 0 & 0 & 0 & 0 & 0 & -1 & 0 & 0 & +1 \end{bmatrix} \begin{bmatrix} \varphi_1 \\ \varphi_2 \\ \varphi_3 \\ \varphi_4 \\ \varphi_5 \\ \varphi_6 \\ \varphi_7 \\ \varphi_8 \\ \varphi_9 \end{bmatrix}. \qquad (5.1.14)$$

In order to show the relationship between primal and dual coboundary processes in two-dimensional spaces, let us consider now a 1-form V^1, defined on the 12 dual sides of Fig. 5.2:

$$\mathbf{V}^1 = \begin{bmatrix} v_1 & v_2 & v_3 & v_4 & v_5 & v_6 & v_7 & v_8 & v_9 & v_{10} & v_{11} & v_{12} \end{bmatrix}^{\mathrm{T}}. \qquad (5.1.15)$$

By the coboundary process on the 1-form V^1, the 12 elements of V^1, $\left[v_1, v_2, ..., v_{12} \right]$, are spread on the nine dual areas, as shown in Fig. 5.4. This generates a 2-form, P^2, on the dual areas:

$$P^2 = \delta V^1, \qquad (5.1.16)$$

where δ denotes the coboundary operator.

As discussed in Section 4.3, for consistency between the two-dimensional and three-dimensional problems, one should consider that plane domains are provided with unit thickness. Consequently, V^1 and P^2 are, more properly, a 2-form and a 3-form, respectively. Moreover, δ

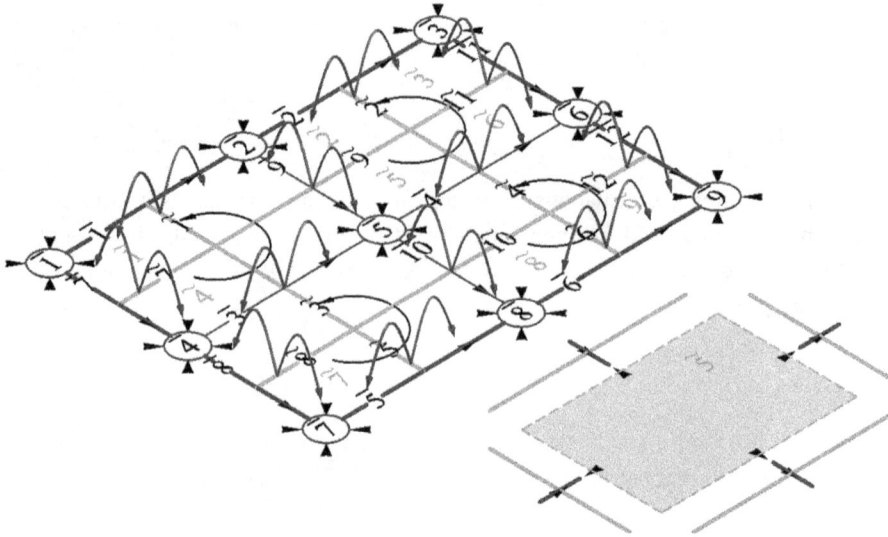

Figure 5.4. Coboundary process performed on the dual cell complex.

can be expressed by the incidence matrix $\widetilde{\mathbf{D}}$, the matrix of the incidence numbers between dual 3-cells and dual 2-cells. In the following, we will continue to call V^1 a 1-form and P^2 a 2-form, but, in order to preserve the relationship between tensor and their adjoints, we will express the dual coboundary process by means of the incidence matrix $\widetilde{\mathbf{D}}$. Denoted by $\mathbf{P}^2$ the vector of the 2-form P^2, we can therefore write

$$\mathbf{P}^2 = \widetilde{\mathbf{D}}\mathbf{V}^1; \tag{5.1.17}$$

$$
\begin{bmatrix} p_1 \\ p_2 \\ p_3 \\ p_4 \\ p_5 \\ p_6 \\ p_7 \\ p_8 \\ p_9 \end{bmatrix} =
\begin{bmatrix}
+1 & 0 & 0 & 0 & 0 & 0 & +1 & 0 & 0 & 0 & 0 & 0 \\
-1 & +1 & 0 & 0 & 0 & 0 & 0 & 0 & +1 & 0 & 0 & 0 \\
0 & -1 & 0 & 0 & 0 & 0 & 0 & 0 & 0 & 0 & +1 & 0 \\
0 & 0 & +1 & 0 & 0 & 0 & -1 & +1 & 0 & 0 & 0 & 0 \\
0 & 0 & -1 & +1 & 0 & 0 & 0 & 0 & -1 & +1 & 0 & 0 \\
0 & 0 & 0 & -1 & 0 & 0 & 0 & 0 & 0 & 0 & -1 & +1 \\
0 & 0 & 0 & 0 & +1 & 0 & 0 & -1 & 0 & 0 & 0 & 0 \\
0 & 0 & 0 & 0 & -1 & +1 & 0 & 0 & 0 & -1 & 0 & 0 \\
0 & 0 & 0 & 0 & 0 & -1 & 0 & 0 & 0 & 0 & 0 & -1
\end{bmatrix}
\begin{bmatrix} v_1 \\ v_2 \\ v_3 \\ v_4 \\ v_5 \\ v_6 \\ v_7 \\ v_8 \\ v_9 \\ v_{10} \\ v_{11} \\ v_{12} \end{bmatrix}, \tag{5.1.18}
$$

where, since the dual areas have an outward outer orientation (Fig. 5.4) while the primal nodes are sinks:

$$\widetilde{\mathbf{D}} = -\overline{\mathbf{G}}^{\mathrm{T}}. \tag{5.1.19}$$

As a final remark, note that, in space/time cell complexes, the time instants are the faces of the time intervals, both in primal and in dual cell complexes. Therefore, the time instants define the boundary on the time intervals, both in primal and in dual space/time cell complexes. Consequently, even the processes for the formation of the mean velocities and mean accelerations, described in Section 4.4, are coboundary processes.

5.1.1 Performing the Coboundary Process on Discrete 0-forms in Space Domain: Analogies Between Algebraic and Differential Operators

When performed for space elements of $\mathbb{R}^3$ or $\mathbb{R}^2$, the coboundary process on a discrete 0-form is the algebraic counterpart of finding the gradient of a scalar field.

In fact, both the gradient and the discrete 1-form generated by the coboundary process on a discrete 0-form comes from scalar quantities associated with points. More precisely, the gradient at the point $\mathbf{P}$ of a scalar valued function $\phi(\mathbf{P})$ is the vector valued function $\mathbf{u}(\mathbf{P})$ generated by the following five-step process:

1. Evaluating the increments of $\phi(\mathbf{P})$ along various oriented directions outgoing from the point $\mathbf{P}$, by forming the differences $\phi(\mathbf{Q}) - \phi(\mathbf{P})$, where $\mathbf{Q}$ is a point in the neighborhood of $\mathbf{P}$.
2. Evaluating the ratios of these increments to the distances between $\mathbf{Q}$ and $\mathbf{P}$.
3. Evaluating the limit for these ratios when the distances go to zero.
4. Finding the direction for which this limit is a maximum and considering this as the privileged direction.
5. Introducing a vector with origin at the point $\mathbf{P}$, with modulus equal to the maximum ratio found, arranged along the privileged direction. Such a vector valued function, $\mathbf{u}(\mathbf{P})$, is defined as the gradient of the scalar valued function ϕ at the point $\mathbf{P}$:

$$\mathbf{u}(\mathbf{P}) = \operatorname{grad} \phi(\mathbf{P}) = \nabla \phi(\mathbf{P}); \tag{5.1.20}$$

or, in Cartesian coordinates:

$$\mathbf{u}(x, y, z) = \operatorname{grad} \phi(x, y, z) = \nabla \phi(x, y, z). \tag{5.1.21}$$

The algebraic counterpart of this five-step process is evaluating the normalized increments of the discrete 0-form $\Phi^0 = \left[\phi_1, \phi_2, ..., \phi_{n_0} \right]$, defined for the n_0 vertices of the cell complex, along the edges connected with the point $\mathbf{P}$. Since evaluating the normalized increments of $\Phi^0(\mathbf{P})$ involves taking into account the incidence numbers between $\mathbf{P}$ and its cofaces, we can operatively divide this unique step into the two steps, spreading and collecting, shown in Fig. 5.5:

1. Spreading the value ϕ_i of the i-th 0-cell, $\mathbf{P}_i$, to all the cofaces, $\mathbf{L}_j$, of the 0-cell, each multiplied by the mutual incidence number between $\mathbf{P}_i$ and $\mathbf{L}_j$ and divided by L_j, the length of $\mathbf{L}_j$.
2. Adding the two values that have been spread on the same 1-cell, $\mathbf{L}_j$.

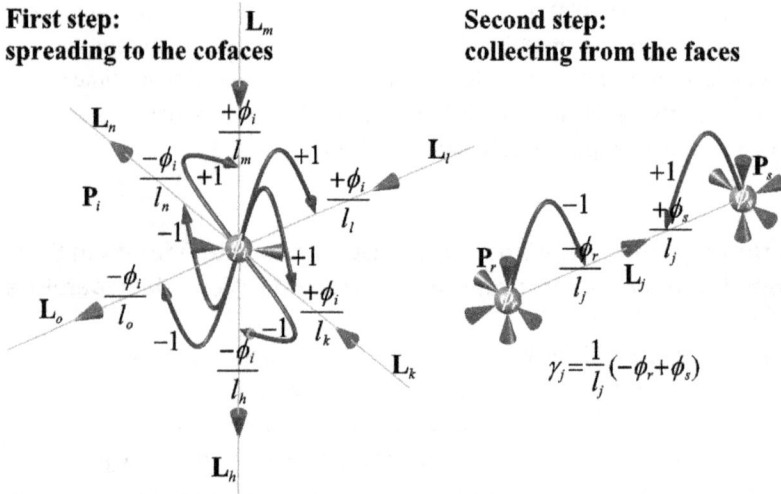

Figure 5.5. The coboundary process on a discrete 0-form defined on the nodes of a space a cell complex.

The discrete 1-form $\Gamma^1 = \left[\gamma_1, \gamma_2, ..., \gamma_{n_1} \right]$, generated on the n_1 1-cells of the space cell complex by the coboundary process on the discrete 0-form, is the coboundary of the discrete 0-form:

$$\Gamma^1 = \delta \Phi^0. \tag{5.1.22}$$

In the general case, where the 1-cells are not of unit length, γ_j, the j-th element of the discrete 1-form Γ^1, is given by the summation on the two 0-cells connected to the j-th 1-cell:

$$\gamma_j = \frac{1}{l_j} \sum_{i=1}^{2} g_{ji} \phi_i, \tag{5.1.23}$$

where g_{ji} is the element in row j and column i of the incidence matrix $\mathbf{G}$, and l_j is the length of the j-th 1-cell, $\mathbf{L}_j$, on which we are spreading the value ϕ_i of the i-th 0-cell, $\mathbf{P}_i$.

The elements of the matrix $\mathbf{T}^G$, which represents the coboundary operator for the given labeling of the cell complex, have the form:

$$T_{ji}^G = \frac{1}{l_j} g_{ji}. \tag{5.1.24}$$

$\mathbf{T}^G$ may be obtained as the sum of expanded local matrices. More precisely, let n be the number of 1-cells and let m be the number of 0-cells, then each local matrix is the expansion to the $n \times m$ dimensions of the 1×1 matrix $\mathbf{t}_j^G$ (a scalar), whose unique element is $1/l_j$:

$$\mathbf{t}_j^G = \left[\frac{1}{l_j} \right]. \tag{5.1.25}$$

The number of local matrices equals the number of 1-cells of the cell complex, that is, the number of times we must perform the second step of the coboundary process.

The assembling procedure for building $\mathbf{T}^G$ generalizes the assembling procedure used for the stiffness matrix to the case of non-symmetric matrices. The generalization is achieved by defining two collocation vectors, $\mathbf{l}_j$ (where "**l**" stands for left) and $\mathbf{r}_j$ (where "**r**" stands for right), according to the following twofold criterion:

- $\mathbf{l}_j$ is an $n \times 1$ Boolean vector, whose unique nonzero value is the j-th element:

$$\left(\mathbf{l}_j \right)_h = \delta_{jh}, \; h = 1, 2, ..., n, \tag{5.1.26}$$

where δ_{jh} is the Kronecker delta symbol.
- $\mathbf{r}$ is the $m \times 1$ vector such that $\mathbf{r}^{\mathrm{T}}$ is the j-th row of the incidence matrix $\mathbf{G}$:

$$\left(\mathbf{r}_j \right)_k = g_{jk}, \; k = 1, 2, ..., m. \tag{5.1.27}$$

In these assumptions, the expansion of the j-th local matrix, $\mathbf{t}_j^{\mathrm{G}}$, is given by the matrix product $\mathbf{l}_j \mathbf{t}_j^{\mathrm{G}} \mathbf{r}_j^{\mathrm{T}}$, and $\mathbf{T}^{\mathrm{G}}$ is provided by the sum:

$$\mathbf{T}^{\mathrm{G}} = \sum_{i=1}^{n} \mathbf{l}_j \mathbf{t}_j^{\mathrm{G}} \mathbf{r}_j^{\mathrm{T}}. \tag{5.1.28}$$

Γ^1, the vector of the 1-form Γ^1 can then be obtained as the product:

$$\Gamma^1 = \mathbf{T}^{\mathrm{G}} \Phi^0. \tag{5.1.29}$$

By way of example, if the primal 1-cells in Fig. 5.2 were not of unit length, the 5-th local matrix of the primal cell complex would be

$$\overline{\mathbf{t}}_5^{\mathrm{G}} = \left[\frac{1}{\overline{l}_5} \right], \tag{5.1.30}$$

and its expansion to the 12×9 dimensions of the problem would be provided by

$$\overline{\mathbf{l}}_5 \overline{\mathbf{t}}_5^{\mathrm{G}} \overline{\mathbf{r}}_5^{\mathrm{T}} = \begin{bmatrix} 0 \\ 0 \\ 0 \\ 0 \\ 0 \\ 1 \\ 0 \\ 0 \\ 0 \\ 0 \\ 0 \\ 0 \end{bmatrix} \left[\frac{1}{\overline{l}_5} \right] \begin{bmatrix} 0 & 0 & 0 & 0 & 0 & 0 & -1 & +1 & 0 \end{bmatrix} = \begin{bmatrix} 0 & 0 & 0 & 0 & 0 & 0 & 0 & 0 & 0 \\ 0 & 0 & 0 & 0 & 0 & 0 & 0 & 0 & 0 \\ 0 & 0 & 0 & 0 & 0 & 0 & 0 & 0 & 0 \\ 0 & 0 & 0 & 0 & 0 & 0 & 0 & 0 & 0 \\ 0 & 0 & 0 & 0 & 0 & 0 & \frac{-1}{\overline{l}_5} & \frac{+1}{\overline{l}_5} & 0 \\ 0 & 0 & 0 & 0 & 0 & 0 & 0 & 0 & 0 \\ 0 & 0 & 0 & 0 & 0 & 0 & 0 & 0 & 0 \\ 0 & 0 & 0 & 0 & 0 & 0 & 0 & 0 & 0 \\ 0 & 0 & 0 & 0 & 0 & 0 & 0 & 0 & 0 \\ 0 & 0 & 0 & 0 & 0 & 0 & 0 & 0 & 0 \\ 0 & 0 & 0 & 0 & 0 & 0 & 0 & 0 & 0 \\ 0 & 0 & 0 & 0 & 0 & 0 & 0 & 0 & 0 \end{bmatrix}. \tag{5.1.31}$$

After sum over the index j, the primal matrix $\overline{\mathbf{T}}^G$ would be

$$\overline{\mathbf{T}}^G = \sum_{j=1}^{12} \overline{\mathbf{l}}_j \overline{\mathbf{t}}_j^G \overline{\mathbf{r}}_j^T = \begin{bmatrix} -1/\overline{l}_1 & +1/\overline{l}_1 & 0 & 0 & 0 & 0 & 0 & 0 & 0 \\ 0 & -1/\overline{l}_2 & +1/\overline{l}_2 & 0 & 0 & 0 & 0 & 0 & 0 \\ 0 & 0 & 0 & -1/\overline{l}_3 & +1/\overline{l}_3 & 0 & 0 & 0 & 0 \\ 0 & 0 & 0 & 0 & -1/\overline{l}_4 & +1/\overline{l}_4 & 0 & 0 & 0 \\ 0 & 0 & 0 & 0 & 0 & 0 & -1/\overline{l}_5 & +1/\overline{l}_5 & 0 \\ 0 & 0 & 0 & 0 & 0 & 0 & 0 & -1/\overline{l}_6 & +1/\overline{l}_6 \\ -1/\overline{l}_7 & 0 & 0 & +1/\overline{l}_7 & 0 & 0 & 0 & 0 & 0 \\ 0 & 0 & 0 & -1/\overline{l}_8 & 0 & 0 & +1/\overline{l}_8 & 0 & 0 \\ 0 & -1/\overline{l}_9 & 0 & 0 & +1/\overline{l}_9 & 0 & 0 & 0 & 0 \\ 0 & 0 & 0 & 0 & -1/\overline{l}_{10} & 0 & 0 & +1/\overline{l}_{10} & 0 \\ 0 & 0 & -1/\overline{l}_{11} & 0 & 0 & +1/\overline{l}_{11} & 0 & 0 & 0 \\ 0 & 0 & 0 & 0 & 0 & -1/\overline{l}_{12} & 0 & 0 & +1/\overline{l}_{12} \end{bmatrix},$$

(5.1.32)

and the coboundary process on the 0-form Φ^0 in Eq. (5.1.12) would be expressed as

$$\mathbf{U}^1 = \overline{\mathbf{T}}^G \mathbf{\Phi}^0. \tag{5.1.33}$$

Note that the matrix $\overline{\mathbf{T}}^G$ becomes equal to the incidence matrix $\mathbf{G}$ when all the lengths are equal to 1. Note also that the matrix $\overline{\mathbf{T}}^G$ does not contain any material parameter, but only metric notions, such as must be for a topological operator.

Finally, in order to derive a geometrical representation of the coboundary process on the discrete 0-forms, we can recall that, for the properties of a Banach space,[79] there exists an isomorphism between the orthogonal space and the dual space. This allows us to extend the geometric interpretation of the linear functional to the coboundary process performed on the discrete 0-forms. In general, we can view the action of the coboundary operators on the discrete p-forms in terms of hyperplanes, as shown in Fig. 2.2.

On the other hand, we have also seen that, when the scalar valued potential function $\phi(\mathbf{P})$ in the neighborhood of the point $\mathbf{P}$ is approximated with an affine scalar field, ϕ, the equipotential surfaces are parallel planes (Section 4.2, Fig. 5.6). In this case, let $\mathbf{Q}$ and $\mathbf{R}$ be two points of the neighborhood of $\mathbf{P}$, arbitrarily chosen on two equipotential planes, then the ratio:

$$\frac{\phi_R - \phi_Q}{\|\mathbf{R} - \mathbf{Q}\|}, \tag{5.1.34}$$

where we used the notation of Grassmann[111] for denoting vectors, provides the same value for any couple of intersection points between the equipotential planes and a straight line parallel to t, which is the line that passes through $\mathbf{Q}$ and $\mathbf{R}$ (Fig. 5.6). Therefore, the ratio in Eq. (5.1.34) depends on the direction of t, but does not depend on the point in which it is evaluated.

In differential formulation, the ratio in Eq. (5.1.34) is called the gradient in the direction t and is denoted by G_t. The maximum value of G_t is attained when the distance between two

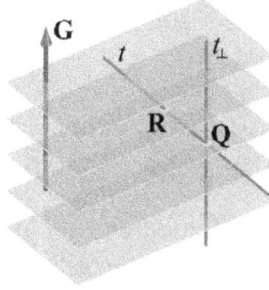

Figure 5.6. Relation of orthogonality between the gradient of an affine function and the equipotential planes of the affine function.

successive intersection points is minimal, that is, when **R** is placed on the perpendicular to the equipotential plane passing through **Q** (the straight line t is perpendicular to the equipotential planes, as the line $t_\perp$ in Fig. 5.6).

Since it does not depend on the evaluation point, the maximum value of G_t can be represented by a free vector,[73] which is denoted by **G** and called the gradient of the scalar valued affine function, ϕ. The sense of **G** is the one along which the potential increases and the modulus of **G** is equal to the maximum value of G_t.

According to the Riesz representation theorem in Euclidean n-space, expressed by Eq. (2.2.15), **G** is the unique vector that represent the linear form "grad":

$$grad(\phi) = \mathbf{G} \cdot (\mathbf{R} - \mathbf{Q}). \tag{5.1.35}$$

Therefore, the same term, "gradient", applies both to the linear form (which is a covector) and to the vector that represents it.

Analogously, the coboundary process on $\Phi^0(\mathbf{P})$, the affine scalar field ϕ that approximates the potential in the neighborhood of the point **P**, produces the normalized increments of $\Phi^0(\mathbf{P})$ along the directions of the edges connected with **P**. These normalized increments depend on the edge direction, while they do not vary if **P** is substituted with another point of the neighborhood in which the affine scalar field has been evaluated.

We will denote by δ_t^G the algebraic equivalent of G_t and by δ_G the algebraic equivalent of **G**, where δ_G is a uniform[153] vector field in the neighborhood of the point **P**. As shown in Fig. 5.7, δ_G is orthogonal to the equipotential parallel planes of $\Phi^0(\mathbf{P})$:

$$\delta^G(\Phi^0) = \delta_G \cdot (\mathbf{R} - \mathbf{Q}). \tag{5.1.36}$$

5.1.2 Performing the Coboundary Process on Discrete 0-forms in Time Domain: Analogies Between Algebraic and Differential Operators

As we have discussed in Section 3.8, when our goal is that of studying how a given phenomenon evolves in time, we must add a time axis to the three-dimensional and two-dimensional

---

[153] A uniform vector field is a vector field that does not depend on the point.

space cell complexes of $\mathbb{R}^3$ and $\mathbb{R}^2$, thus originating four-dimensional and three-dimensional space/time cell complexes, respectively.

Since the elements of a space/time cell complex are of different nature, some p-cells are associated with a variation of the space variables, some other p-cells are associated with a variation of the time variables, and some other p-cells are associated with a variation of both the space and time variables.

Now, since a discrete 0-form is defined on each zero-dimensional element of a cell complex, the coboundary process on the discrete 0-form naturally extends to all the zero-dimensional elements of the space/time cell complex, that is, both to the 0-cells associated with a given time instant and to the 0-cells associated with adjacent time instants. Thus, if we start from space cell complexes in dimension 3, adding the time axis will cause the discrete 0-form to be defined on all the 0-cells of a tesseract (Fig. 5.8).

Moreover, the coboundary process on the discrete 0-form generates discrete 1-forms on all the one-dimensional elements of the tesseract, that is, on both the one-dimensional elements

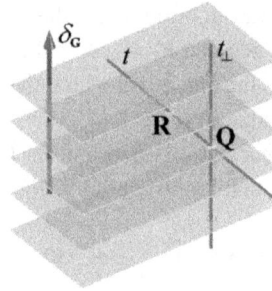

Figure 5.7. Relation of orthogonality between δ_G and the equipotential planes of $\Phi^0(\mathbf{P})$.

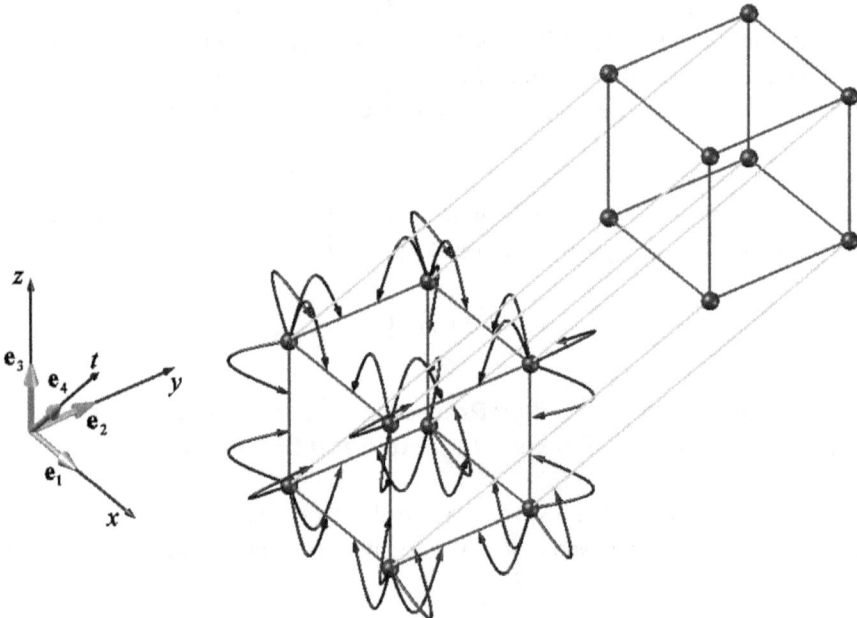

Figure 5.8. The coboundary process on a discrete 0-form defined for the nodes $t = t_1$ of a tesseract.

of the kind "space" (the edges of the space elements) and the one-dimensional elements of the kind "time" (the time intervals). This results in two different kinds of discrete 1-forms, generated by the same discrete 0-form (Fig. 5.8):

- Discrete 1-forms of the kind "space": the discrete 1-forms generated on the 1-cells of the kind "space".
- Discrete 1-forms of the kind "time": the discrete 1-forms generated on the 1-cells of the kind "time".

The process for the generation of the 1-forms of the kind "space" is, even in this second case, the algebraic counterpart of finding the gradient of a scalar valued function of the point, while the process for the generation of the 1-forms of the kind "time" has a different nature. This second process is the algebraic version of finding the derivative of a function of one variable.

For the sake of completeness, let us recall that the derivative of a function is a process which can be divided into three steps:

- Forming the increment in time of the function of one variable.
- Forming the ratio of this increment for the time interval.
- Forming the limit process for the time interval going to zero.

Therefore, the same coboundary process performed on a discrete 0-form in space/time gives rise to two operators at the same time, the algebraic version of the gradient and the algebraic version of the derivative of a function of one variable. This allows us to say that finding the increments in space and finding the increments in time are two sides of the same coin, that is, the coboundary process on discrete 0-forms.

By extending these results to the differential formulation, the gradient and the derivative of a function of one variable can now be seen as two sides of the same coin, that is, the exterior differentiation on exterior differential forms. In effect, this conclusion is implicit in the possibility of decomposing the differential in the product of the derivative times the differential of the independent variable.

The coboundary process on a discrete 0-form in space/time can also be seen as the algebraic counterpart of the spacetime gradient in spacetime algebra,[154] which is built up from combinations of one time-like basis vector, γ_0, and three orthogonal space-like vectors, $\{\gamma_1, \gamma_2, \gamma_3\}$.

Associated with the orthogonal basis, $\{\gamma_\mu\}$, is the reciprocal basis:

$$\gamma^\mu = \frac{1}{\gamma_\mu}, \tag{5.1.37}$$

---

[154] In mathematical physics, **spacetime algebra** (STA) is a name for the Clifford algebra $C\ell_{1,3}(\mathbb{R})$,[57] or equivalently the geometric algebra $G_4 = G(M4)$ (Section 2.2), which can be particularly closely associated with the geometry of special relativity and relativistic spacetime. It is a linear algebra allowing not just vectors, but also directed quantities associated with particular planes (for example: areas, or rotations) or associated with particular (hyper-)volumes to be combined, as well as rotated, reflected, or Lorentz boosted. It is also the natural parent algebra of spinors in special relativity. These properties allow many of the most important equations in physics to be expressed in particularly simple forms and can be very helpful towards a more geometric understanding of their meanings.

for all $\mu = 0,...,3$, satisfying the relation:

$$\gamma_\mu \cdot \gamma^\nu = \delta_\mu^\nu, \tag{5.1.38}$$

where δ_μ^ν is the Kronecker delta symbol.

The spacetime gradient, like the gradient in a Euclidean space, is defined such that the directional derivative relationship is satisfied:

$$a \cdot \nabla F(x) = \lim_{\tau \to 0} \frac{F(x + a\tau) - F(x)}{\tau}. \tag{5.1.39}$$

This requires the definition of the gradient to be

$$\nabla = \gamma^\mu \frac{\partial}{\partial x^\mu} = \gamma^\mu \partial_\mu. \tag{5.1.40}$$

Written out explicitly with

$$x = ct\gamma_0 + x^k \gamma_k, \tag{5.1.41}$$

these partials are

$$\partial_0 = \frac{1}{c} \frac{\partial}{\partial t}; \tag{5.1.42}$$

$$\partial_k = \frac{\partial}{\partial x^k}. \tag{5.1.43}$$

5.1.3 Performing the Coboundary Process on Discrete 1-forms in Space/Time Domain: Analogies Between Algebraic and Differential Operators

When performed for space elements of $\mathbb{R}^3$ or $\mathbb{R}^2$, the coboundary process on a discrete 1-form is the algebraic counterpart of finding the curl of a vector field.

Both the curl and the discrete 2-form generated by the coboundary process on a discrete 1-form come from vector quantities associated with lines. More precisely, the curl at the point **P** of a vector valued function $\mathbf{u}(\mathbf{P})$ is the vector valued function $\mathbf{v}(\mathbf{P})$ generated by the following five-step process:

1. Evaluating the line integral of the vector $\mathbf{u}(\mathbf{P})$ along the boundary of a small plane surface centered at the point **P**.
2. Evaluating the ratios between such circulation and the area of the surface.
3. Performing the limit of this ratio when the surface shrinks to the point **P**.
4. Looking for the attitude of the plane surface on which this limit is maximum and considering this as the privileged attitude.
5. Introducing a vector with origin at the point **P**, with modulus equal to the maximum ratio found, disposed normal to the privileged attitude and oriented in such a way that the orientation of the closed line (the one used to evaluate the circulation) and the vector

form a clockwise screw. This vector, $\mathbf{v}(\mathbf{P})$, is defined as the curl of the vector field $\mathbf{u}$ at the point $\mathbf{P}$:

$$\mathbf{v}(\mathbf{P}) = \operatorname{curl} \mathbf{u}(\mathbf{P}) = \nabla \times \mathbf{u}(\mathbf{P}); \tag{5.1.44}$$

or, in Cartesian coordinates:

$$\mathbf{v}(x, y, z) = \operatorname{curl} \mathbf{u}(x, y, z) = \nabla \times \mathbf{u}(x, y, z). \tag{5.1.45}$$

Let $\Gamma^1 = \left[\gamma_1, \gamma_2, \dots, \gamma_{n_1} \right]$ be the discrete 1-form of the line integrals along the n_1 edges, $\mathbf{L}_i$, of the space cell complex; then the algebraic counterpart of the five-step process for finding the curl at the point $\mathbf{P}$ is the coboundary process on Γ^1, which can be described by the following two-step process (Fig. 5.9):

1. Spreading the value γ_i of the i-th 1-cell, $\mathbf{L}_i$, to all the cofaces, $\mathbf{S}_j$, of the 1-cell, each multiplied by the mutual incidence number between $\mathbf{L}_i$ and $\mathbf{S}_j$ and divided by S_j, the area of $\mathbf{S}_j$.
2. Adding all the values that have been spread on the same 2-cell, $\mathbf{S}_j$.

The discrete 2-form $\Psi^2 = \left[\psi_1, \psi_2, \dots, \psi_{n_2} \right]$, generated on the n_2 2-cells of the space cell complex by the coboundary process on the discrete 1-form, is the coboundary of the discrete 1-form:

$$\Psi^2 = \delta\Gamma^1. \tag{5.1.46}$$

The j-th element of the discrete 2-form Ψ^2 is given by the summation on the 1-cells connected to the j-th 2-cell:

$$\psi_j = \frac{1}{S_j} \sum_i c_{ji} \gamma_i, \tag{5.1.47}$$

where c_{ji} is the element in row j and column i of the incidence matrix $\mathbf{C}$, and S_j is the area of the j-th 2-cell, $\mathbf{S}_j$, on which we are spreading the value γ_i of the i-th 1-cell, $\mathbf{L}_i$.

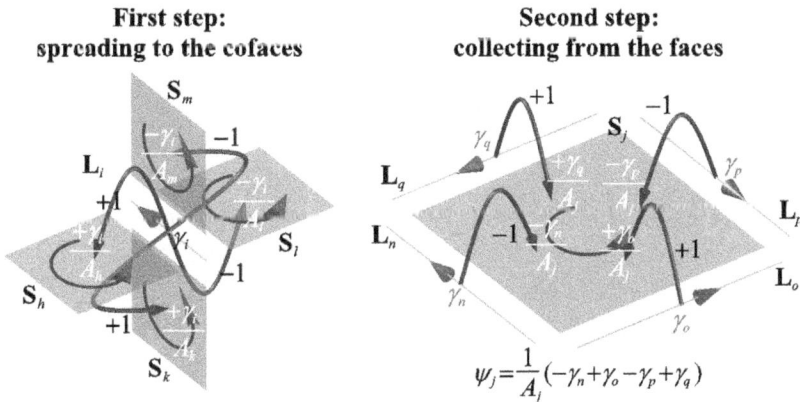

Figure 5.9. The coboundary process on a discrete 1-form defined on the edges of a space cell complex.

As for the coboundary process on 0-forms, even in this case we can obtain the matrix $\mathbf{T}^C$, which expresses the coboundary operator for the given labeling of the cell complex, as a sum of expanded local matrices. The elements of $\mathbf{T}^C$ have the form

$$T^C_{ji} = \frac{1}{S_j} c_{ji}. \tag{5.1.48}$$

Let n be the number of 2-cells and let m be the number of 1-cells; then each local matrix is the expansion to the $n \times m$ dimensions of the 1×1 matrix $\mathbf{t}^C_j$ (a scalar), whose unique element is $1/S_j$:

$$\mathbf{t}^C_j = \left[\frac{1}{S_j} \right]. \tag{5.1.49}$$

The number of local matrices equals the number of 2-cells of the cell complex, that is, the number of times we must perform the second step of the coboundary process.

The two collocation vectors, $\mathbf{l}_j$ and $\mathbf{r}_j$, are defined according to the following twofold criterion:

- $\mathbf{l}_j$ is an $n \times 1$ Boolean vector, whose unique nonzero value is the j-th element:

$$\left(\mathbf{l}_j \right)_h = \delta_{jh}, \; h = 1, 2, ..., n, \tag{5.1.50}$$

where δ_{jh} is the Kronecker delta symbol.
- $\mathbf{r}$ is the $m \times 1$ vector such that $\mathbf{r}^T$ is the j-th row of the incidence matrix $\mathbf{C}$:

$$\left(\mathbf{r}_j \right)_k = c_{jk}, \; k = 1, 2, ..., m. \tag{5.1.51}$$

The expansion of the j-th local matrix, $\mathbf{t}^C_j$, is given by the matrix product $\mathbf{l}_j \mathbf{t}^C_j \mathbf{r}^T_j$, and $\mathbf{T}^C$ is provided by the sum:

$$\mathbf{T}^C = \sum_{i=1}^n \mathbf{l}_j \mathbf{t}^C_j \mathbf{r}^T_j. \tag{5.1.52}$$

The 2-form Ψ^2 can then be obtained as the product:

$$\Psi^2 = \mathbf{T}^C \Gamma^1. \tag{5.1.53}$$

In the case where the primal 1-cells in Fig. (5.2) are not of unit length, the discrete 2-form defined on the dual 1-cells gives rise to local matrices for the dual cell complex. By way of example, the 5-th local matrix of the dual cell complex is

$$\tilde{\mathbf{t}}^C_5 = \left[\frac{1}{\tilde{S}_5} \right], \tag{5.1.54}$$

and the expanded form of the 5-th local matrix is provided by the collocation dual vectors $\tilde{\mathbf{l}}_5$ and $\tilde{\mathbf{r}}$:

$$\tilde{\mathbf{l}}_5\tilde{\mathbf{t}}_5^C\tilde{\mathbf{r}}_5^T = \begin{bmatrix} 0 \\ 0 \\ 0 \\ 0 \\ 1 \\ 0 \\ 0 \\ 0 \\ 0 \end{bmatrix} \begin{bmatrix} \dfrac{1}{\tilde{S}_5} \end{bmatrix} \begin{bmatrix} 0 & 0 & -1 & +1 & 0 & 0 & 0 & 0 & -1 & +1 & 0 & 0 \end{bmatrix}, \qquad (5.1.55)$$

providing the result:

$$\tilde{\mathbf{l}}_5\tilde{\mathbf{t}}_5^C\tilde{\mathbf{r}}_5^T = \begin{bmatrix} 0 & 0 & 0 & 0 & 0 & 0 & 0 & 0 & 0 & 0 & 0 & 0 \\ 0 & 0 & 0 & 0 & 0 & 0 & 0 & 0 & 0 & 0 & 0 & 0 \\ 0 & 0 & 0 & 0 & 0 & 0 & 0 & 0 & 0 & 0 & 0 & 0 \\ 0 & 0 & 0 & 0 & 0 & 0 & 0 & 0 & 0 & 0 & 0 & 0 \\ 0 & 0 & \dfrac{-1}{\tilde{S}_5} & \dfrac{+1}{\tilde{S}_5} & 0 & 0 & 0 & 0 & \dfrac{-1}{\tilde{S}_5} & \dfrac{+1}{\tilde{S}_5} & 0 & 0 \\ 0 & 0 & 0 & 0 & 0 & 0 & 0 & 0 & 0 & 0 & 0 & 0 \\ 0 & 0 & 0 & 0 & 0 & 0 & 0 & 0 & 0 & 0 & 0 & 0 \\ 0 & 0 & 0 & 0 & 0 & 0 & 0 & 0 & 0 & 0 & 0 & 0 \\ 0 & 0 & 0 & 0 & 0 & 0 & 0 & 0 & 0 & 0 & 0 & 0 \end{bmatrix}. \qquad (5.1.56)$$

The matrix $\tilde{\mathbf{T}}^C$ then follows from the assembling procedure:

$$\tilde{\mathbf{T}}^C = \sum_{j=1}^{9} \tilde{\mathbf{l}}_j\tilde{\mathbf{t}}_j^C\tilde{\mathbf{r}}_j^T = \begin{bmatrix} \dfrac{+1}{\tilde{S}_1} & 0 & 0 & 0 & 0 & 0 & \dfrac{+1}{\tilde{S}_1} & 0 & 0 & 0 & 0 & 0 \\ \dfrac{-1}{\tilde{S}_2} & \dfrac{+1}{\tilde{S}_2} & 0 & 0 & 0 & 0 & 0 & 0 & \dfrac{+1}{\tilde{S}_2} & 0 & 0 & 0 \\ 0 & \dfrac{-1}{\tilde{S}_3} & 0 & 0 & 0 & 0 & 0 & 0 & 0 & 0 & \dfrac{+1}{\tilde{S}_3} & 0 \\ 0 & 0 & \dfrac{+1}{\tilde{S}_4} & 0 & 0 & 0 & \dfrac{-1}{\tilde{S}_4} & \dfrac{+1}{\tilde{S}_4} & 0 & 0 & 0 & 0 \\ 0 & 0 & \dfrac{-1}{\tilde{S}_5} & \dfrac{+1}{\tilde{S}_5} & 0 & 0 & 0 & 0 & \dfrac{-1}{\tilde{S}_5} & \dfrac{+1}{\tilde{S}_5} & 0 & 0 \\ 0 & 0 & 0 & \dfrac{-1}{\tilde{S}_6} & 0 & 0 & 0 & 0 & 0 & 0 & \dfrac{-1}{\tilde{S}_6} & \dfrac{+1}{\tilde{S}_6} \\ 0 & 0 & 0 & 0 & \dfrac{+1}{\tilde{S}_7} & 0 & 0 & \dfrac{-1}{\tilde{S}_7} & 0 & 0 & 0 & 0 \\ 0 & 0 & 0 & 0 & \dfrac{-1}{\tilde{S}_8} & \dfrac{+1}{\tilde{S}_8} & 0 & 0 & 0 & \dfrac{-1}{\tilde{S}_8} & 0 & 0 \\ 0 & 0 & 0 & 0 & 0 & \dfrac{-1}{\tilde{S}_9} & 0 & 0 & 0 & 0 & 0 & \dfrac{-1}{\tilde{S}_9} \end{bmatrix}. \qquad (5.1.57)$$

As discussed in Section 4.3, if we provide a plane domain with a unit thickness, the incidence matrix $\tilde{\mathbf{C}}$ plays the role of the incidence matrix $\tilde{\mathbf{D}}$ of the equivalent three-dimensional problem. In this case, $\tilde{\mathbf{T}}^C$ is equal to the matrix $\tilde{\mathbf{T}}^D$ of the equivalent three-dimensional problem, and we can write

$$\mathbf{P}^2 = \tilde{\mathbf{T}}^D \mathbf{V}^1, \tag{5.1.58}$$

where $\tilde{\mathbf{T}}^D$ becomes the incidence matrix $\tilde{\mathbf{D}}$ in the special case where all the primal 1-cells have unit lengths.

When the vector valued potential function $\mathbf{u}(\mathbf{P})$ in the neighborhood of the point $\mathbf{P}$ is approximated with an affine vector field and the plane domain is made of triangles, an interesting property can be derived from the theorems of the vector affine fields (Section 4.2). Let $\mathbf{Q}$, $\mathbf{R}$, and $\mathbf{S}$ be the vertices of a triangle and let $\mathbf{u}_Q$, $\mathbf{u}_R$, and $\mathbf{u}_S$ be three vectors defined at the vertices $\mathbf{Q}$, $\mathbf{R}$, and $\mathbf{S}$. The vector field for the points of the triangle can be approximated by the two affine functions:

$$u_x = a + bx + cy; \tag{5.1.59}$$

$$u_y = d + ex + fy. \tag{5.1.60}$$

Let us denote the three oriented edges of the triangle by $\mathbf{L}_1$, $\mathbf{L}_2$, and $\mathbf{L}_3$. For the inner orientations in Fig. 5.10, it follows:

$$\mathbf{L}_1 = \mathbf{R} - \mathbf{Q}; \tag{5.1.61}$$

$$\mathbf{L}_2 = \mathbf{S} - \mathbf{R}; \tag{5.1.62}$$

$$\mathbf{L}_3 = \mathbf{Q} - \mathbf{S}; \tag{5.1.63}$$

$$\mathbf{L}_1 + \mathbf{L}_2 + \mathbf{L}_3 = \mathbf{0}. \tag{5.1.64}$$

Due to Eqs. (4.2.22) and (4.2.23), we can compute the circulation, Γ, along the boundary of the triangle $\mathbf{QRS}$ as

$$
\begin{aligned}
\Gamma &= \frac{1}{2}\left[\left(\mathbf{u}_Q + \mathbf{u}_R\right)\cdot\mathbf{L}_1 + \left(\mathbf{u}_R + \mathbf{u}_S\right)\cdot\mathbf{L}_2 + \left(\mathbf{u}_S + \mathbf{u}_Q\right)\cdot\mathbf{L}_3\right] = \\
&= \frac{1}{2}\left[\mathbf{u}_Q\cdot\left(\mathbf{L}_1 + \mathbf{L}_3\right) + \mathbf{u}_R\cdot\left(\mathbf{L}_1 + \mathbf{L}_2\right) + \mathbf{u}_S\cdot\left(\mathbf{L}_2 + \mathbf{L}_3\right)\right] = \\
&= -\frac{1}{2}\left[\mathbf{u}_Q\cdot\mathbf{L}_2 + \mathbf{u}_R\cdot\mathbf{L}_3 + \mathbf{u}_S\cdot\mathbf{L}_1\right],
\end{aligned}
\tag{5.1.65}
$$

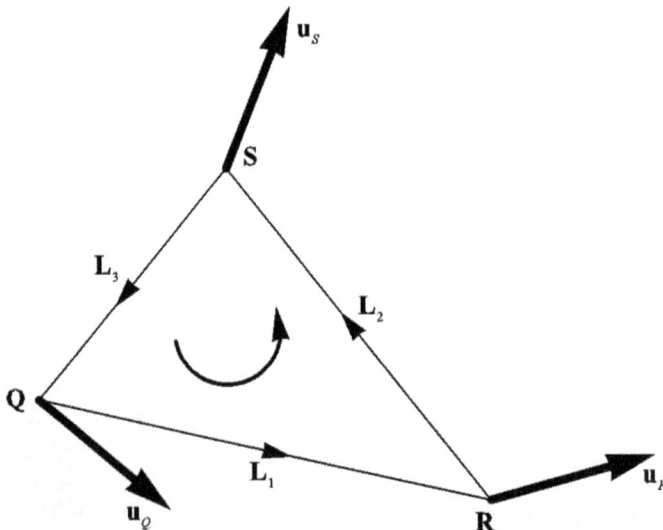

Figure 5.10. Approximation of a vector field in a triangular region with an affine vector field.

where Eq. (5.1.64) has been used for simplifying the calculations. By using Eqs. (5.1.59) and (5.1.60) for expressing the components of $\mathbf{u}_Q$, $\mathbf{u}_R$, and $\mathbf{u}_S$, Eq. (5.1.65) can be re-written in scalar form as

$$-2\Gamma = \begin{bmatrix} a + bx_Q + cy_Q \\ d + ex_Q + fy_Q \end{bmatrix} \cdot \begin{bmatrix} x_S - x_R \\ y_S - y_R \end{bmatrix} + \begin{bmatrix} a + bx_R + cy_R \\ d + ex_R + fy_R \end{bmatrix} \cdot \begin{bmatrix} x_Q - x_S \\ y_Q - y_S \end{bmatrix} + \begin{bmatrix} a + bx_S + cy_S \\ d + ex_S + fy_S \end{bmatrix} \cdot \begin{bmatrix} x_R - x_Q \\ y_R - y_Q \end{bmatrix},$$

$$\text{(5.1.66)}$$

which, after simplification, provides

$$-2\Gamma = (c - e)\left(x_S y_Q - x_R y_Q + x_Q y_R - x_S y_R + x_R y_S - x_Q y_S \right). \tag{5.1.67}$$

According to Eq. (4.2.24), Γ must be proportional to the area, A, of the triangle. In particular, by comparing Eq. (5.1.67) with twice the area:

$$2A = \begin{vmatrix} 1 & x_Q & y_Q \\ 1 & x_R & y_R \\ 1 & x_S & y_S \end{vmatrix} = \left(x_S y_Q - x_R y_Q + x_Q y_R - x_S y_R + x_R y_S - x_Q y_S \right), \tag{5.1.68}$$

we find that the ratio between the circulation and the area is a constant that is characteristic of the affine field, that is, is a characteristic of the neighborhood:

$$\frac{\Gamma}{A} = e - c. \tag{5.1.69}$$

In a three-dimensional affine vector field, the ratio between Γ and A depends on the inclination of the plane space element, while it does not depend on the point of the neighborhood.

The vector whose direction is perpendicular to the plane for which we have the maximum value of this ratio, whose modulus is the value of this maximum ratio and whose orientation is the one obtained by applying the screw rule to the circuit along which the circulation is evaluated is called the curl of the affine vector field. In particular, the curl of a plane affine vector field is orthogonal to the plane.

As discussed in Section 3.8, when we add a time axis to the three-dimensional and two-dimensional space cell complexes of $\mathbb{R}^3$ and $\mathbb{R}^2$, we obtain space/time cell complexes with p-cells of different kinds ("space", "time", and "space/time").

Since the discrete p-forms are defined on each p-dimensional element of the space/time cell complex, the coboundary process on the discrete 1-form naturally extends both to the 1-cells associated with a given time instant and to the 1-cells associated with adjacent time instants. Starting from space cell complexes in dimension 3, this implies that the discrete 1-form in space/time is defined on all the 1-cells of a tesseract (Fig. 5.11).

Consequently, the coboundary process on the discrete 1-form generates discrete 2-forms for all the two-dimensional elements of the tesseract, that is, both for the two-dimensional elements of the kind "space" and for the two-dimensional elements of the kind "time". This results in two different kinds of discrete 2-forms, generated by the same discrete 1-form (Fig. 5.11):

- Discrete 2-forms of the kind "space": the discrete 2-forms generated on the 2-cells of the kind "space".
- Discrete 2-forms of the kind "space/time": the discrete 2-forms generated on the 2-cells of the kind "space/time".

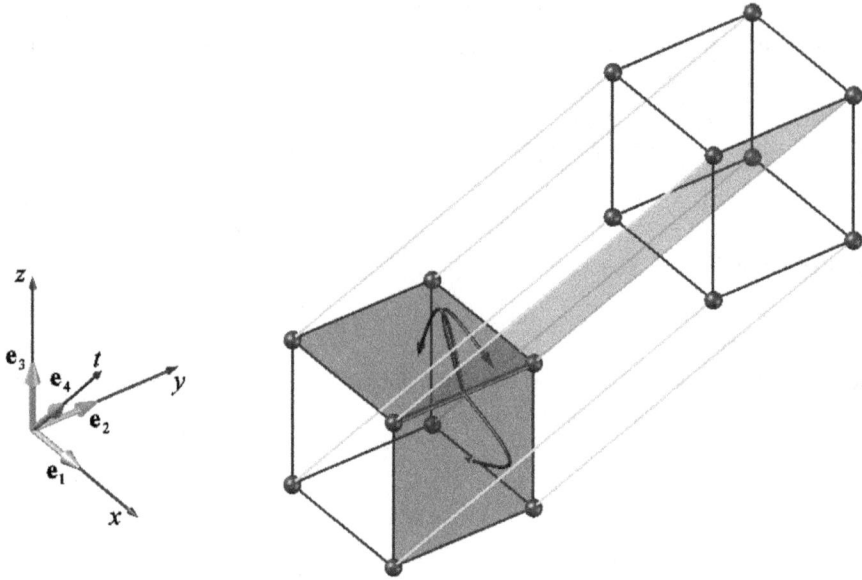

Figure 5.11. The coboundary process on a discrete 1-form defined for one edge $t = t_1$ of a tesseract.

The process for the generation of the 2-forms of the kind "space" is the algebraic counterpart of finding the curl of a vector valued function of the point, while the process for the generation of the 2-forms of the kind "space/time" is the algebraic version of finding the derivative of a function of two variables.

In conclusion, the coboundary process performed on a discrete 1-form in space/time gives rise to two operators at the same time, the algebraic version of the curl and the algebraic version of the derivative of a function of two variables. This unify the two related processes as two sides of the same coin.

5.1.4 Performing the Coboundary Process on Discrete 2-forms in Space/Time Domain: Analogies Between Algebraic and Differential Operators

When performed for space elements of $\mathbb{R}^3$ or $\mathbb{R}^2$, the coboundary process on a discrete 2-form is the algebraic counterpart of finding the divergence of a vector field.

Both the divergence and the discrete 3-form generated by the coboundary process on a discrete 2-form come from vector quantities associated with surfaces. More precisely, the divergence at the point $\mathbf{P}$ of a vector valued function $\mathbf{v}(\mathbf{P})$ is the scalar valued function $\psi(\mathbf{P})$ generated by the following three-step process:

1. Evaluating the flux of the vector $\mathbf{v}(\mathbf{P})$ across the boundary of a small space region centered at the point $\mathbf{P}$.
2. Evaluating the ratio between such flux and the volume of the small region.
3. Performing the limit of this ratio when the region contracts to the point $\mathbf{P}$. This scalar valued function, $\psi(\mathbf{P})$, is called the divergence of the vector valued function $\mathbf{v}$ at the point $\mathbf{P}$:

$$\psi(\mathbf{P}) = \operatorname{div} \mathbf{v}(\mathbf{P}) = \nabla \cdot \mathbf{v}(\mathbf{P}), \tag{5.1.70}$$

or, in Cartesian coordinates:

$$\psi(x, y, z) = \operatorname{div} \mathbf{v}(x, y, z) = \nabla \cdot \mathbf{v}(x, y, z). \tag{5.1.71}$$

Let $\Psi^2 = \left[\psi_1, \psi_2, \ldots, \psi_{n_2}\right]$ be the discrete 2-form of the fluxes through the n_2 2-cells, S_i, of the space cell complex; then the algebraic counterpart of the three-step process for finding the divergence at the point $\mathbf{P}$ is the coboundary process on Ψ^2, which can be described by the following two-step process (Fig. 5.12):

1. Spreading the value ψ_i of the i-th 2-cell, S_i, to all the cofaces, V_j, of the 2-cell, each multiplied by the mutual incidence number between S_i and V_j and divided by V_j, the area of V_j.
2. Adding all the values that have been spread on the same 3-cell, V_j.

The discrete 3-form $W^3 = \left[w_1, w_2, \ldots, w_{n_3}\right]$, generated on the n_3 3-cells of the space cell complex by the coboundary process on the discrete 2-form, is the coboundary of the discrete 2-form:

$$W^3 = \delta\Psi^2. \tag{5.1.72}$$

The j-th element of the discrete 3-form W^3 is given by the summation on the 2-cells connected to the j-th 3-cell:

$$w_j = \frac{1}{V_j}\sum_i d_{ji}\psi_i, \tag{5.1.73}$$

where d_{ji} is the element in row j and column i of the incidence matrix $\mathbf{D}$, and V_j is the volume of the j-th 3-cell, V_j, on which we are spreading the value ψ_i of the i-th 2-cell, S_i.

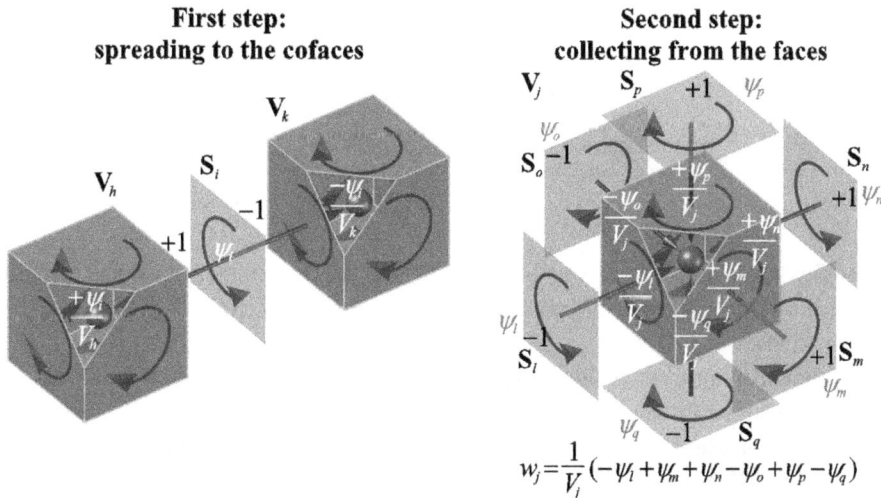

Figure 5.12. The coboundary process on a discrete 2-form defined on the surfaces of a space cell complex.

We can obtain the matrix $\mathbf{T}^D$, which expresses the coboundary operator for the given labeling of the cell complex, as a sum of expanded local matrices. The elements of $\mathbf{T}^D$ have the form:

$$T_{ji}^D = \frac{1}{V_j} d_{ji}. \tag{5.1.74}$$

Let n be the number of 3-cells and let m be the number of 2-cells; then each local matrix is the expansion to the $n \times m$ dimensions of the 1×1 matrix $\mathbf{t}_j^D$ (a scalar), whose unique element is $1/V_j$:

$$\mathbf{t}_j^D = \left[\frac{1}{V_j} \right]. \tag{5.1.75}$$

The number of local matrices equals the number of 3-cells of the cell complex, that is, the number of times we must perform the second step of the coboundary process.

The two collocation vectors, $\mathbf{l}_j$ and $\mathbf{r}_j$, are defined according to the following twofold criterion:

- $\mathbf{l}_j$ is an $n \times 1$ Boolean vector, whose unique nonzero value is the j-th element:

$$\left(\mathbf{l}_j \right)_h = \delta_{jh}, \; h = 1, 2, ..., n, \tag{5.1.76}$$

where δ_{jh} is the Kronecker delta symbol.
- $\mathbf{r}$ is the $m \times 1$ vector such that $\mathbf{r}^T$ is the j-th row of the incidence matrix $\mathbf{D}$:

$$\left(\mathbf{r}_j \right)_k = d_{jk}, \; k = 1, 2, ..., m. \tag{5.1.77}$$

The expansion of the j-th local matrix, $\mathbf{t}_j^D$, is given by the matrix product $\mathbf{l}_j \mathbf{t}_j^D \mathbf{r}_j^T$, and $\mathbf{T}^D$ is provided by the sum:

$$\mathbf{T}^D = \sum_{i=1}^{n} \mathbf{l}_j \mathbf{t}_j^D \mathbf{r}_j^T. \tag{5.1.78}$$

The 3-form W^3 can then be obtained as the product:

$$\mathbf{W}^3 = \mathbf{T}^D \mathbf{\Psi}^2. \tag{5.1.79}$$

Even this latter time, when we add a time axis to the three-dimensional and two-dimensional space cell complexes of $\mathbb{R}^3$ and $\mathbb{R}^2$, we obtain space/time cell complexes with p-cells of different kinds.

The coboundary process on the discrete 2-form naturally extends both to the 2-cells associated with a given time instant and to the 2-cells associated with adjacent time instants. Starting from space cell complexes in dimension 3, this implies that the discrete 2-form in space/time is defined on all the 2-cells of a tesseract (Fig. 5.13).

The coboundary process on the discrete 2-form generates discrete 3-forms for all the three-dimensional elements of the tesseract, that is, both for the three-dimensional elements of the

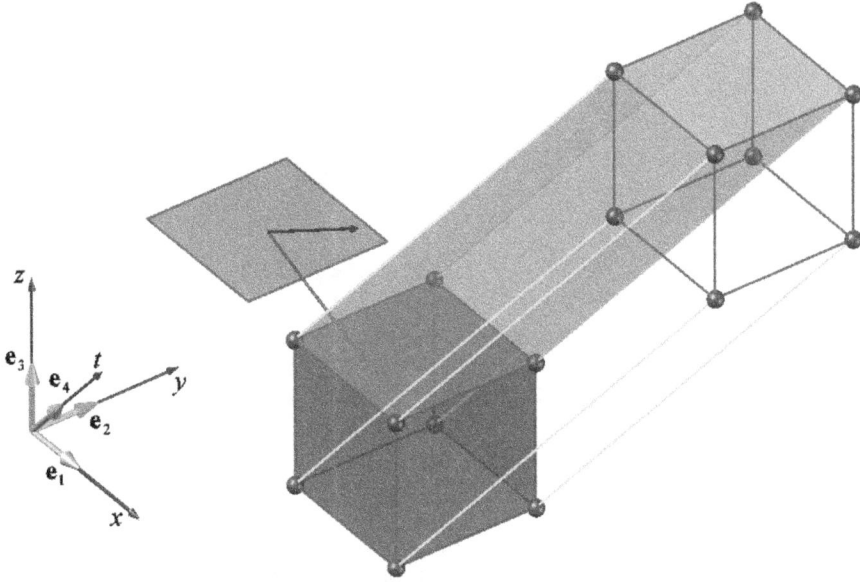

Figure 5.13. The coboundary process on a discrete 2-form defined for one face $t = t_1$ of a tesseract.

kind "space" and for the three-dimensional elements of the kind "time". This results in two different kinds of discrete 3-forms, generated by the same discrete 2-form (Fig. 5.13):

- Discrete 3-forms of the kind "space": the discrete 3-forms generated on the 3-cells of the kind "space".
- Discrete 3-forms of the kind "space/time": the discrete 3-forms generated on the 3-cells of the kind "space/time".

The process for the generation of the 3-forms of the kind "space" is the algebraic counterpart of finding the divergence of a vector valued function of the point, while the process for the generation of the 3-forms of the kind "space/time" is the algebraic version of finding the derivative of a function of three variables.

In conclusion, the coboundary process performed on a discrete 2-form in space/time gives rise to two operators at the same time, the algebraic version of the divergence and the algebraic version of the derivative of a function of three variables. This unifies the two related processes as two sides of the same coin.

5.2 How to Compose the Fundamental Equation of a Physical Theory

The fundamental equation describes mathematically the fundamental problem of the theory (Section 4.1), that is, the problem of finding the configuration of a system, once the sources are assigned.

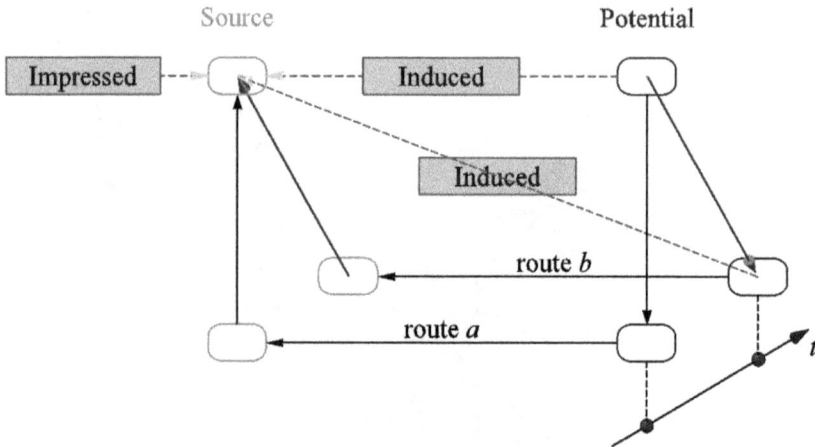

Figure 5.14. Building the fundamental equation.

As far as the sources are concerned, we have seen in Section 4.4 that there are two kinds of sources, impressed and induced sources, which are assigned by the problem and induced by the second Newton's law or dissipative laws on the configuration variables, respectively.

Consequently, we have two ways for relating source and configuration variables. The first way is that of following the topological equations reversely, from the sources of the problem to the main configuration variables, by using the constitutive equations for passing from the column of the source variables to the column of the configuration variables (Fig. 5.14).

Following this path, we insert the variables of the path in the main topological equations of the path (those contained in the first square box of the path, starting from the main sources), by successive substitutions.

The second way is that of expressing the sources of the main topological equations in terms of configuration variables by using the second Newton's law and the dissipative laws.

In doing so, we obtain one or more topological equations of the column of the source variables (balance equations, circuital equations, or space differences) written in terms of the configuration variables, which are the unknowns of the fundamental problem.

5.3   Analogies in Physics

When a relation or a statement is invariant in different physical theories under the exchanges of the elements involved in them, the corresponding variables of two different theories are called homologous[155] and the relationship established between the theories by the homologous variables is called an analogy. In particular, the existence of analogies is made manifest by the recurrent presence in the fundamental equations of the field theories of the same operators, "grad", "curl" and "div", and the same equations between variables, as the equations of Laplace, Poisson, and d'Alembert.

---

[155] In an analogy between two physical phenomena the homologous variables differ in many senses: they have different physical meaning, different physical dimensions and, in general, different mathematical nature.

Analogies are well known in physics and played an essential role in the development of new disciplines. They allow us, in fact, to explore new fields by using the established knowledge of other fields, through the intermediation of the homologous elements. This has ever been a common practice in physics,[156] despite it was never provided a compelling reason for the existence of analogies between different disciplines. Over the centuries, the consolidated custom of using analogies systematically led to the construction of many mathematical formalisms, such as

- the formalism of dynamical systems,
- the formalism of generalized network theory,
- the formalism of irreversible thermodynamics,
- the formalism of mathematical field theory,
- the formalism of variational principles,
- the formalism of the first quantization,
- the formalism of the second quantization.

The classification diagrams of the global variables, together with the orientations and the relations of the global variables, allow us, now, to provide an explanation to the existence of homologous variables in physical theories. In fact, the classification diagrams highlight a general structure, which is common to all field physical theories of the macrocosm. This general structure characterizes both the algebraic and the differential formulation due to the **inherited association**, that is, the fact that densities and rates (which are field variables) inherit an association with the space and time element to which the related global variable corresponds.[157]

---

[156] In electrostatics, which is deduced from the laws of action at a distance, the idea that the electromagnetic action is also transmitted by contact was introduced by analogy with heat conduction in solids, which is transmitted by contact. Similarly, Siméon Denis Poisson (21 June 1781–25 April 1840) introduced the idea of electric potential by analogy with the notion of temperature in a thermal field, a subject previously treated by Jean Baptiste Joseph Fourier (21 March 1768–16 May 1830) in his book on heat conduction. Another analogy arises from the comparison between the propagation of waves in a material continuum (solid or fluid) and the propagation of electromagnetic waves in free space, since both physical phenomena are governed by reflection, refraction, interference, diffraction, and polarization.

[157] By way of example of inherited associations:

- Mass density, ρ, inherits from the mass content, M^c, the association with time instants endowed with inner orientation, $\bar{\mathrm{I}}$, and volumes endowed with outer orientation, $\widetilde{\mathbf{V}}$:

$$M^c\left[\bar{\mathrm{I}}, \widetilde{\mathbf{V}}\right] \Rightarrow \rho\left[\bar{\mathrm{I}}, \widetilde{\mathbf{V}}\right].$$

- Electric field strength, $\mathbf{E}$, inherits from the electric voltage, E, the association with time intervals endowed with outer orientation, $\widetilde{\mathrm{T}}$, and lines endowed with inner orientation, $\bar{\mathrm{L}}$:

$$E\left[\widetilde{\mathrm{T}}, \bar{\mathrm{L}}\right] \Rightarrow \mathbf{E}\left[\widetilde{\mathrm{T}}, \bar{\mathrm{L}}\right].$$

- Heat current density, $\mathbf{q}$, inherits from the heat, Q, the association with time intervals endowed with inner orientation, $\bar{\mathrm{T}}$, and surfaces endowed with outer orientation, $\widetilde{\mathbf{S}}$:

$$Q\left[\bar{\mathrm{T}}, \widetilde{\mathbf{S}}\right] = \mathbf{q}\left[\bar{\mathrm{T}}, \widetilde{\mathbf{S}}\right].$$

(continues next page)

As a consequence, we can indicate the associated space and time element next to the variables in Figs. 5.15–5.17, despite they are the point-wise variables of the differential formulation.

In particular, Fig. 5.15 shows the classification diagram for elastostatics of deformable bodies, where the strains ε_i are related to the displacements along the directions of the x, y, z axes, u, v, w, respectively, in the assumption of linear relationships between the components of strain and the displacement derivatives of the first order (theory of the first order for small strains). The congruence equations, $\varepsilon = \mathbf{D}u$, are therefore expressed as

$$
\begin{cases}
\varepsilon_x = \dfrac{\partial u}{\partial x} \\[2mm]
\varepsilon_y = \dfrac{\partial v}{\partial y} \\[2mm]
\varepsilon_z = \dfrac{\partial w}{\partial z} \\[2mm]
\gamma_{yz} = \dfrac{\partial v}{\partial z} + \dfrac{\partial w}{\partial y} \\[2mm]
\gamma_{xz} = \dfrac{\partial u}{\partial z} + \dfrac{\partial w}{\partial x} \\[2mm]
\gamma_{xy} = \dfrac{\partial u}{\partial y} + \dfrac{\partial v}{\partial x}
\end{cases}
. \tag{5.3.1}
$$

The constitutive equations of the primal cycle, $\sigma = \mathbf{C}\varepsilon$, relate the normal stresses σ_i to the normal strains ε_i in the $i = x, y, z$ directions, and the shear stresses τ_{ij} to the shear strains γ_{ij} in the $i, j = x, y, z$ directions, with $i \neq j$. They are expressed by Hooke's well-known generalized laws:

$$
\begin{cases}
\sigma_x = \lambda I_{1\varepsilon} + 2\mu\varepsilon_x \\
\sigma_y = \lambda I_{1\varepsilon} + 2\mu\varepsilon_y \\
\sigma_z = \lambda I_{1\varepsilon} + 2\mu\varepsilon_z \\
\tau_{yz} = \mu\gamma_{yz} \\
\tau_{xz} = \mu\gamma_{xz} \\
\tau_{xy} = \mu\gamma_{xy}
\end{cases}
, \tag{5.3.2}
$$

(*continues from previous page*)

- Volume forces and surface forces, $\mathbf{F}$ and $\mathbf{T}$, respectively, inherit from the impulse, $\mathbf{J}$, the association with time intervals endowed with outer orientation, $\widetilde{\mathbf{T}}$, because the force is the time rate of impulse:

$$
\mathbf{J}^V\left[\widetilde{\mathbf{T}}, \overline{\mathbf{V}}\right] \Rightarrow \mathbf{F}\left[\widetilde{\mathbf{T}}, \overline{\mathbf{V}}\right];
$$

$$
\mathbf{J}^S\left[\widetilde{\mathbf{T}}, \overline{\mathbf{L}}\right] \Rightarrow \mathbf{T}\left[\widetilde{\mathbf{T}}, \overline{\mathbf{L}}\right].
$$

The space global variables which are associated with points, such as temperature, electric potential, chemical potential, displacement, velocity potential, and so forth, are automatically field variables. We will use round brackets for these variables, for example $q(t, P)$ for heat current density, in order to distinguish them from those point-wise field variables that are obtained as densities of the corresponding global variables.

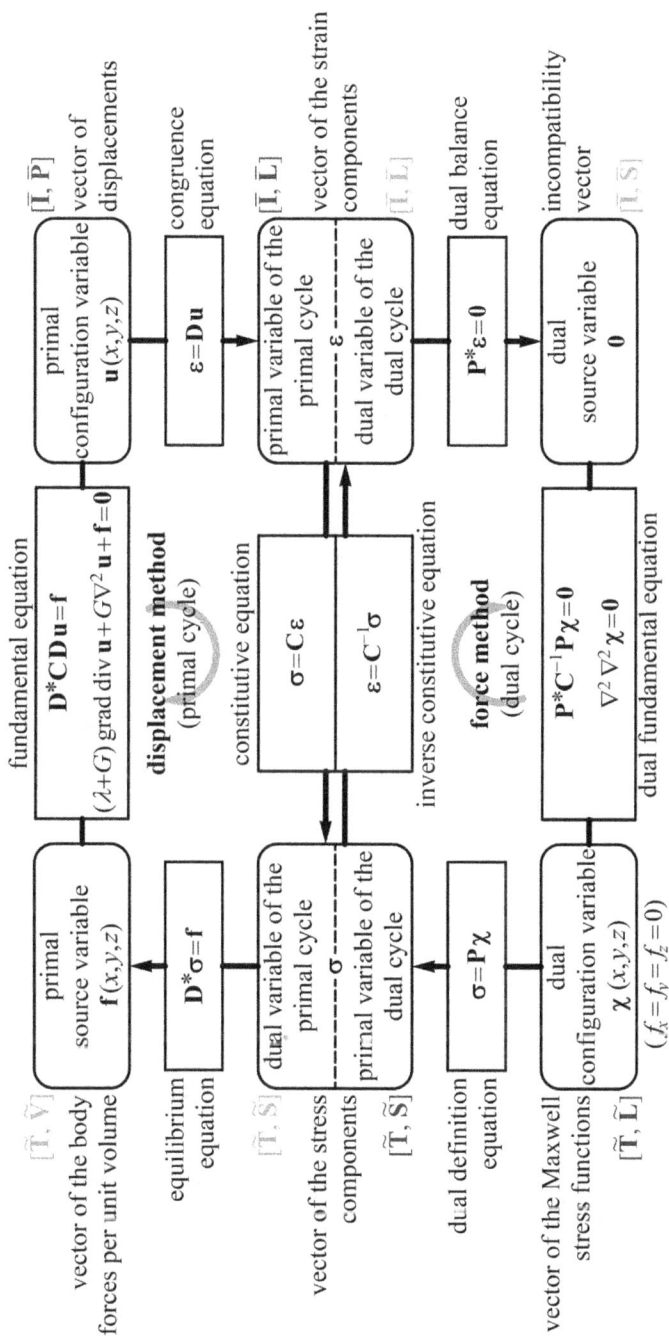

Figure 5.15. Classification diagram for elastostatics of deformable solids.

where $I_{1\varepsilon} = \mathrm{tr}(\varepsilon)$ is the first invariant of strain, giving the bulk (volumetric) strain, $\Delta V/V$, in the first order theory:

$$I_{1\varepsilon} = \varepsilon_x + \varepsilon_y + \varepsilon_z = \frac{\partial u}{\partial x} + \frac{\partial v}{\partial y} + \frac{\partial w}{\partial z} = \frac{\Delta V}{V}, \tag{5.3.3}$$

and λ and μ are the two parameters of Lamé, expressed in functions of E and υ, which are the longitudinal modulus of elasticity, or Young's modulus, and Poisson's ratio, respectively:

$$\lambda = E\frac{\upsilon}{(1+\upsilon)(1-2\upsilon)}; \tag{5.3.4}$$

$$\mu = G = \frac{E}{2(1+\upsilon)}. \tag{5.3.5}$$

The governing equations in the partial differential forms of the primal cycle are obtained starting from the last topological equations of the primal cycle, that is, the equations of equilibrium $\mathbf{D^*\sigma = f}$:

$$\begin{cases} \dfrac{\partial \sigma_x}{\partial x} + \dfrac{\partial \tau_{xy}}{\partial y} + \dfrac{\partial \tau_{xz}}{\partial z} + f_x = 0 \\[2mm] \dfrac{\partial \tau_{xy}}{\partial x} + \dfrac{\partial \sigma_y}{\partial y} + \dfrac{\partial \tau_{yz}}{\partial z} + f_y = 0, \\[2mm] \dfrac{\partial \tau_{xz}}{\partial x} + \dfrac{\partial \tau_{yz}}{\partial y} + \dfrac{\partial \sigma_z}{\partial z} + f_z = 0 \end{cases} \tag{5.3.6}$$

where f_x, f_y, f_z are the body forces per unit volume ΔV, along the x, y, and z axes, respectively:

$$\begin{cases} f_x = \lim\limits_{\Delta V \to 0} \dfrac{\Delta F_x}{\Delta V} = \dfrac{dF_x}{dV} \\[2mm] f_y = \lim\limits_{\Delta V \to 0} \dfrac{\Delta F_y}{\Delta V} = \dfrac{dF_y}{dV}. \\[2mm] f_z = \lim\limits_{\Delta V \to 0} \dfrac{\Delta F_z}{\Delta V} = \dfrac{dF_z}{dV} \end{cases} \tag{5.3.7}$$

By following the primal cycle reversely, at first we substitute Hooke's generalized laws into the equilibrium equations:

$$\mathbf{D^*C\varepsilon = f}; \tag{5.3.8}$$

then we use the congruence equations for expressing the equations of equilibrium in terms of displacements:

$$\mathbf{D^*CDu = f}. \tag{5.3.9}$$

By making use of the Laplace operator, ∇^2:

$$\nabla^2 = \frac{\partial}{\partial x^2} + \frac{\partial}{\partial y^2} + \frac{\partial}{\partial z^2}, \tag{5.3.10}$$

we can finally derive the fundamental equations in the scalar form:

$$\begin{cases} (\lambda + \mu)\dfrac{\partial I_{1\varepsilon}}{\partial x} + \mu\nabla^2 u + f_x = 0 \\[2mm] (\lambda + \mu)\dfrac{\partial I_{1\varepsilon}}{\partial y} + \mu\nabla^2 v + f_y = 0 \\[2mm] (\lambda + \mu)\dfrac{\partial I_{1\varepsilon}}{\partial z} + \mu\nabla^2 w + f_z = 0 \end{cases} \tag{5.3.11}$$

Analogously, in the dual cycle we start from the equations of dual balance (the compatibility equations), $\mathbf{P}^*\varepsilon = \mathbf{0}$:

$$\begin{bmatrix} \dfrac{\partial^2}{\partial y^2} & \dfrac{\partial^2}{\partial x^2} & 0 & 0 & 0 & -\dfrac{\partial^2}{\partial x \partial y} \\[3mm] \dfrac{\partial^2}{\partial z^2} & 0 & \dfrac{\partial^2}{\partial x^2} & 0 & -\dfrac{\partial^2}{\partial x \partial z} & 0 \\[3mm] 0 & \dfrac{\partial^2}{\partial z^2} & \dfrac{\partial^2}{\partial y^2} & -\dfrac{\partial^2}{\partial y \partial z} & 0 & 0 \\[3mm] 2\dfrac{\partial^2}{\partial y \partial z} & 0 & 0 & \dfrac{\partial^2}{\partial x^2} & -\dfrac{\partial^2}{\partial x \partial y} & -\dfrac{\partial^2}{\partial x \partial z} \\[3mm] 0 & 2\dfrac{\partial^2}{\partial x \partial z} & 0 & -\dfrac{\partial^2}{\partial x \partial y} & \dfrac{\partial^2}{\partial y^2} & -\dfrac{\partial^2}{\partial y \partial z} \\[3mm] 0 & 0 & 2\dfrac{\partial^2}{\partial x \partial y} & -\dfrac{\partial^2}{\partial x \partial z} & -\dfrac{\partial^2}{\partial y \partial z} & \dfrac{\partial^2}{\partial z^2} \end{bmatrix} \begin{bmatrix} \varepsilon_x \\ \varepsilon_y \\ \varepsilon_z \\ \gamma_{yz} \\ \gamma_{xz} \\ \gamma_{xy} \end{bmatrix} = \begin{bmatrix} 0 \\ 0 \\ 0 \\ 0 \\ 0 \\ 0 \end{bmatrix}, \tag{5.3.12}$$

and follow the dual cycle reversely, by substituting the inverse equations of Hooke's generalized laws, $\varepsilon = \mathbf{C}^{-1}\sigma$:

$$\begin{cases} \varepsilon_x = \dfrac{1}{E}\left[\sigma_x - \upsilon\left(\sigma_y + \sigma_z\right)\right] \\[2mm] \varepsilon_y = \dfrac{1}{E}\left[\sigma_y - \upsilon\left(\sigma_z + \sigma_x\right)\right] \\[2mm] \varepsilon_z = \dfrac{1}{E}\left[\sigma_z - \upsilon\left(\sigma_x + \sigma_y\right)\right] \\[2mm] \gamma_{yz} = \dfrac{1}{G}\tau_{yz} \\[2mm] \gamma_{xz} = \dfrac{1}{G}\tau_{xz} \\[2mm] \gamma_{xy} = \dfrac{1}{G}\tau_{xy} \end{cases} \tag{5.3.13}$$

and the equations of dual definition, $\sigma = P\chi$, where χ is the vector of the stress functions of Maxwell, which make identically satisfied the equilibrium equations when the body forces are equal to zero:

$$\chi = \begin{bmatrix} \mathcal{U} \\ \mathcal{V} \\ \mathcal{W} \end{bmatrix}; \tag{5.3.14}$$

$$\begin{cases} \sigma_x = \dfrac{\partial^2 \mathcal{W}}{\partial y^2} + \dfrac{\partial^2 \mathcal{V}}{\partial z^2} \\[2mm] \sigma_y = \dfrac{\partial^2 \mathcal{U}}{\partial z^2} + \dfrac{\partial^2 \mathcal{W}}{\partial x^2} \\[2mm] \sigma_z = \dfrac{\partial^2 \mathcal{V}}{\partial x^2} + \dfrac{\partial^2 \mathcal{U}}{\partial y^2} \\[2mm] \tau_{yz} = -\dfrac{\partial^2 \mathcal{U}}{\partial y \partial z} \\[2mm] \tau_{xz} = -\dfrac{\partial^2 \mathcal{V}}{\partial x \partial z} \\[2mm] \tau_{xy} = -\dfrac{\partial^2 \mathcal{W}}{\partial x \partial y} \end{cases} ; \tag{5.3.15}$$

$$\begin{cases} \dfrac{\partial \sigma_x (\mathcal{V}, \mathcal{W})}{\partial x} + \dfrac{\partial \tau_{xy} (\mathcal{W})}{\partial y} + \dfrac{\partial \tau_{xz} (\mathcal{V})}{\partial z} = 0 \\[2mm] \dfrac{\partial \tau_{xy} (\mathcal{W})}{\partial x} + \dfrac{\partial \sigma_y (\mathcal{U}, \mathcal{W})}{\partial y} + \dfrac{\partial \tau_{yz} (\mathcal{U})}{\partial z} = 0 . \\[2mm] \dfrac{\partial \tau_{xz} (\mathcal{V})}{\partial x} + \dfrac{\partial \tau_{yz} (\mathcal{U})}{\partial y} + \dfrac{\partial \sigma_z (\mathcal{U}, \mathcal{V})}{\partial z} = 0 \end{cases} \tag{5.3.16}$$

In vector notation, we obtain the dual fundamental equation of elastostatics with the following three steps:

$$P^* \varepsilon = 0; \tag{5.3.17}$$

$$P^* C^{-1} \sigma = 0; \tag{5.3.18}$$

$$P^* C^{-1} P\chi = 0. \tag{5.3.19}$$

The equations of the primal and dual cycles of elastostatics are shown in Fig. 5.16 for the special case of plane elasticity. By way of example of the common structure exhibited by different physical theories, these latter equations can be compared to the equations of plane motion of a perfect, incompressible fluid, shown in Fig. 5.17. Many further examples of common structure can be found in the papers and books by Tonti.

The existence of an underlying structure, common to different physical theories, is precisely the main responsible for the structural similarities, present in physical theories, commonly called analogies. Moreover, the classification diagrams also allow us to explain the analogies in the light of the association between the global variables and the four space elements, since the

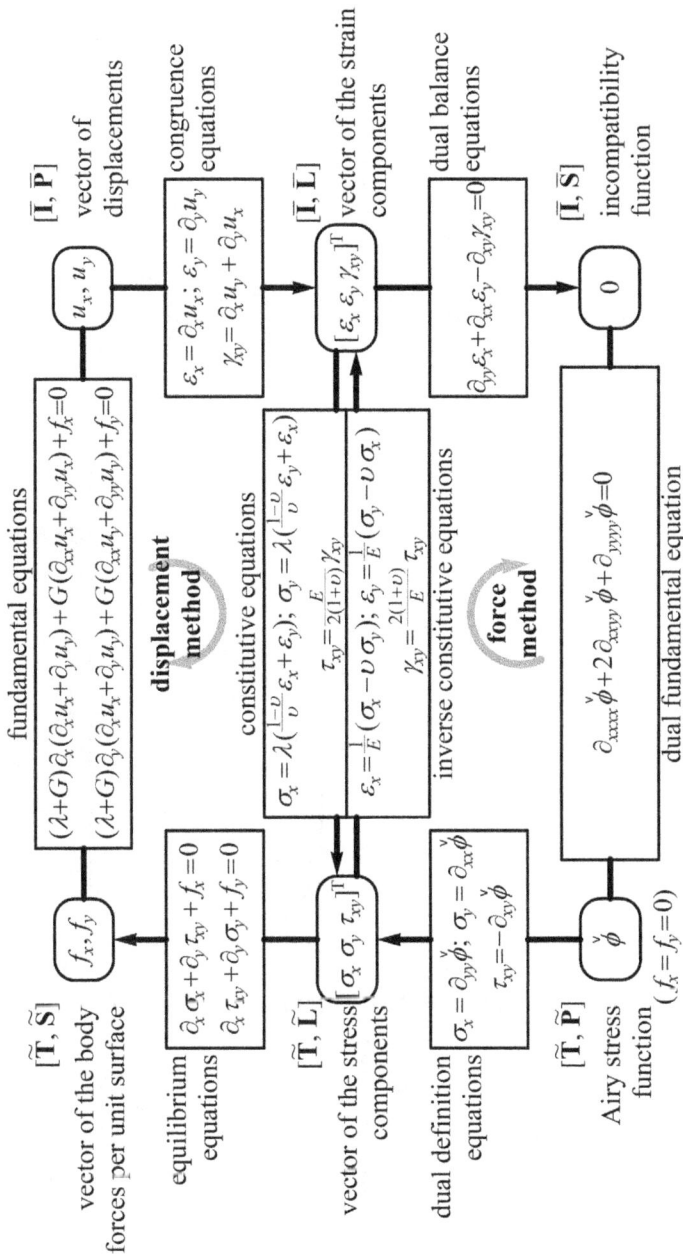

Figure 5.16. Classification diagram for plane elasticity of deformable solids.

$[\bar{\mathbf{I}}, \bar{\mathbf{P}}]$ vector of displacements

u_x, u_y

congruence equations

$\varepsilon_x = \partial_x u_x; \ \varepsilon_y = \partial_y u_y$
$\gamma_{xy} = \partial_x u_y + \partial_y u_x$

$[\bar{\mathbf{I}}, \mathbf{L}]$ vector of the strain components

$[\varepsilon_x \ \varepsilon_y \ \gamma_{xy}]^{\mathrm{T}}$

dual balance equations

$\partial_{yy}\varepsilon_x + \partial_{xx}\varepsilon_y - \partial_{xy}\gamma_{xy} = 0$

$[\bar{\mathbf{I}}, \bar{\mathbf{S}}]$ incompatibility function

0

fundamental equations

$(\lambda + G)\partial_x(\partial_x u_x + \partial_y u_y) + G(\partial_{xx} u_x + \partial_{yy} u_x) + f_x = 0$
$(\lambda + G)\partial_y(\partial_x u_x + \partial_y u_y) + G(\partial_{xx} u_y + \partial_{yy} u_y) + f_y = 0$

displacement method

constitutive equations

$\sigma_x = \lambda\left(\frac{1-\nu}{\nu}\varepsilon_x + \varepsilon_y\right); \ \sigma_y = \lambda\left(\frac{1-\nu}{\nu}\varepsilon_y + \varepsilon_x\right)$
$\tau_{xy} = \frac{E}{2(1+\nu)}\gamma_{xy}$

$\varepsilon_x = \frac{1}{E}(\sigma_x - \nu\sigma_y); \ \varepsilon_y = \frac{1}{E}(\sigma_y - \nu\sigma_x)$
$\gamma_{xy} = \frac{2(1+\nu)}{E}\tau_{xy}$

inverse constitutive equations

force method

$\partial_{xxxx}\overset{\vee}{\phi} + 2\partial_{xxyy}\overset{\vee}{\phi} + \partial_{yyyy}\overset{\vee}{\phi} = 0$

dual fundamental equation

$[\tilde{\mathbf{T}}, \tilde{\mathbf{S}}]$ vector of the body forces per unit surface

f_x, f_y

equilibrium equations

$\partial_x\sigma_x + \partial_y\tau_{xy} + f_x = 0$
$\partial_x\tau_{xy} + \partial_y\sigma_y + f_y = 0$

$[\tilde{\mathbf{T}}, \mathbf{L}]$ vector of the stress components

$[\sigma_x \ \sigma_y \ \tau_{xy}]^{\mathrm{T}}$

dual definition equations

$\sigma_x = \partial_{yy}\overset{\vee}{\phi}; \ \sigma_y = \partial_{xx}\overset{\vee}{\phi}$
$\tau_{xy} = -\partial_{xy}\overset{\vee}{\phi}$

$[\tilde{\mathbf{T}}, \tilde{\mathbf{P}}]$ Airy stress function $(f_x = f_y = 0)$

$\overset{\vee}{\phi}$

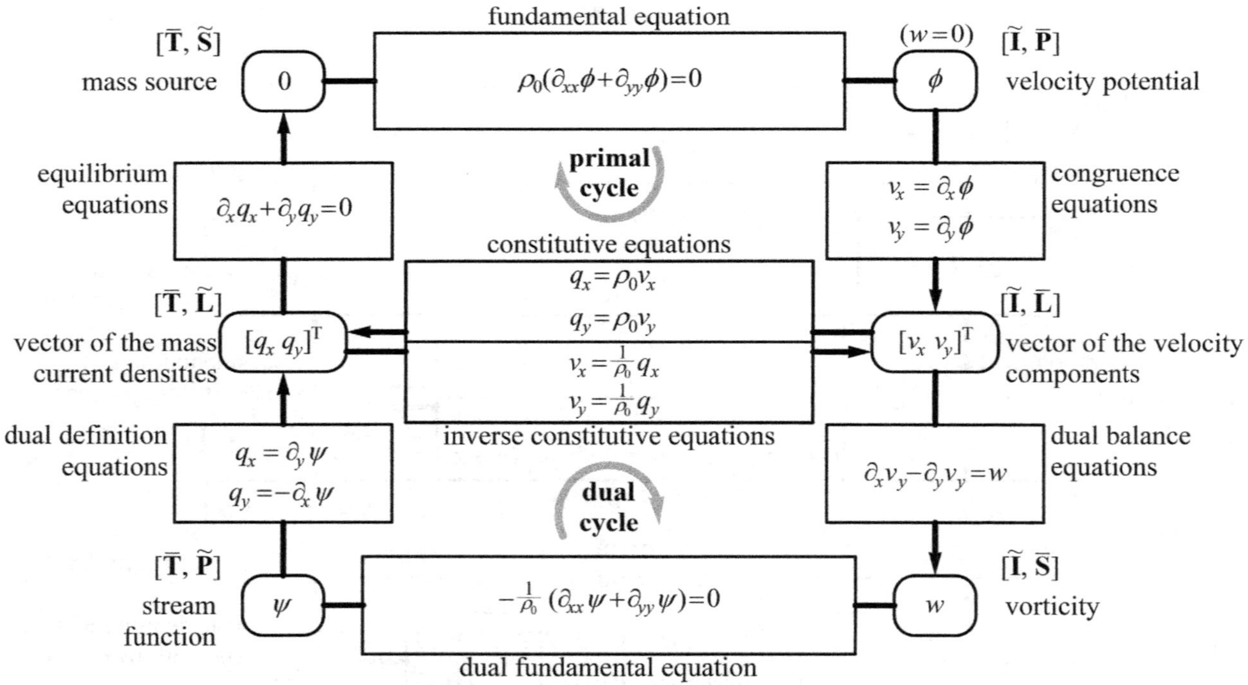

Figure 5.17. Classification diagram for the plane motion of a perfect, incompressible fluid.

homologous global variables of two physical theories are those associated with the same space and time element. In other words, the analogies between physical theories arise from the geometrical structure of the global variables and not from the similarity of the equations that relate variables to each other in different physical theories.

As quoted even in Tonti (2013), some early ideas on the relationship between the physical variables, the space, and the extent of space elements can be found in the works by Feynman[158] and Maxwell.[159]

5.4 Physical Theories with Reversible Constitutive Laws

A constitutive relation is reversible when it connects a variable associated with a time element to another associated with its dual time element:

$$\overline{I} \leftrightarrow \tilde{T}, \tag{5.4.1}$$

---

[158] The question on why the equations from different phenomena are so similar was raised also by the American theoretical physicist Richard Phillips Feynman (May 11, 1918–February 15, 1988). He did not actually answer the question, even if he put the attention on the space:

We might say: "It is the underlying unity of nature." But what does that mean? What could such a statement mean? It could mean simply that the equations are similar for different phenomenon; but then, of course, we have given no explanation. The "underlying unity" might mean that everything is made out of the same stuff, and therefore obeys the same equations. That sounds like a good explanation, but let us think. The electrostatic potential, the diffusion of neutrons, heat flow—are we really dealing with the same stuff? Can we really imagine that the electrostatic potential is physically identical to the temperature, or to the density of particles? Certainly φ is not exactly the same as the thermal energy of particles. The displacement of a membrane is certainly not like a temperature. Why, then, is there "an underlying unity"? A closer look at the physics of the various subjects shows, in fact, that the equations are not really identical. The equation we found for neutron diffusion is only an approximation that is good when the distance over which we are looking is large compared with the mean free path. If we look more closely, we would see the individual neutrons running around. Certainly the motion of an individual neutron is a completely different thing from the smooth variation we get from solving the differential equation. The differential equation is an approximation, because we assume that the neutrons are smoothly distributed in space. Is it possible that this is the clue? That the thing which is common to all the phenomena is the space the framework into which the physics is put?

[159] In 1871, James Clerk Maxwell (13 June 1831–5 November 1879) published an article entitled "Remarks on the Mathematical Classification of Physical Quantities" in which he wrote:

Of the factors which compose it [energy], one is referred to unit of length, and the other to unit of area. This gives what I regard as a very important distinction among vector quantities

Speaking about analogies Maxwell said:

But it is evident that all analogies of this kind depend on principles of a more fundamental nature; and that, if we had a true mathematical classification of quantities, we should be able at once to detect the analogy between any system of quantities presented to us and other systems of quantities in known sciences, so that we should lose no time in availing ourselves of the mathematical labours of those who had already solved problems essentially the same. [...] At the same time, I think that the progress of science, both in the way of discovery, and in the way of diffusion, would be greatly aided if more attention were paid in a direct way to the classification of quantities.

or

$$\bar{\mathbf{T}} \leftrightarrow \tilde{\mathbf{I}}. \tag{5.4.2}$$

In the classification diagram, the reversible constitutive relations are represented by links between left and right column in the front or in the rear. As a consequence, the reversible relations connect two variables whose product is an energy or an energy density.

Conversely, a constitutive relation is irreversible when it connects a variable associated with a time interval to another associated with its dual time element:

$$\bar{\mathbf{T}} \leftrightarrow \tilde{\mathbf{T}}. \tag{5.4.3}$$

In this case, the constitutive relation is a link that connects the left column in the front with the right column in the rear (and there are not other existing connections). Therefore, the irreversible relations connect two variables whose product is a power.

The reversible constitutive equations of physical theories share the common properties of being usually linear and having symmetric operators, often represented by matrixes. In the rare cases in which the operators are nonlinear, the derivatives of the operators are symmetric. This implies some interesting mathematical properties:

- The constitutive equations can be derived from the condition of stationarity of a function, usually the potential energy but in some case the kinetic energy. Thus, the potential and kinetic energies play the role of potentials of the constitutive equations. In general, the stationary value is a minimum.
- The stationary property, combined with the property of adjointness of the operators, gives rise to a reciprocity principle and to a variational principle for the fundamental equation.

5.5 The Choice of Primal and Dual Cell Complexes in Computation

As we have already pointed out in Section 3.1, the cell complexes usually employed in numerical modeling are simplicial complexes (triangles in two-dimensional spaces and tetrahedra in three-dimensional spaces). Even the cell method uses simplexes for defining the primal mesh. The reason for this choice is that, as discussed in Section 4.2, each scalar or vector field in the neighborhood of every point, in a region of regularity, can be approximated by an affine field. Now, simplexes are compatible with the affine description of the field (Fig. 5.18), while cells of arbitrary number of sides are not compatible.

In fact, in a three-dimensional Cartesian coordinate system, the 12 components of the four vectors at the four vertices of the tetrahedron are in number strictly necessary for evaluating the 12 unknowns of the affine vector field described by Eqs. (4.2.20). Moreover, since the unknowns of the affine vector field reduce to six in a two-dimensional space, a_x, a_y, h_{xx}, h_{xy}, h_{yx}, and h_{yy}, in this second case the triangle is the compatible cell, because the components of the three vectors at its vertices are six, that is, in number strictly necessary for evaluating the six unknown of the affine vector field.

From the point of view of algebraic topology, the reason for which a vector field in the neighborhood of a point in the three-dimensional space is compatible with tetrahedra, and not with 3-cells with a greater number of sides, can be found in the fact that the tetrahedron

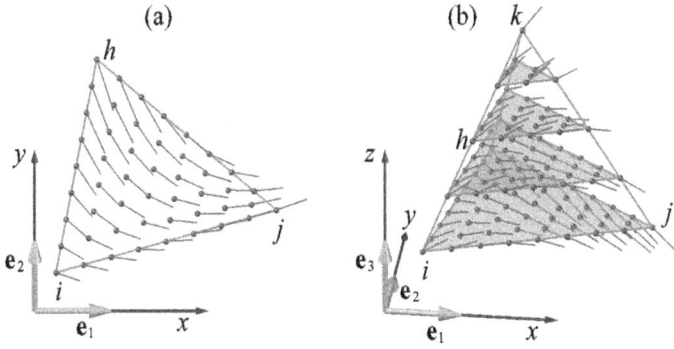

Figure 5.18. Approximation of a vector field by an affine vector field defined by the vectors at the vertices of a triangle, in two-dimensional space (a), and of a tetrahedron, in three-dimensional space (b).

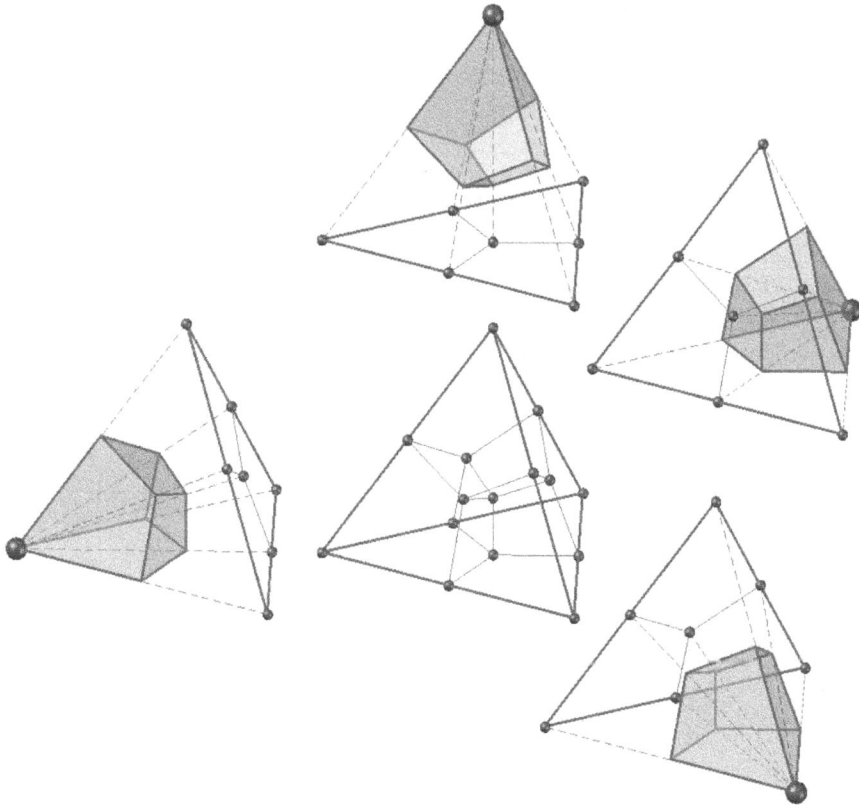

Figure 5.19. Vertex-centered central projection of a tesseract: detail of the projection of 4 of 8 cubic cells.

forms the convex hull[98] of the tesseract's vertex-centered central projection to three dimensions (Fig. 5.19). In fact, the central projection from a vertex gives a three-dimensional shadow, with a tetrahedral convex hull.

By recalling that the tesseract is the suitable 4-cell for analysis in space/time (Section 3.8), the vertex-centered central projection can be seen as the projection of a given physical phenomenon from the four-dimensional space to the three-dimensional space of the three spatial dimensions. In other words, performing a vertex-centered central projection is equivalent to fixing a time instant and analyzing the given physical phenomenon in that instant. In this sense, the global physical variables of a physical problem of statics can be associated with the space elements of a tetrahedron.

For the same reason, the suitable 2-cell for static analysis in a two-dimensional space is the triangle, since it is the two-dimensional shadow of the tetrahedron.

As we have discussed in Section 4.1, there are many ways for building a dual cell complex, once a primal cell complex has been provided. The two easier constructions of the dual cell complex are represented by

- The **barycentric dual cell complex**. In two-dimensional domains (Fig. 5.20a), the dual polygons are obtained by connecting the barycenter of every triangle with the mid-points of the edges of the triangle. In doing so, the dual of each primal 1-cell (a primal side) is not a straight line. Analogously, in three-dimensional domains (Fig. 5.21a), the dual of a primal 1-cell (a primal edge) is composed of many flat faces.
- The **Voronoi dual cell complex**. In two-dimensional domains (Fig. 5.20b), the mesh is formed by the polygons whose vertexes are at the circumcenters of the primal mesh. The sides of the dual polygons are the axes of the edges of the primal complex. The Voronoi diagram has the property that for each side every point in the region around that side is closer to that side than to any of the other sides. In three-dimensional domains (Fig. 5.21b), the dual node is the spherocenter, that is, the center of the sphere whose surface passes through the four vertices of a tetrahedron.

Of these two possible dual meshes, the one usually employed in numerical analyses is the Voronoi mesh, because, in simplicial complexes, the line connecting the circumcenters of two adjacent cells is orthogonal to their common face and the surface connecting the spherocenters of four adjacent tetrahedra is orthogonal to their common edge. This condition of orthogonality is very useful in numerical analysis, but the Voronoi mesh also has a disadvantage. In fact, while

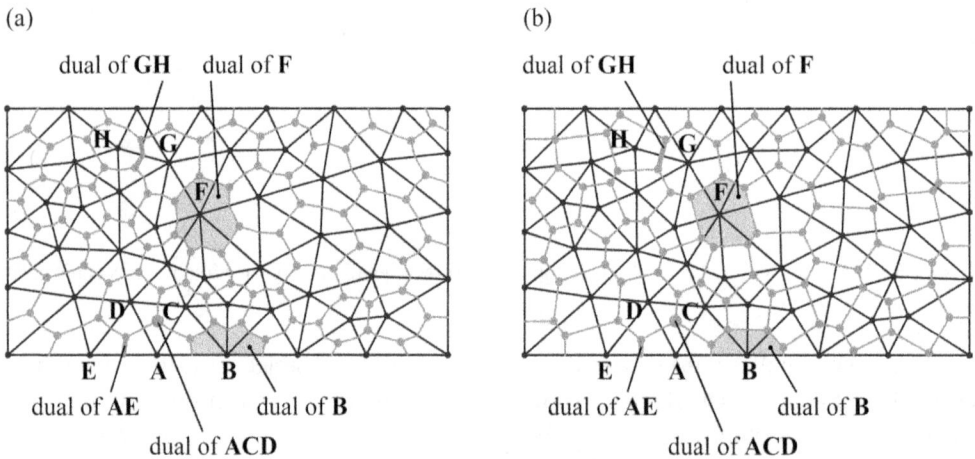

Figure 5.20. The barycentric dual mesh (a) and the dual mesh of Voronoi (b) in two-dimensional domains.

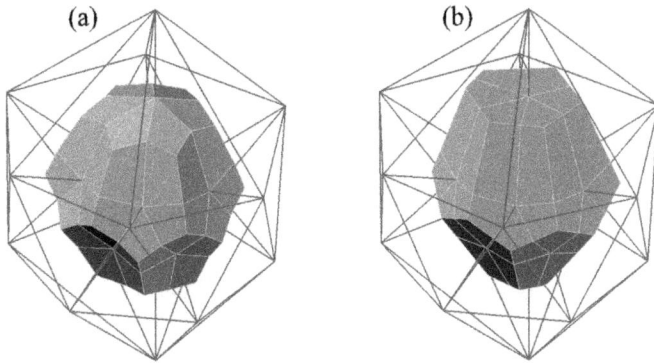

Figure 5.21. The barycentric dual mesh (a) and the dual mesh of Voronoi (b) three-dimensional domains.

the barycentric dual mesh does not involve any restriction on the primal mesh, the dual mesh of Voronoi requires an appropriate choice of the primal mesh, in order to avoid that some simplex has a circumcenter, or a spherocenter, that lie outside the simplex itself. In a two-dimensional domain, this happens whenever there are triangles with an angle greater than $\pi/2$.

Avoiding that some circumcenters lie outside the corresponding triangles is essential for the accurateness of the numerical result, since these circumcenters generate numerical errors. This is the reason why, in plane domains, the dual mesh of Voronoi is used together with a primal mesh of Delaunay, which is a triangulation such that the circumcenter of each triangle lies inside the triangle. A primal mesh of Delaunay, together with its dual mesh of Voronoi, is denoted as the Voronoi-Delaunay complex.

CHAPTER 6

THE PROBLEM OF THE SPURIOUS
SOLUTIONS IN COMPUTATIONAL PHYSICS

In Section 6.1, we analyze the mathematical structure of the fundamental equations, which can be classified as elliptic, parabolic, and hyperbolic fundamental equations. The solution of the fundamental equation is then discussed with regard to the exact closed form and the numerical solutions. A brief discussion on the stability of some numerical methods of the first- and the second-orders is also presented, in order to compare the CM with the most commonly used numerical techniques. In particular, the coboundary process in time-domain, used by the CM, is found to be very similar to the algebraic version of the leapfrog integration, which is a second-order method for solving dynamical systems of classical mechanics. The similarity between the time-marching schemes of the CM and leapfrog integration establishes a strict relationship between the CM and the Finite-difference time-domain method (FDTD), a differential time-domain numerical modeling method that uses the leapfrog integration, together with grid staggering (Yee lattice), for applications in computational electromagnetics. Therefore, the CM can be considered a generalization of the FDTD to space/time-domain numerical modeling, which also allows us to achieve a fourth-order convergence in space-domain.

In Sections 6.2, 6.3, and 6.4, we introduce a discussion on non-local models, which are needed both in the microcosm (quantum non-locality) and the macrocosm. Non-locality is also discussed with reference to the numerical modeling. In particular, the motivation for using non-locality in modeling the physical theories of the macrocosm lies in the regularization effect of the non-local numerical methods, which restore the well-posedness of the boundary value problem in differential formulation. In differential formulation, non-locality is usually achieved by enriching the material description with an internal length scale. The CM does not need this enrichment in order to provide a non-local description of physics. In fact, the CM born as a non-local method, due to three reasons:

- It replaces the field variables of the differential formulation with the global variables, which are associated with geometrical objects provided with an extent (we could say, with internal length scales in dimension 1, 2, and 3).
- It uses two staggered meshes, both in space and in time. Since the configuration variables are associated with the space elements of the primal mesh and the source variables are

associated with the space elements of the dual (staggered) mesh, the algebraic constitutive relations are not established in the point, but within a volume surrounding the point. The non-degenerate dimension of this volume is also the reason why the CM does not present problems of localization with zero dissipated energy.

- It obtains the algebraic topological equations (that is, both balance and kinematic equations) by means of coboundary processes. The typical two-step procedure of any coboundary process establishes a relationship between the p-cells and their cofaces, which also takes into account the extent of the p-cells. Therefore, the equations are not established in a point but within a volume, once again.

This three-fold motivation provides a non-local nature both to the algebraic variables and the governing equations of the CM.

In the final part of Section 6.4, the CM geometrical approach for treating space and time global variables is compared to the Minkowski spacetime, which, just as the CM, is based on a geometrical view of space-time. Since Minkowski spacetime is the mathematical space setting in which Einstein's theory of special relativity is most conveniently formulated, we can reasonably view the CM as a possible contribution for finding the basis of the purely algebraic theory, theorized by Einstein during the last decades of his life, for the representation of reality through a purely algebraic unifying gravitational theory.

6.1 Stability and Instability of the Numerical Solution

In differential formulation, the fundamental equations of any physical problem are expressed by partial differential equations (PDEs) of second order. Depending on the physical theory involved, the particular path followed for putting in relationship the configuration with the source variables may result in either elliptic (Fig. 6.1), or parabolic (Fig. 6.2), and hyperbolic equations (Fig. 6.3).

The reason for the terms "elliptic", "parabolic", and "hyperbolic" is the general form assumed by the second order PDE in two independent variables. Let A, B, and C be three coefficients that may depend upon x and y, and satisfy the condition:

$$A^2 + B^2 + C^2 > 0; \qquad (6.1.1)$$

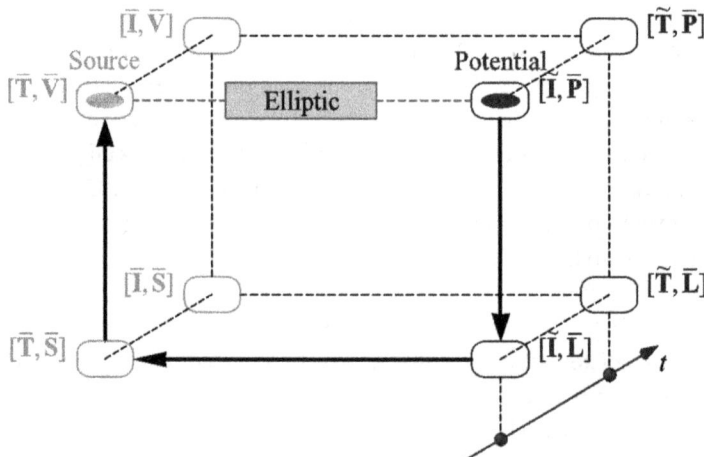

Figure 6.1. Elliptic equations in the classification diagram.

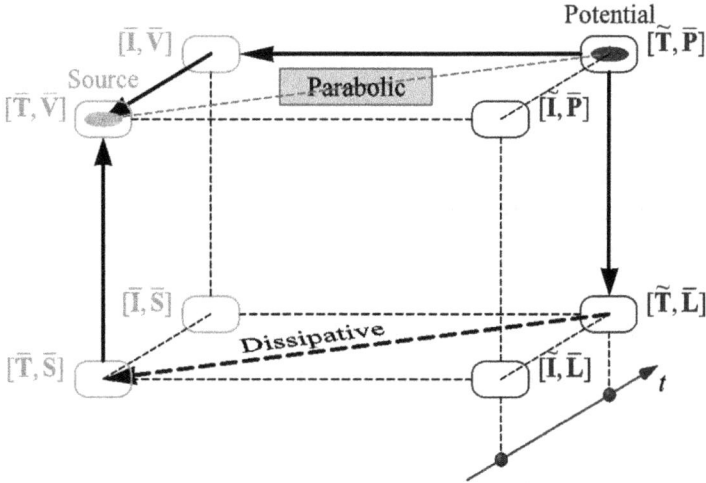

Figure 6.2. Parabolic equations in the classification diagram.

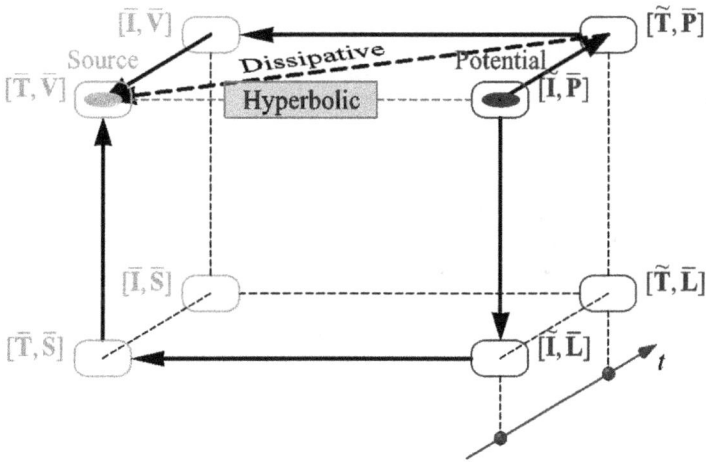

Figure 6.3. Hyperbolic equations in the classification diagram.

over a region of the xy plane, then the general second order PDE form in that region is

$$Au_{xx} + 2Bu_{xy} + Cu_{yy} + \dots \text{(lower-order terms)} = 0, \qquad (6.1.2)$$

where we have assumed:

$$u_{xy} = u_{yx}. \qquad (6.1.3)$$

Replacing ∂_x by x and ∂_y by y (formally this is done by a Fourier transform[160]) converts a constant-coefficient PDE into a polynomial of the same degree. Therefore, the general second-order PDE form in two independent variables is formally analogous to the equation for a conic section:

$$Ax^2 + 2Bxy + Cy^2 + \dots \text{(lower-order terms)} = 0. \qquad (6.1.4)$$

---

[160] The **Fourier transform** is named after Jean Baptiste Joseph Fourier (21 March 1768–16 May 1830).

Just as one classifies conic sections and quadratic forms into elliptic, parabolic, and hyperbolic based on the discriminant:

$$\Delta = B^2 - 4AC, \tag{6.1.5}$$

the same can be done for a second-order PDE at a given point. Due to the convention of the xy term in a PDE being $2B$ rather than B, the discriminant of the associated quadratic form is

$$\Delta = (2B)^2 - 4AC = 4(B^2 - AC). \tag{6.1.6}$$

Consequently, the discriminant in a PDE is given by $B^2 - AC$, with the factor of four dropped for simplicity. The following three cases may occur:

1. $B^2 - AC < 0$: the equation is elliptic. The solutions of elliptic PDEs are as smooth as the coefficients allow, within the interior of the region, where the equation and solutions are defined. For example, solutions of Laplace's equation[161] are analytic within the domain, where they are defined, but solutions may assume boundary values that are not smooth.

---

[161] In three dimensions, **Laplace's equation** is often written as:

$$\Delta\varphi = 0 \quad \text{or} \quad \nabla^2\varphi = 0,$$

where $\Delta = \nabla^2 = \nabla \cdot \nabla = \text{div grad}$ is the Laplace operator, or Laplacian (Eq. (5.3.10)), named after Pierre-Simon de Laplace (23 March 1749–5 March 1827) and φ is a twice-differentiable real-valued function of real variables x, y, and z. The Laplace operator can be generalized to operate on functions defined on surfaces in Euclidean space and, more generally, on Riemannian and pseudo-Riemannian manifolds. This more general operator goes by the name **Laplace–Beltrami operator**, after Laplace and Eugenio Beltrami (November 16, 1835, Cremona–June 4, 1899, Rome). Like the Laplacian, the Laplace–Beltrami operator is defined as the divergence of the gradient and is a linear operator taking functions into functions. The operator can be extended to operate on tensors as the divergence of the covariant derivative. Alternatively, the operator can be generalized to operate on differential forms using the divergence and exterior derivative. The resulting operator is called the **Laplace–de Rham operator**, named after Georges de Rham (10 September 1903–9 October 1990). On a Riemannian manifold it is an elliptic operator, while on a Lorentzian manifold it is hyperbolic. The Laplace–de Rham operator is defined by

$$\Delta = d\delta + \delta d = (d + \delta)^2.$$

d is the exterior derivative, or differential, and δ is the codifferential, acting as $(-1)^{kn+n+1} \star d \star$ on k-forms, where $\star$ is the Hodge star operator, or Hodge dual, that is, a one-to-one mapping from the space of k-vectors and the space of $(n-k)$-vectors, where n is the dimensionality of the oriented inner product space and k is an integer such that $0 \le k \le n$. When computing Δf for a scalar function f, we have $\delta f = 0$, so that

$$\Delta f = \delta df.$$

Up to an overall sign, The Laplace–de Rham operator is equivalent to the Laplace–Beltrami operator when acting on a scalar function. On functions, the Laplace–de Rham operator is actually the negative of the Laplace–Beltrami operator, as the conventional normalization of the codifferential assures that the Laplace–de Rham operator is (formally) positive definite, whereas the Laplace–Beltrami operator is typically negative. The sign is a pure convention, however, and both are common in the literature. The Laplace–de Rham operator differs more significantly from the

(continues next page)

The motion of a fluid at subsonic speeds can be approximated with elliptic PDEs, and the Euler–Tricomi equation:[162]

$$u_{xx} = xu_{yy}, \tag{6.1.7}$$

is elliptic, where $x < 0$.

2. $B^2 - AC = 0$: the equation is parabolic. It can be transformed into a form analogous to the heat equation[163] by a change of independent variables. Solutions smooth out as the

---

(*continues from previous page*)

tensor Laplacian restricted to act on skew-symmetric tensors. Apart from the incidental sign, the two operators differ by a Weitzenböck identity that explicitly involves the Ricci curvature tensor.

Laplace first applied the operator ∇^2 to the study of celestial mechanics, where the operator gives a constant multiple of the mass density when it is applied to a given gravitational potential. Solutions of the Laplace's equation are the so-called harmonic functions and represent the possible gravitational fields in free space. The Laplacian occurs in differential equations that describe many physical phenomena, such as electric and gravitational potentials, the diffusion equation for heat and fluid flow, wave propagation, and quantum mechanics. The Laplacian represents the flux density of the gradient flow of a function. For instance, the net rate at which a chemical dissolved in a fluid moves toward or away from some point is proportional to the Laplacian of the chemical concentration at that point. Expressed symbolically, the resulting equation is the diffusion equation. For these reasons, it is extensively used in the sciences for modelling all kinds of physical phenomena. In image processing and computer vision, the Laplacian operator has been used for various tasks such as blob and edge detection.

[162] The Euler–Tricomi equation is used in the investigation of transonic flow. It is named after Leonhard Euler (15 April 1707–18 September 1783) and Francesco Giacomo Tricomi (5 May 1897–21 November 1978). Its characteristics are

$$xdx^2 = dy^2,$$

which have the integral

$$y \pm \frac{2}{3}\sqrt{x^3} = C,$$

where C is a constant of integration. The characteristics thus comprise two families of semicubical parabolas, with cusps on the line $x = 0$, the curves lying on the right-hand side of the y-axis.

[163] For a function $u(x, y, z, t)$ of three spatial variables (x, y, z) and the time variable t, the **heat equation** is

$$\alpha \left(\frac{\partial^2 u}{\partial x^2} + \frac{\partial^2 u}{\partial y^2} + \frac{\partial^2 u}{\partial z^2} \right) = \frac{\partial u}{\partial t}.$$

More generally, in any coordinate system:

$$\alpha \nabla^2 u = \frac{\partial u}{\partial t},$$

where α is a positive constant, and ∇^2 denotes the Laplace operator (Eq. (5.3.10)). In the physical problem of temperature variation, $u(x, y, z, t)$ is the temperature and α is the thermal diffusivity. For the mathematical treatment it is sufficient to consider the case $\alpha = 1$. The heat equation is of fundamental importance in diverse scientific fields. In mathematics, it is the prototypical parabolic partial differential equation. In probability theory, the heat equation is connected with the study of Brownian motion via the Fokker–Planck equation. In financial mathematics it is used to solve the Black–Scholes partial differential equation. It is also important in Riemannian geometry and thus topology:

(*continues next page*)

transformed time variable increases. The Euler–Tricomi equation has parabolic type on the line, where $x = 0$.

3. $B^2 - AC > 0$: the equation is hyperbolic. It retains any discontinuities of functions or derivatives in the initial data. An example is the wave equation.[164] Moreover, the motion of a fluid at supersonic speeds can be approximated with hyperbolic PDEs, and the Euler–Tricomi equation is hyperbolic where $x > 0$.

(*continues from previous page*)

it was adapted by Richard Streit Hamilton (born 1943) when he defined the Ricci flow that was later used by Grigori Yakovlevich Perelman (born 13 June 1966) to solve the topological Poincaré conjecture. The diffusion equation, a more general version of the heat equation, arises in connection with the study of chemical diffusion and other related processes.

[164] In its simplest form, the wave equation concerns a time variable t, one or more spatial variables $x_1, x_2, \ldots, x_n$, and a scalar function $u = u(x_1, x_2, \ldots, x_n, t)$, whose values could model the displacement of a wave. The wave equation for u is

$$c^2 \nabla^2 u = \frac{\partial^2 u}{\partial t^2},$$

where ∇^2 is the (spatial) Laplacian (Eq. (5.3.10)) and where c is a fixed constant. The constant c is identified with the propagation speed of the wave. This equation is linear, as the sum of any two solutions is again a solution (superposition principle). The equation alone does not specify a solution. A unique solution is usually obtained by setting a problem with further conditions, such as initial conditions, which prescribe the value and velocity of the wave. Another important class of problems specifies boundary conditions, for which the solutions represent standing waves, or harmonics, analogous to the harmonics of musical instruments.

To model dispersive wave phenomena, those in which the speed of wave propagation varies with the frequency of the wave, the constant c is replaced by the phase velocity:

$$v_p = \frac{\omega}{k}.$$

The elastic wave equation in three dimensions describes the propagation of waves in an isotropic homogeneous elastic medium. Most solid materials are elastic, so this equation describes such phenomena as seismic waves in the Earth and ultrasonic waves used to detect flaws in materials. While linear, this equation has a more complex form than the equations given above, as it must account for both longitudinal and transverse motions:

$$(\lambda + 2\mu)\nabla(\nabla \cdot \mathbf{u}) - \mu\nabla \times (\nabla \times \mathbf{u}) + \mathbf{f} = \rho\ddot{\mathbf{u}},$$

where

- λ and μ are the so-called Lamé parameters (Eqs. (5.3.4) and (5.3.5)), describing the elastic properties of the medium.
- ρ is the density.
- $\mathbf{f}$ is the source variable (driving force).
- $\mathbf{u}$ is the configuration variable (displacement vector).

Since both force and displacement are vector quantities, this equation is sometimes known as the vector wave equation.

If there are n independent variables $x_1, x_2, \ldots, x_n$, a general linear partial differential equation of second order has the form:

$$Lu = \sum_{i=1}^{n} \sum_{j=1}^{n} a_{i,j} \frac{\partial^2 u}{\partial x_i \partial x_j} + \ldots \text{(lower-order terms)} = 0. \qquad (6.1.8)$$

The classification of PDEs then depends upon the signature of the eigenvalues of the coefficient matrix $a_{i,j}$.

1. Elliptic equation: the eigenvalues are all positive or all negative.
2. Parabolic equation: the eigenvalues are all positive or all negative, save one that is zero.
3. Hyperbolic equation: there is only one negative eigenvalue and all the rest are positive, or there is only one positive eigenvalue and all the rest are negative.
4. Ultra-hyperbolic equation: there is more than one positive eigenvalue and more than one negative Eigenvalue, and there are no zero eigenvalues.

In several real-world problems, it is not possible to derive closed form solutions of the fundamental equations, for the multitude of irregular geometries, various constitutive relations[128] of media, and boundary conditions. Computational numerical techniques can overcome this inability, providing us with important tools for design and modeling. To achieve this, time and space are divided into a discrete grid and the continuous differential equations are discretized. In general, the simulated system behaves differently than the intended physical system. The amount and character of the difference depends on the system being simulated and the type of discretization that is used.

Choosing the right numerical technique for solving a problem is important, as choosing the wrong one can either result in incorrect results, or results which take excessively long time to compute. In particular, the equation which approximates the equation to be studied is probable to become unstable, meaning that errors in the input data and intermediate calculations can be magnified in the limit, instead of damped, causing the error to grow exponentially.

Form the numerical point of view, an unstable solution occurs in differential formulation whenever the algebraic system of discretized equations derived from an elliptic equation ceases to be elliptic. The same occurs when the algebraic systems of a parabolic or hyperbolic equation are not parabolic or hyperbolic, respectively. The causes for this are several. In some cases, they consist in the integration method adopted. In particular, it is important to use a stable method whenever we want to solve a stiff equation,[165] that is, a differential equation for which certain numerical methods for solving the equation are numerically unstable, unless the step

---

[165] The behavior of numerical methods on stiff problems can be analyzed by applying these methods to the test equation:

$$y' = ky,$$

subject to the initial condition:

$$y(0) = 1 \text{ with } k \in \mathbb{C}.$$

The solution of this equation is

$$y(t) = e^{kt}.$$

(continues next page)

size is taken to be extremely small. A problem is stiff when the step size is forced down to an unacceptably small level in a region where the solution curve is very smooth, while one would expect the requisite step size to be relatively small in a region where the solution curve displays much variation and to be relatively large where the solution curve straightens out to approach a line with slope nearly zero. A method that is stable on stiff problems is called an A-stable method.[166]

A method is A-stable if the region of absolute stability contains the set:

$$\left\{ z \in \mathbb{C} : \mathrm{Re}(z) < 0 \right\}; \qquad (6.1.9)$$

that is, the left half complex plane. By way of example, the **Euler method**[167] with step size h[168] is numerically unstable for the linear stiff equation $y' = ky$ whenever the product hk is outside

---

(*continues from previous page*)

This solution approaches zero as $t \to \infty$ when $\mathrm{Re}(k) < 0$. If the numerical method also exhibits this behavior, then the method is said to be A-stable.

[166] Yet another definition of stability is used in numerical partial differential equations. According to the **Lax equivalence theorem**, an algorithm for solving a linear evolutionary partial differential equation converges if it is consistent and stable, in the sense that the total variation of the numerical solution at a fixed time remains bounded as the step size goes to zero. This theorem is due to Peter David Lax (born 1 May 1926). It is sometimes called the **Lax–Richtmyer theorem**, after Peter Lax and Robert Davis Richtmyer (1910–2003).

[167] The **Euler method** is a numerical procedure for solving ordinary differential equations (ODEs) with a given initial value. It is named after Leonhard Euler, who treated it in his book *Institutionum calculi integralis* (published 1768–1770). The Euler method can be derived in a number of ways. One possibility is to substitute the forward finite difference formula for the derivative:[10]

$$y'(t_0) = \frac{y(t_0 + h) - y(t_0)}{h},$$

in the differential equation $y' = f(t, y)$. This is equivalent to substitute the differential equation $y' = f(t, y)$ in the Taylor expansion of the function y around t_0:

$$y(t_0 + h) = y(t_0) + hy'(t_0) + \frac{1}{2}h^2 y''(t_0) + O(h^3),$$

and ignore the quadratic and higher-order terms. From the geometrical point of view, the Euler method takes a small step along the tangent line at the starting point, A_0, of the ordinary differential equation $y' = f(t, y)$ with initial value $y(t_0) = y_0$, where the function f and the initial data t_0 and y_0 are known, while the function y depends on the real variable t and is unknown. Along this small step, the slope does not change too much, so the end point of the small step, which we denote by A_1, will be close to the curve. If we pretend that A_1 is still on the curve, the same reasoning as for the point A_0 above can be used. After several steps, a polygonal curve $A_0, A_1, A_2, A_3, \ldots$ is computed. In general, this curve does not diverge too far from the original unknown curve, and the error between the two curves can be made small if the step size is small enough and the interval of computation is finite.

[168] A numerical method produces a sequence:

$$y_0, y_1, y_2, \ldots,$$

such that y_k approximates $y(t_0 + kh)$, where h is called the step size.

the (linear) stability region, which, in the complex plane, is given by the disk with radius one centered at $(-1,0)$ (Fig. 6.4):

$$\{z \in \mathbb{C} : |z+1| \le 1\}. \qquad (6.1.10)$$

Therefore, the Euler method is not A-stable. A simple modification of the Euler method that eliminates the stability problems is the **backward Euler method**, which differs from the (standard, or forward) Euler method in that the function f is evaluated at the end point of the step, instead of the starting point:

$$y_{n+1} = y_n + hf\left(t_{n+1}, y_{n+1}\right). \qquad (6.1.11)$$

The region of absolute stability for the backward Euler method is the complement, in the complex plane, of the disk with radius one centered at $(+1, 0)$ (Fig. 6.5). This includes the whole

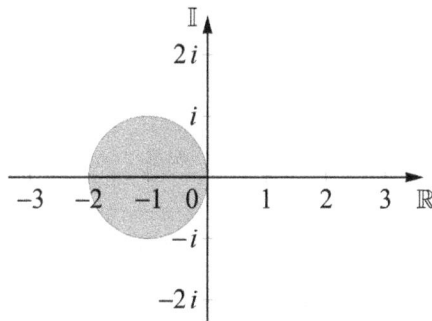

Figure 6.4. The gray disk shows the stability region for the Euler method.

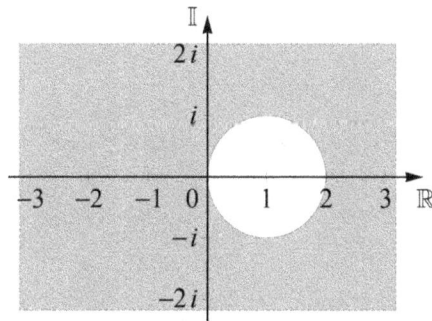

Figure 6.5. The gray region outside the disk shows the stability region of the backward Euler method.

left half of the complex plane, so the backward Euler method is A-stable,[169] making it suitable for the solution of stiff equations.

Since the formula for the backward Euler method has y_{n+1} on both sides, the backward Euler method is an implicit method.[170] This makes the implementation more costly.

Both the forward and the backward Euler methods are first-order methods, which means that the local error (error per step) is proportional to the square of the step size, and the global error (error at a given time) is proportional to the step size. Using the big O notation,[171] this means that the local truncation error (the error made in one step) is $O(h^2)$, and the error at a specific time t is $O(h)$.

---

[169] The backward Euler method is even **L-stable**, meaning that it is A-stable and the stability function ϕ[173] vanishes at infinity:

$$|\phi(z)| \to 0 \quad \text{as } z \to \infty.$$

The definition of L-stability, or Lascal (*La-skal*) stability, is due to Ehle (1969).

[170] **Explicit and implicit methods** are approaches used in numerical analysis for obtaining numerical solutions of time-dependent ordinary and partial differential equations. Explicit methods calculate the state of a system at a later time from the state of the system at the current time, while implicit methods find a solution by solving an equation involving both the current state of the system and the later one. Implicit methods require an extra computation, and they can be much harder to implement. Implicit methods are used because many problems arising in practice are stiff, for which the use of an explicit method requires impractically small time steps Δt to keep the error in the result bounded. In the vast majority of cases, the equation to be solved when using an implicit scheme has not an exact solution. Then one uses root-finding algorithms, such as Newton's method.

[171] **Big O notation** describes the limiting behavior of a function when the argument tends towards a particular value or infinity. It is a member of a larger family of notations that is called **Landau notation, Bachmann–Landau notation** (after Edmund Georg Hermann Landau, 14 February 1877–19 February 1938, and Paul Gustav Heinrich Bachmann, 22 June 1837–31 March 1920), or **asymptotic notation**. Big O notation characterizes functions according to their growth rates: different functions with the same growth rate may be represented using the same O notation. The letter O is used because the growth rate of a function is also referred to as order of the function. Let $f(x)$ and $g(x)$ be two functions defined on some subset of the real numbers. One writes

$$f(x) = O(g(x)) \quad \text{as } x \to \infty,$$

if and only if there exists a positive real number M such that, for all sufficiently large values of x, $f(x)$ is at most M multiplied by $g(x)$, in the absolute value:

$$|f(x)| \le M|g(x)| \quad \text{for all } x > x_0,$$

where x_0 is a real number. The notation can also be used to describe the behavior of $f(x)$ near some real number a (often, $a = 0$):

$$f(x) = O(g(x)) \quad \text{as } x \to a,$$

if and only if there exist positive numbers δ and M such that

$$|f(x)| \le M|g(x)| \quad \text{for all } |x - a| > \delta.$$

More complicated methods can achieve a higher order (and more accuracy). One possibility[172] is to use more function evaluations, leading to the family of Runge–Kutta methods.[173] This is illustrated by the **midpoint method**:[174]

$$y_{n+1} = y_n + hf\left(t_n + \frac{1}{2}h, y_n + \frac{1}{2}hf\left(t_n, y_n\right)\right),$$ (6.1.12)

---

[172] The other possibility is to use more past values, as illustrated by the two-step **Adams–Bashforth method**:

$$y_{n+1} = y_n + \frac{3}{2}hf\left(t_n, y_n\right) - \frac{1}{2}hf\left(t_{n-1}, y_{n-1}\right).$$

This leads to the family of **linear multistep methods**.

[173] The **Runge–Kutta methods** are an important family of implicit and explicit iterative methods for the approximation of solutions of ordinary differential equations, which take the form:

$$y_{n+1} = y_n + h\sum_{i=1}^{s} b_i k_i \, ;$$

$$k_i = f\left(t_n + c_i h, y_n + h\sum_{j=1}^{s} a_{ij} k_j\right).$$

Runge–Kutta methods applied to the test equation $y' = ky$ take the form:

$$y_{n+1} = \phi\left(hk\right) y_n,$$

and, by induction:

$$y_n = \left[\phi\left(hk\right)\right]^n y_0.$$

The function ϕ is called the **stability function**. These techniques were developed around 1900 by the German mathematicians Carl David Tolmé Runge (1856–1927) and Martin Wilhelm Kutta (November 3, 1867–December 25, 1944). The methods are each defined by its **Butcher tableau**, which puts the coefficients of the method in a table as follows:

$$
\begin{array}{c|cccc}
c_1 & a_{11} & a_{12} & \cdots & a_{1s} \\
c_2 & a_{21} & a_{22} & \cdots & a_{2s} \\
\vdots & \vdots & \vdots & \ddots & \vdots \\
c_s & a_{s1} & a_{s2} & \cdots & a_{ss} \\
\hline
 & b_1 & b_2 & \cdots & b_s
\end{array}
$$

The explicit methods are those where the matrix $\left[a_{ij}\right]$ is a lower triangular. They include, for example, Forward Euler method, Generic second-order method, Kutta's third-order method, Classic fourth-order method, and 3/8-rule fourth-order method. Examples of implicit methods include Backward Euler method, Implicit midpoint method, and Lobatto methods (Lobatto IIIA methods, Lobatto IIIB methods, Lobatto IIIC methods). Moreover, the embedded methods are designed to produce an estimate of the local truncation error of a single Runge-Kutta step, and as result, allow to control the error with adaptive step-size. This is done by having two methods in the tableau, one with order p and one with order $p-1$. Examples of embedded methods include Heun–Euler method, Bogacki–Shampine method, Fehlberg method, Cash-Karp method, and Dormand–Prince method.

[174] The midpoint method is also known as the **modified Euler method**.

where the error is roughly proportional to the square of the step size.[175] For this reason, the midpoint method is said to be a second-order method.

Leapfrog integration is another second-order method, which is used for numerically integrating differential equations of the form:

$$\ddot{x} = -\nabla V(x), \tag{6.1.13}$$

where V is the potential energy of the system, particularly in the case of a dynamical system of classical mechanics.

Leapfrog integration is equivalent to updating positions $x(t)$ and velocities $v(t) = \dot{x}(t)$ at interleaved time points, staggered in such a way that they "leapfrog" over each other. For example, the position is updated at integer time steps and the velocity is updated at integer-plus-a-half time steps. Unlike Euler integration, it is stable for oscillatory motion, as long as the time-step Δt is constant, and

$$\Delta t \leq \frac{2}{\omega}, \tag{6.1.14}$$

where ω is the angular frequency (measured in radians per second). In leapfrog integration, the equations for updating position and velocity are

$$x_i = x_{i-1} + v_{i-\frac{1}{2}}\Delta t, \tag{6.1.15}$$

$$a_i = F(x_i), \tag{6.1.16}$$

$$v_{i+\frac{1}{2}} = v_{i-\frac{1}{2}} + a_i\Delta t, \tag{6.1.17}$$

where x_i is the position at step i, $v_{i+\frac{1}{2}}$ is the velocity, or first derivative of x, at step $i+\frac{1}{2}$, $a_i = F(x_i)$ is the acceleration, or second derivative of x, at step i, and Δt is the size of each time step.

By comparison between Eq. (6.1.15) and (4.4.4), which can be put in the form:

$$\overline{x}_2 = \overline{x}_1 + \overline{v}_{x1}\Delta t = \overline{x}_1 + \tilde{v}_{x1}\Delta t, \tag{6.1.18}$$

where $\overline{x}_1$ is the position associated with the first primal time instant $\overline{\mathbf{I}}^1$ (first primal step), $\overline{v}_{x1}$ is the velocity associated with the first primal time interval $\overline{\mathbf{T}}^1$, $\tilde{v}_{x1}$ is the velocity associated with the first dual time instant $\tilde{\mathbf{I}}^1$ (first dual step), and primal and dual time instants (or time steps) are staggered for a-half time step, as shown in Fig. 4.13, we can observe that the leapfrog integration uses the same explicit[170] time-marching scheme of the CM. We may come to the same conclusion by comparing Eq. (6.1.17) with Eq. (4.4.10), which can be put in the form:

$$\tilde{v}_{x2} = \tilde{v}_{x1} + \tilde{a}_{x1}\Delta t = \tilde{v}_{x1} + \overline{a}_{x2}\Delta t, \tag{6.1.19}$$

where $\tilde{v}_{x1}$ is the velocity associated with the first dual time instant $\tilde{\mathbf{I}}^1$ (first dual step), $\tilde{a}_{x1}$ is the acceleration associated with the first dual time interval $\tilde{\mathbf{T}}^1$, $\overline{a}_{x2}$ is the acceleration associated with the second primal time instant $\overline{\mathbf{I}}^2$ (second primal step), and primal and dual time instants

---

[175] The local error at each step of the midpoint method is of order $O(h^3)$, giving a global error of order $O(h^2)$.

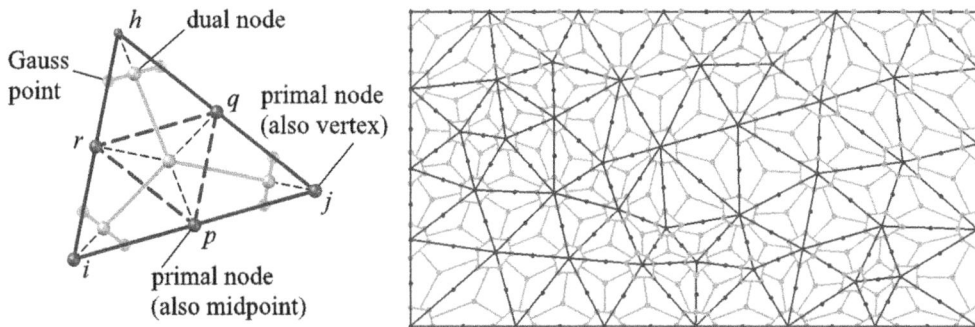

Figure 6.6. Construction of the dual polygons (a), and primal (dark gray) and dual (light gray) mesh for a plane domain (b).

are interleaved as shown in Fig. 4.13. Consequently, the time-marching scheme of the CM can be viewed as the algebraic version of the leapfrog integration in the differential formulation. This allows us to state that the leapfrog integration is equivalent to a coboundary process performed for the scalar variables of the 0-form $\overline{\Phi}_1^0 = \left[\overline{x}_i \right]$ on the primal time instants, which generates the 0-form $\widetilde{\Phi}^0 = \left[\tilde{v}_{xi} \right]$ on the dual time instants and a second 0-form, $\overline{\Phi}_2^0 = \left[\overline{a}_{x(i+1)} \right]$, on the primal time instants (Section 4.4).

The most important consequence of the similarity between the leapfrog integration and the time-marching scheme of the CM is that also the CM is stable for oscillatory motion. The similarity then extends to the convergence order, since even the CM, in its original formulation with barycentric or circumcentric dual polygons in space (Section 5.5), is a second-order method, both in space and in time. Nevertheless, by modifying the shape of the dual polygons in space, it is possible to achieve higher convergence orders for the CM. In particular, we attain a fourth-order convergence[176] in space by choosing Gauss points, besides the primal barycenters and the midpoints of the dual sides, for building the dual polygons around the primal nodes (Figs. 6.6, 6.7).

Attaining a fourth-order convergence with the CM is all the more relevant as it was not possible to attain convergence greater than second-order for any of the methods which are similar to the CM, such as the direct or physical approach of the FEM, the vertex-based scheme of the FVM, and the FDM.

In order for the leapfrog integration used to solve the partial differential equations does not become unstable, space and time steps must satisfy the **Courant–Friedrichs–Lewy condition**[177] (CFL condition), which is a necessary condition for the convergence of explicit[170] time-marching schemes while solving certain partial differential equations (usually hyperbolic PDEs) numerically by the method of finite differences. According to the CFL condition, the time step must be less than a certain time in many explicit time-marching computer simulations; otherwise the simulation will produce incorrect results. In essence, the CFL condition states that the numerical domain of dependence of any point in space and time (which data values in the initial conditions affect the numerical computed value at that point) must include the analytical

---

[176] The CM was implemented for solid mechanics with quadratic interpolation of the displacements in two-dimensional domains by Cosmi (2000), and three-dimensional domains by Pani and Taddei (2013).

[177] The condition is named after Richard Courant (January 8, 1888–January 27, 1972), Kurt Otto Friedrichs (September 28, 1901–December 31, 1982), and Hans Lewy (20 October 1904–23 August 1988) who described it in their 1928 paper.

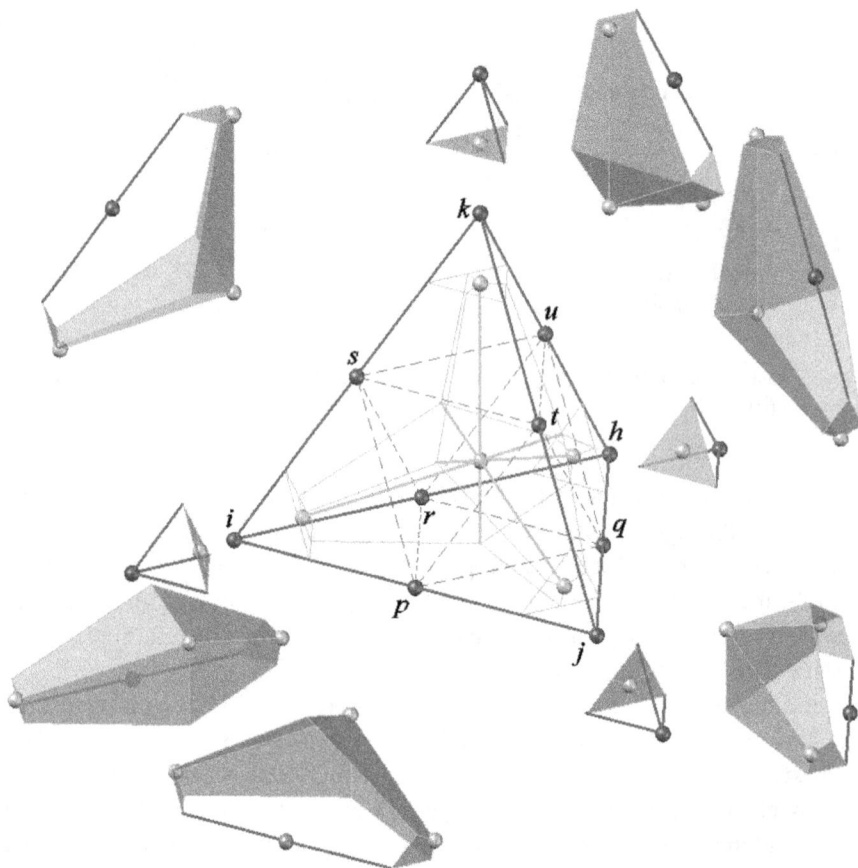

Figure 6.7. Construction of the dual polygons for a three-dimensional domain.

domain of dependence (where in the initial conditions has an effect on the exact value of the solution at that point) in order to assure that the scheme can access the information required to form the solution. For example, if a wave is moving across a discrete spatial grid and we want to compute its amplitude at discrete time steps of equal length, then this length must be less than the time for the wave to travel to adjacent grid points.

The CFL condition can be a very limiting constraint on the time step Δt. For example, in the finite-difference approximation of certain fourth-order nonlinear partial differential equations, it can have the following form:

$$\frac{\Delta t}{(\Delta x)^4} < Cu, \qquad (6.1.20)$$

meaning that a decrease in the length interval Δx requires a fourth-order decrease in the time step Δt for the condition to be fulfilled. Moreover, since the CFL condition is a necessary condition, but may not be sufficient for the convergence of the finite-difference approximation of a given numerical problem, in order to establish the convergence of the finite-difference approximation, it is necessary to use other methods, which, in turn, could imply further limitations on the length of the time step and/or the lengths of the spatial intervals. Therefore, when solving particularly stiff problems, efforts are often made to avoid the CFL condition, for example by using implicit methods.

Leapfrog integration is used in the **Finite-difference time-domain method** (FDTD),[178] which is a numerical analysis technique for modeling computational electromagnetics (CEM).[179] CEM typically solves the problem of computing the E (electric) and H (magnetic) fields across the problem domain. It typically involves using computationally efficient approximations to Maxwell's equations, which can be formulated as a hyperbolic system of partial differential equations.

The FDTD method belongs in the general class of grid-based differential time-domain numerical modeling methods (finite difference methods). Since it is a time-domain method, FDTD solutions can cover a wide frequency range with a single simulation run and treat non-linear material properties in a natural way. The time-dependent Maxwell's equations (in partial differential form) are discretized using central-difference[12] approximations to the space and time partial derivatives.

The equations are solved in a cyclic manner: the electric field vector components in a volume of space are solved at a given instant in time, then the magnetic field vector components in the same spatial volume are solved at the next instant in time, and the process is repeated over and over again until the desired transient or steady-state electromagnetic field behavior is fully evolved. Since the change in the E-field in time (the time derivative) is dependent on the change in the H-field across space (the curl), at any point in space, the updated value of the E-field in time is dependent on the stored value of the E-field and the numerical curl of the local distribution of the H-field in space. Analogously, at any point in space, the updated value of the H-field in time is dependent on the stored value of the H-field and the numerical curl of the local distribution of the E-field in space. Iterating the E-field and H-field updates the results in a marching-in-time process wherein sampled-data analogs of the continuous electromagnetic waves under consideration propagate in a numerical grid stored in the computer memory.

This description holds true for 1-D, 2-D, and 3-D FDTD techniques. When multiple dimensions are considered, calculating the numerical curl can become complicated. Kane Yee's seminal 1966 paper proposed spatially staggering the vector components of the E-field and H-field about rectangular unit cells of a Cartesian computational grid so that each E-field vector component is located midway between a pair of H-field vector components, and conversely. This scheme, now known as a **Yee lattice**, has proven to be very robust and remains at the core of many current FDTD software constructs. Furthermore, Yee proposed a leapfrog scheme for marching in time wherein the E-field and H-field updates are staggered so that E-field updates are conducted midway during each time-step between successive H field updates, and

---

[178] The basic FDTD algorithm traces back to a seminal 1966 paper by Kane Yee in IEEE Transactions on Antennas and Propagation. Allen Taflove originated the descriptor "Finite-difference time-domain" and its corresponding "FDTD" acronym in a 1980 paper in IEEE Transactions on Electromagnetic Compatibility. Since about 1990, FDTD techniques have emerged as the primary means to model many scientific and engineering problems addressing electromagnetic wave interactions with material structures. An effective technique based on a time-domain finite-volume discretization procedure was introduced by Mohammadian et al. in 1991. Current FDTD modeling applications range from near-DC (ultralow-frequency geophysics involving the entire Earth-ionosphere waveguide) through microwaves (radar signature technology, antennas, wireless communications devices, digital interconnects, biomedical imaging/treatment) to visible light (photonic crystals, nanoplasmonics, solitons, and biophotonics).

[179] **Computational electromagnetics**, **computational electrodynamics** or **electromagnetic modeling** is the process of modeling the interaction of electromagnetic fields with physical objects and the environment.

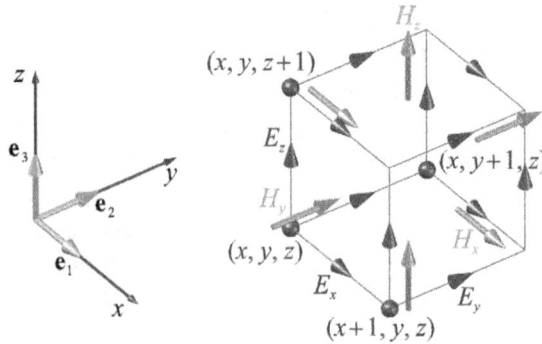

Figure 6.8. Illustration of a standard Cartesian Yee cell used for FDTD, about which electric and magnetic field vector components are distributed.

conversely. On the plus side, this explicit time-stepping scheme avoids the need to solve simultaneous equations and furthermore yields dissipation-free numerical wave propagation. On the minus side, this scheme mandates an upper bound on the time-step to ensure numerical stability. This allows us to avoid spurious solutions, that is, to avoid a numerical drawback.

If visualized as a cubic voxel,[180] the electric field components form the edges of the cube, and the staggered magnetic field components form the normals to the faces of the cube (Fig. 6.8). A three-dimensional space lattice consists of a multiplicity of such Yee cells,[181] leading to a scheme analogous to the CM scheme with primal and dual cells (Fig. 4.3). Therefore, the Yee lattice can be considered the particularization of the primal and dual cell complexes of the CM, when a differential formulation is derived from the algebraic formulation. Moreover, we may generalize the notion of inherited association (Section 5.3) to the stability of the numerical solution, by assuming that the numerical stability is inherited by the Yee lattice from the CM cell complexes, when the field variables are derived from the global variables.

6.2 The Need for Non-Local Models in Quantum Physics

Non-locality is one of the remarkable features of the microscopic world prescribed by quantum theory. In simple terms, non-locality is the idea that objects can instantaneously know about each other's state, even when separated by large distances. Non-locality occurs due to the phenomenon of entanglement,[182] whereby particles that have been interacted with each other

---

[180] A **voxel** (volumetric pixel or Volumetric Picture Element) is a volume element, representing a value on a regular grid in three-dimensional spaces. This is analogous to a texel, which represents 2D image data in a bitmap (which is sometimes referred to as a pixmap).

[181] An electromagnetic wave interaction structure is mapped into the space lattice by assigning appropriate values of permittivity to each electric field component, and permeability to each magnetic field component.

[182] **Entanglement** is usually created by direct interactions between subatomic particles. These interactions can take numerous forms. One of the most commonly used methods is spontaneous parametric down-conversion to generate a

(continues next page)

become permanently correlated, to the extent that they effectively lose their individuality and in many ways behave as a single entity. The idea of instantaneous action at a distance[183] or transfer of information, as if the universe were able to anticipate future events, contradicts the relativistic upper limit on speed of propagation of information in special relativity, which states that information can never be transmitted faster than the speed of light without violating causality. It is generally believed that any theory which violates causality would also be internally inconsistent, and thus useless. Non-locality also violates the "principle of local action"[184] of Einstein,[185]

(continues from previous page)

pair of photons entangled in polarization. Other methods include the use of a fiber coupler to confine and mix photons, the use of quantum dots to trap electrons until decay occurs, the use of the Hong-Ou-Mandel effect, and so forth. In the earliest tests of Bell's theorem,[191] the entangled particles were generated using atomic cascades. As far as the measurement of entanglement is concerned, entropy provides one tool that can be used to quantify entanglement, although other entanglement measures exist. If the overall system is pure, the entropy of one subsystem can be used to measure its degree of entanglement with the other subsystems.

[183] In physics, **action at a distance** is the non-local interaction of objects that are separated in space. Efforts to account for action at a distance in the theory of electromagnetism led to the development of the concept of a field, which mediated interactions between currents and charges across empty space. According to field theory, we account for the Coulomb (electrostatic) interaction between charged particles through the fact that charges produce around themselves an electric field, which can be felt by other charges as a force. The concept of the field was elevated to fundamental importance in Maxwell's equations, which used the field to elegantly account for all electromagnetic interactions, as well as light (which, until then, had been a completely unrelated phenomenon). In Maxwell's theory, the field is its own physical entity, carrying momenta and energy across space, and action at a distance is only the apparent effect of local interactions of charges with their surrounding field. Electrodynamics can be described without fields (in Minkowski space[61]) as the direct interaction of particles with light-like[61] separation vectors. This results in the Fokker-Tetrode-Schwarzschild action integral. This kind of electrodynamic theory is often called "direct interaction", to distinguish it from field theories where action at a distance is mediated by a localized field (localized in the sense that its dynamics are determined by the nearby field parameters). This description of electrodynamics, in contrast with Maxwell's theory, explains an apparent action at a distance not by postulating a mediating entity (the field) but by appealing to the natural geometry of special relativity.

[184] In pre-quantum mechanics, Newton's law of universal gravitation (17th Century) was formulated in terms of action at a distance, 183 thereby violating the principle of locality. Coulomb's law of electric forces was initially also formulated as instantaneous action at a distance, but was later superseded by Maxwell's Equations of electromagnetism, which obey locality. Coherently with his postulate that no material or energy can travel faster than the speed of light, Einstein sought to reformulate physical laws in a way which obeyed the principle of locality. He later succeeded in producing an alternative theory of gravitation, General Relativity, which obeys the principle of locality. Einstein assumed that the principle of locality was necessary, and that there could be no violations of it. He said:

> "(...) The following idea characterizes the relative independence of objects far apart in space, A and B: external influence on A has no direct influence on B; this is known as the **Principle of Local Action**, which is used consistently only in field theory. If this axiom were to be completely abolished, the idea of the existence of quasienclosed systems, and thereby the postulation of laws which can be checked empirically in the accepted sense, would become impossible. (...)"

[185] **Albert Einstein** was born in Ulm, in the Kingdom of Württemberg of the German Empire, on 14 March 1879, and died in Princeton, New Jersey, on 18 April 1955. **General relativity**, or the **general theory of relativity**, is the geometric theory of gravitation published by Albert Einstein in 1916 and the current description of gravitation in modern physics. General relativity generalizes special relativity and Newton's law of universal gravitation, providing a unified description of gravity as a geometric property of space and time, or spacetime. The **Einstein field equations**, presented to the Prussian Academy of Science in November 1915, specify how the geometry of space and time is influenced by whatever matter and radiation are present, and form the core of Einstein's general theory of relativity.

that is, the relative independence of objects far apart in space. For these reasons, in 1935 Albert Einstein and his colleagues Boris Podolsky[186] and Nathan Rosen[187] (known collectively as EPR) heavily criticized non-locality and quantum theory in their paper, known as EPR paper.[188] The authors claim that, given a specific experiment, in which the outcome of a measurement is known before the measurement takes place, there must exist something in the real world, an "element of reality", that determines the measurement outcome. They postulate that these elements of reality are local, in the sense that each belongs to a certain point in spacetime.[61] Each element may only be influenced by events which are located in the backward light cone[61] of its point in spacetime (that is, the past). These claims are founded on assumptions about nature that constitute what is now known as local realism.[189] Using the principle of locality, Einstein, Podolsky, and Rosen designed a thought experiment, sometimes referred to as the EPR paradox, intended to

---

[186] **Boris Yakovlevich Podolsky** (1896–1966) was an American physicist of Russian Jewish descent. He was born in Taganrog, in what was then the Russian Empire, and moved to the United States in 1913. After receiving a Bachelor of Science degree in Electrical Engineering from the University of Southern California in 1918, he served in the US Army and then worked at the Los Angeles Bureau of Power and Light. In 1926, he obtained an MS in Mathematics from the University of Southern California. In 1928, he received a PhD in Theoretical Physics. He worked with Albert Einstein and Nathan Rosen on entangled wave functions and the EPR paradox.

[187] **Nathan Rosen** (March 22, 1909–December 18, 1995) was an American-Israeli physicist. He was born in Brooklyn, New York. He attended MIT during the Great Depression, where he received a bachelor's degree in electromechanical engineering and, later, a masters and a doctorate in physics. As a student he published several papers of note, one being "The Neutron," which attempted to explain the structure of the atomic nucleus a year before their discovery by Sir James Chadwick (20 October 1891–24 July 1974). He also developed an interest in wave functions, and later, gravitation, when he worked as a fellow at the University of Michigan and Princeton University. In 1935, he became Albert Einstein's assistant at The Institute for Advanced Study in Princeton, New Jersey. Einstein and Rosen discovered the mathematical solution for a type of wormhole connecting distant areas in space. Dubbed an Einstein-Rosen bridge, or Schwarzschild Wormhole, the solution was found by merging the mathematical models of a black hole and a white hole (a theoretical black hole moving backward in time), using Einstein's field equations. Einstein-Rosen Bridges are purely theoretical. It was shown in a 1962 paper by theoretical physicists John Archibald Wheeler (July 9, 1911–April 13, 2008) and Robert W. Fuller (born 1936) that these types of wormholes are unstable.

[188] At the time the EPR paper "Can Quantum-Mechanical Description of Physical Reality Be Considered Complete?" was written, it was known from experiments that the outcome of an experiment sometimes cannot be uniquely predicted. An example of such indeterminacy can be seen when a beam of light is incident on a half-silvered mirror. One half of the beam will reflect, and the other will pass. If the intensity of the beam is reduced to so that only one photon is in transit at any time, whether that photon will reflect or transmit cannot be predicted quantum mechanically. The routine explanation of this effect was, at that time, provided by **Heisenberg's uncertainty principle**, named after Werner Karl Heisenberg (5 December 1901–1 February 1976). According to this principle, physical quantities come in pairs that are called conjugate quantities. Examples of such conjugate pairs are position and momentum of a particle and components of spin measured around different axes. When one quantity was measured and became determined, the conjugated quantity became indeterminate. Heisenberg explained this as a disturbance caused by measurement. Heisenberg's principle was an attempt to provide a classical explanation of non-locality. The EPR paper was intended to illustrate that Heisenberg explanation is inadequate.

[189] In the vernacular of Einstein, realism meant the moon is there even when not being observed. **Local realism** is the combination of the principle of locality[184] with the "realistic" assumption that all objects must objectively have a pre-existing value for any possible measurement before the measurement is made.

reveal what they believed to be inadequacies of quantum mechanics. The original EPR paradox challenges the prediction of quantum mechanics that it is impossible to know both the position and the momentum of a quantum particle. This challenge can be extended to other pairs of physical properties.

The EPR paradox showed that quantum mechanics predicts non-locality unless position and momentum were simultaneous "real" properties of a particle. According to EPR, there were two possible explanations to the EPR paradox. Either there was some interaction between the particles, even though they were separated, or the information about the outcome of all possible measurements was already present in both particles. Since the first explanation, that an effect propagated instantly, across a distance, is in conflict with the theory of relativity, the EPR authors preferred the second explanation, according to which that information was encoded in some "hidden parameters". The conclusion the EPR drew was that quantum mechanics is physically incomplete[190] and logically unsatisfactory since, in its formalism, there was no space for such hidden parameters. In particular, Einstein never accepted the idea of non-locality, which he derided and called "spooky actions at a distance", sometimes also referred as "ghostly action at a distance".

Despite EPR's misgivings about entanglement and non-locality, Bell's theorem,[191] published by John Bell[192] in 1964, effectively showed that the results predicted by quantum mechanics could not be explained by any theory which preserved locality. The subsequent

---

[190] A **complete theory** would provide descriptive categories to account for all observable behavior and thus avoid any indeterminism. The existence of indeterminacy for some measurements is a characteristic of prevalent interpretations of quantum mechanics. Bounds for indeterminacy can be expressed in a quantitative form by the Heisenberg uncertainty principle.[188] Under the orthodox Copenhagen interpretation, quantum mechanics is nondeterministic, meaning that it generally does not predict the outcome of any measurement with certainty. Instead, it indicates what the probabilities of the outcomes are, with the indeterminism of observable quantities constrained by the uncertainty principle. Albert Einstein objected to the fundamentally probabilistic nature of quantum mechanics, and famously declared "I am convinced God does not play dice". Einstein, Podolsky, and Rosen had proposed their definition of a complete description as one that uniquely determines the values of all its measurable properties.

[191] In its simplest form, **Bell's theorem** states: "No physical theory of local hidden variables can ever reproduce all of the predictions of quantum mechanics". The theorem shows that, if local hidden variables exist, certain experiments could be performed involving quantum entanglement, where the result would satisfy a **Bell's inequality**. If, on the other hand, statistical correlations resulting from quantum entanglement could not be explained by local hidden variables, Bell's inequality would be violated. Any theory, such as quantum mechanics, that violates Bell's inequalities must abandon either locality[184] or realism.[189] Some physicists dispute that experiments have demonstrated Bell's violations, on the grounds that the sub-class of inhomogeneous Bell's inequalities has not been tested or due to experimental limitations in the tests

[192] **John Stewart Bell** (28 June 1928–1 October 1990) was an Northern Irish physicist. In 1964, he wrote a paper entitled "On the Einstein-Podolsky-Rosen Paradox". In this work, he showed that carrying forward EPR's analysis permits one to derive the famous Bell's theorem.[191] The resultant inequality, derived from certain assumptions, is violated by quantum theory. Bell held that not only local hidden variables, but any and all local theoretical explanations must conflict with the predictions of quantum theory. According to an alternative interpretation, not all local theories in general, but only local hidden variables theories (or local realist theories) have shown to be incompatible with the predictions of quantum theory.

practical experiments by John Clauser[193] and Stuart Freedman in 1972 seem (despite Clauser's initial espousal of Einstein's position) to definitively show that the effects of non-locality are real, and that "spooky actions at a distance" are indeed possible. In effect, non-locality and the related phenomenon of entanglement have been repeatedly demonstrated in laboratory experiments.[194] In 1981, Alain Aspect[195] and Paul Kwiat have performed experiments that have found violations of Bell's inequality[191] up to 242 standard deviations (excellent scientific certainty). This rules out local hidden variable theories, but does not rule out non-local ones.

It is worth noting that entanglement of a two-party state is necessary but not sufficient for that state to be non-local. It is important to recognize that entanglement is more commonly viewed as an algebraic concept, noted for being a precedent to non-locality as well as to quantum teleportation and to superdense coding, whereas non-locality is defined according to experimental statistics and is much more involved with the foundations and interpretations of quantum mechanics.

---

[193] **John Francis Clauser** (born 1 December 1942, Pasadena, California) is an American theoretical and experimental physicist, known for contributions to the foundations of quantum mechanics, in particular the Clauser-Horne-Shimony-Holt (CHSH) inequality, which can be used in the proof of Bell's theorem. The CHSH inequality states that certain consequences of entanglement in quantum mechanics cannot be reproduced by local hidden variable theories. Experimental verification of violation of the CHSH inequality is seen as experimental confirmation that nature cannot be described by local hidden variables theories. In 1972, working with Stuart Freedman, Clauser carried out the first experimental test of the CHSH-Bell's theorem predictions. This was the world's first observation of quantum entanglement and was the first experimental observation of a violation of a Bell inequality. In 1974, working with Michael Horne, he first showed that a generalization of Bell's Theorem provides severe constraints for all local realistic theories of nature (a.k.a. objective local theories). That work introduced the Clauser–Horne (CH) inequality as the first fully general experimental requirement set by local realism. It also introduced the "CH no-enhancement assumption", whereupon the CH inequality reduces to the CHSH inequality, and whereupon associated experimental tests also constrain local realism. In 1976 he carried out the world's second experimental test of the CHSH-Bell's Theorem predictions. John Clauser was awarded the Wolf Prize in Physics in 2010, together with Alain Aspect[195] (born 15 June 1947, in France) and Anton Zeilinger (born 20 May 1945, in Austria).

[194] One of the most typical experiments is that of creating a pair of electrons. Since the total spin of a quantum system must always cancel out to zero, if the two electrons are created together, one will have clockwise spin and the other will have anticlockwise spin. Under quantum theory, a superposition is also possible, so that the two electrons can be considered to simultaneously have spins of clockwise-anticlockwise and anticlockwise-clockwise respectively. If the pair are then separated by any distance (without observing and thereby decohering them) and then later checked, the second particle can be seen to instantaneously take the opposite spin to the first so that the pair maintains its zero total spin, no matter how far apart they may be.

[195] **Alain Aspect** (born 15 June 1947) is a French physicist, noted for his experimental work on quantum entanglement. In the early 1980s, while working on his PhD thesis from the lesser academic rank of lecturer, he performed the elusive Bell test experiments, which showed that Albert Einstein, Boris Podolsky and Nathan Rosen's reductio ad absurdum of quantum mechanics, namely that it implied "ghostly action at a distance", did in fact appear to be realized when two particles were separated by an arbitrarily large distance. A correlation between their wave functions remained, as they were once part of the same wave function, which was not disturbed before one of the child particles was measured.

6.3 Non-Local Computational Models in Differential Formulation

Non-local approaches were employed in various branches of physical sciences, for example, in optimization of slider bearings, or in modeling of liquid crystals, radiative transfer, electric wave phenomena in the cortex, and continuum mechanics. As far as the last branch is concerned, continuum mechanics, there exist two kinds of problem motivated by the need to improve the classical (local) continuum description in order to achieve non-locality: those with strain-softening[196] and those with no strain-softening at all. They all share the common need of modeling the size-effect,[197] which is impossible in the context of the classical plasticity. Only discrete numerical simulations, such as the random particle and lattice models, have succeeded in bringing to light the existence of a non-statistical size-effect.

Earlier studies on non-local elasticity were addressed to problems in which the size-effect is not caused by material softening. For the most part, these studies were motivated by homogenization of the atomic theory of Bravais.[198] They aimed at a better description of phenomena taking place in crystals, on a scale comparable to the range of interatomic forces. These studies showed that non-local continuum models approximately reproduce the dispersion of short elastic waves and improve the description of interactions between crystal defects such as vacancies, interstitial atoms, and dislocations.

The term "non-local" has in the past been used with two meanings, one narrow and one broad. In the narrow sense, it refers strictly to the models with an averaging integral. In the broad sense, it refers to all the constitutive models[128] that involve a characteristic length (material length), which also include the gradient models.

Generally speaking, integral-type non-local models replace one or more state variables by their non-local counterparts, obtained by weighted averaging over a spatial neighborhood of each point under consideration. This leads to an abandonment of the principle of local action.[184]

In gradient-type non-local models, the principle of local action is preserved and the field in the immediate vicinity of the point is taken into account by enriching the local constitutive relations with the first or higher gradients of some state variables or thermodynamic forces.

An internal length scale is incorporated into the material description of both the integral- and gradient-type non-local models.

---

[196] With reference to a stress-strain diagram, **strain-softening** is the decrease of stress[199] at increasing strains.[203]

[197] In classical Fracture Mechanics, the failure stress is considered as a material property. Consequently, geometrically similar structures of different shapes and sizes made from the same material, with the same microstructure, should fail at the same stress. In practice, however, structures of different sizes fail at different stresses. This behavior is called the **size effect**. By way of example, a larger beam will fail at a lower stress than a smaller beam, if they are made of the same material. The size effect can have two causes: statistical, due to material strength randomness, and energetic (non-statistical), due to energy release when a large crack or a large fracture process zone (FPZ) containing damaged material develops before the maximum load is reached.

[198] **Auguste Bravais** (23 August 1811, Annonay, Ardèche–30 March 1863, Le Chesnay, France) was a French physicist known for his work in crystallography, the conception of Bravais lattices, and the formulation of Bravais law. Bravais also studied magnetism, the northern lights, meteorology, geobotany, phyllotaxis, astronomy, and hydrography.

6.3.1 Continuum Mechanics

The classical continuum mechanics were developed together with local material laws, where the stress[199] at a given point uniquely depends on the current values, and also the previous history, of deformation[200] at that point only.

    The classical local continuum concept, leading to constitutive models falling within the category of simple non-polar materials,[201] does not seem to be adequate for modeling heterogeneous materials[202] in the context of the classical differential formulation. The reason for this was found to lie just in the local nature of the constitutive relations between stress and strain tensors.[203] It was argued that the local constitutive relations are not adequate for describing the mechanical behavior of solids since no material is an ideal continuum, decomposable into a set of infinitesimal material volumes, each of which can be described independently. In effect, all materials, natural and man-made, are characterized by microstructural details whose size ranges over many orders of magnitude. In constructing a material model, one must select a certain resolution level below which the microstructural details are not explicitly visible. Instead of refining the explicit resolution level, it is often more effective to use various forms of generalized continuum formulation, dealing with material that are non-simple or polar, or both.

---

[199] The **stress** across a surface element, S, is the force that the material on one side exerts on the material on the other side, divided by the area of the surface. The direction and magnitude of the stress at a point generally depend on the orientation of S. Thus, the stress state of the material must be described by a tensor,[42] called the (Cauchy) stress tensor, which is a linear function that relates the normal vector, n, of the surface S to the stress across S. Stress inside a body may arise by various mechanisms, such as the reaction to external forces applied to the bulk material (like gravity) or to its surface (like contact forces, external pressure, or friction). Stress may exist in the absence of external forces, leading to a **built-in stress**. This stress is important, for example, in prestressed concrete and tempered glass. Stress may also be imposed on a material without the application of net forces, for example by changes in temperature or chemical composition, or by external electromagnetic fields (as in piezoelectric and magnetostrictive materials). Stress that exceeds certain strength limits of the material will result in permanent deformation (such as plastic flow, fracture, cavitation) or even change its crystal structure and chemical composition.

[200] **Deformation** in continuum mechanics is the transformation of a body from a *reference* configuration to a *current* configuration. A configuration is a set containing the positions of all particles of the body. A deformation may be caused by external loads, body forces (such as gravity or electromagnetic forces), or temperature changes within the body. Deformations which are recovered after the stress field[199] has been removed are called **elastic deformations**. In this case, the continuum completely recovers its original configuration. On the other hand, irreversible deformations remain even after stresses have been removed. One type of irreversible deformation is **plastic deformation**, which occurs in material bodies after stresses have attained a certain threshold value, known as the elastic limit or yield stress, and is the result of slip, or dislocation mechanisms at the atomic level. Another type of irreversible deformation is **viscous deformation**, which is the irreversible part of viscoelastic deformation.

[201] Materials that exhibit body couples and couple stresses in addition to moments produced exclusively by forces are called **polar materials**. **Non-polar materials** are then those materials with only moments of forces. In the classical branches of continuum mechanics the development of the theory of stresses is based on non-polar materials.

[202] A **heterogeneous material** is composed of dissimilar parts; hence the constituents are of a different kind and obey to different constitutive relations.

[203] **Strain** is a description of deformation[200] in terms of *relative* displacement of particles in the body that excludes rigid-body motions. More precisely, a strain is a normalized measure of deformation, representing the displacement between particles in the body relative to a reference length. Hence strains are dimensionless and are usually expressed as a decimal fraction, a percentage or in parts-per notation. A strain is in general a tensor quantity.[42] Different equivalent choices may be made for the expression of a strain field depending on whether it is defined with respect to the initial or the final configuration of the body and on whether the metric tensor or its dual is considered.

Some preliminary ideas on non-local elasticity can be traced back to the late 19th century. Beginning with Krumhansl (1965), Rogula (1965), Eringen (1966), Kunin (1966), and Kröner (1968), the idea was promulgated that heterogeneous materials should properly be modeled by some type of non-local continuum. Non-local continua are continua in which the stress at a certain point is not a function of the strain at the same point, but a function of the strain distribution over a certain representative volume of the material centered at that point. Therefore, non-locality is tantamount to an abandonment of the principle of the local action of classical continuum mechanics.

In computational continuum mechanics, non-local models are differential numerical models that take into account possible interactions between the given point and other material points. Theoretically, the stress at a point can depend on the strain history in the entire body, but the long range interactions certainly diminish with increasing distance, and can be neglected when the distance exceeds a certain limit called the interaction radius R. The interval, circle or sphere, of radius R is called the domain of influence.

Physical justifications of the non-locality well-posedness[204] may be summarized as follows:

1. Homogenization of the heterogeneous microstructure on a scale sufficiently small for it to be impossible to consider the smoothed strain field as uniform.
2. Homogenization of regular or statistically regular lattices or frames.
3. Need to capture the size-effects observed in experiments and in discrete simulations.
4. Impossibility to simulate numerically the observed distributed cracking with local continuum models.
5. Dependence of the microcrack growth on the average deformation of a finite volume surrounding the whole microcrack, and not on the local stress or strain tensor at the point corresponding to the microcrack center.
6. Microcrack interaction, leading to either amplification of the stress intensity factor or crack shielding depending on the orientations of the microcracks, the orientation of the vectors joining the centers, and the size of the microcracks.

---

[204] The mathematical term **well-posed problem** stems from a definition given by the French mathematician Jacques Salomon Hadamard (8 December 1865–17 October 1963). He believed that mathematical models of physical phenomena should have the properties that:

1. A solution exists.
2. The solution is unique.
3. The solution's behavior hardly changes when there's a slight change in the initial condition (topology).

Problems that are not well-posed in the sense of Hadamard are termed **ill-posed**. Inverse problems are often ill-posed. Even if a problem is well-posed, it may still be **ill-conditioned**, meaning that a small error in the initial data can result in much larger errors in the answers. An ill-conditioned problem is indicated by a large **condition number**, where the condition number of a function with respect to an argument measures how much the output value of the function can change for a small change in the input argument. If the problem is well-posed, then it stands a good chance of solution on a computer using a stable algorithm.[165] If it is not well-posed, it needs to be re-formulated for numerical treatment. Typically this involves including additional assumptions, such as smoothness of solution. This process is known as **regularization**.

7. Density of geometrically necessary dislocations in metals, whose effect, after continuum smoothing, naturally leads to a first-gradient model (metal plasticity).

8. Paradoxical situations or incorrect predictions arising from a Wiebull-type weakest link theory of quasi-brittle structural failure[205] on the assumption that the failure probability at a point of a material depends on the continuum stress at the point, rather than on the average strain from a finite neighborhood of that point.

Some studies have also been made to justify the characteristic length in the non-local approach by microstructure.

The enrichment of the classical continuum by incorporating non-local effects into the constitutive equations has the great advantage to avoid the ill-posedness[204] of boundary value problems with strain-softening constitutive models. Strain-softening is one of the most remarkable causes of spurious solutions in modeling heterogeneous materials, since, when the material tangent stiffness matrix ceases to be positive definite, the governing differential equations may lose ellipticity (Fig. 6.1, Section 6.1). Finite element solutions of such problems exhibit a pathological sensitivity to the element size and do not converge to physically meaningful solutions as the mesh is refined. In fact, the boundary value problem does not have a unique solution with continuous dependence on the given data.

From experimental tests on heterogeneous brittle materials[206] with traditional identification process, it appears that the strain-softening zone is of finite size and dissipates a finite amount of energy. However, when strain-softening is applied in conjunction with the classical local continuum concept and the differential formulation, the strain-softening zone is found to localize, in those simple cases for which exact solutions have been found, into a zone of zero thickness. Thus, the numerical solution by finite element converges with mesh refinement to a strain-softening zone of zero thickness and to zero energy dissipated by failure. Strain localizes into a narrow band whose width depends on the element size and tends to zero as the mesh is refined. This is not a realistic result. The corresponding load–displacement diagram always exhibits snapback for a sufficiently fine mesh, independent of the size of the structure and of the ductility of the material. To remedy the loss of ellipticity, a length scale is often incorporated, implicitly, or explicitly, into the material description or the formulation of the boundary value problem.

Incorporating a length scale remedies the loss of ellipticity because the actual width of the zone of localized plastic strain is related to the heterogeneous material microstructure and can be correctly predicted only by models having a parameter with the dimension of length. A properly formulated enhancement has a regularizing effect[204] in differential formulations, since it acts as a localization limiter restoring the well-posedness of the boundary value problem. As far as the reasons for the absence of a length scale from standard theories of elasticity or plasticity are concerned, it has been discussed in Section 1.1 how they follow directly from performing

---

[205] **Quasi-brittle failure** is undergone by **quasi-brittle materials**, that is, materials that possess considerable hardness, which is similar to ceramic hardness, so often it is called ceramic hardness. In modern fracture mechanics, concrete is considered as a quasi-brittle material.

[206] A material is **brittle** if, when subjected to stress,[199] it breaks without significant deformation.[200] Brittle materials include most ceramics and glasses (which do not deform plastically) and some polymers, such as PMMA and polystyrene. Many steels become brittle at low temperatures, depending on their composition and processing.

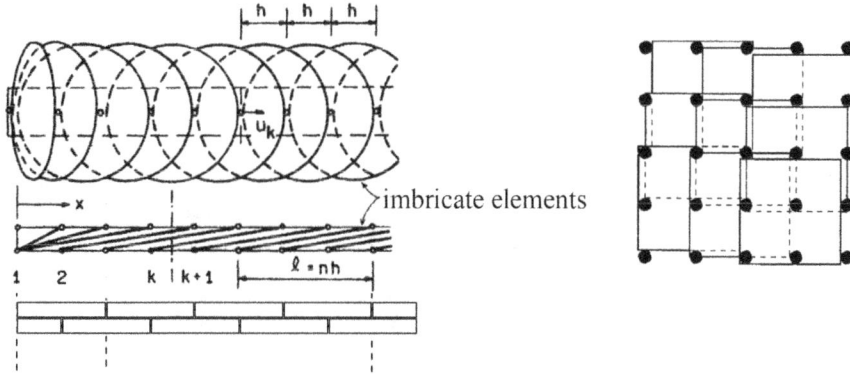

Figure 6.9. Two examples of imbrication (overlapping) of finite elements. (Bažant et al., 1984)

the limit process. The lack of a length scale is thus directly bonded to the use of the differential formulation, not to the physical problem in itself.

Early extensions of the non-local concept to strain-softening material are due to Bažant,[207] in 1984. He applied the idea of using the staggered elements, originally introduced by Yee for avoiding spurious solutions (Section 6.1), in the regularization of material instability problems of strain-softening materials, leading to the so-called imbricate continuum. Imbricate elements overlap like tiles on a roof (Fig. 6.9). In the FEM, imbrication is a way to provide the continuum with non-locality properties. Imbricate elements were later improved by the non-local damage theory and adapted for concrete.

Non-local formulations were elaborated for a wide spectrum of models, including softening plasticity, hardening crystal plasticity, progressively cavitating porous plastic solids, smeared crack models, and microplane models. Form the purely phenomenological point of view, the choice of the variable to be averaged remains to some extent arbitrary. This leads to a great number of possible non-local formulations. Nevertheless, one must be careful when selecting a certain formulation from the literature, because almost all of them capture the onset of localization properly, but some fail to give physically reasonable results at later stages of the localization process. Moreover, the basic model with damage evolution driven by the damage energy release rate is not suitable for quasi-brittle materials, since it gives the same response in tension and in compression. A number of non-local damage formulations of the simple isotropic damage model with one scalar damage variable appeared during the last decades, aimed at emphasizing the effect of tension on the propagation of cracks. Nevertheless, a unified non-local formulation applicable to any inelastic constitutive model with strain-softening as a reliable localization limiter is not available, at present.

Most non-local damage formulations lead to a progressive shrinking of the zone in which local strains increase. This is not realistic, since the thickness of the zone of increasing damage can never be smaller than the support diameter of the non-local weight function. Numerical

---

[207] **Zdeněk Pavel Bažant** (born December 10, 1937) is McCormick School Professor and Walter P. Murphy Professor of Civil Engineering and Materials Science in the Department of Civil and Environmental Engineering at Northwestern University's Robert R. McCormick School of Engineering and Applied Science.

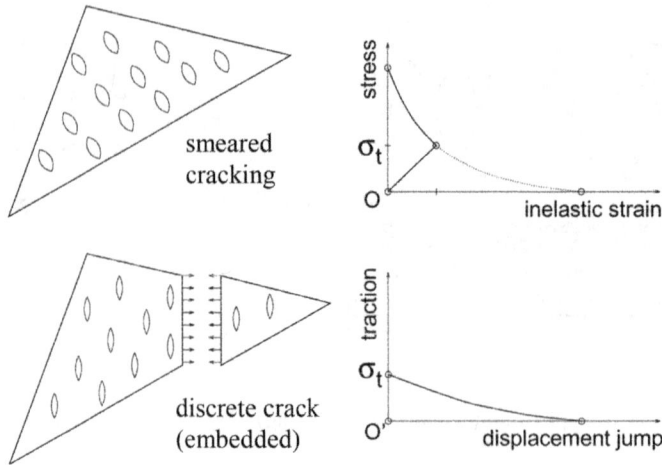

Figure 6.10. Transition from a continuum model to a discontinuity. (Jirásek, 1999)

problems thus occur, when the residual stiffness of the material inside this zone becomes too small. These numerical problems are all the more severe if body forces are present, leading to divergence of the equilibrium iteration process. Transition from highly localized strains to displacement discontinuities embedded in the interior of finite elements (Fig. 6.10) can be used to remedy the loss of convergence when body forces are present.

Finally, in Bažant and Chang (1984) and Jirásek and Rolshoven (2002), it was shown that numerical instabilities do not occur only if softening laws taking into account both the local and non-local effects are used. This means that the principle of the local action of the classical continuum mechanics must somehow be taken into account even in a non-local approach.

6.4 Algebraic Non-Locality of the CM

In Section 1.1, we have already pointed out that, avoiding the limit process and using global variables instead of field variables, the CM preserves the association between physical variables and geometrical objects provided with an extent. In the light of the former discussion on the opportunity of enriching the constitutive relations with internal length parameters (Section 6.3), we can now add that the CM does not need to incorporate any length scale into the constitutive relations, or somewhere else, precisely for the initial choice of avoiding the limit process, a choice that leads to a preservation of the length scales associated with the physical variables. In other words, by simply taking account of the association between global variables and extended space elements, the CM gives a direct non-local description of the continuum, without requiring any sort of enrichment of the constitutive laws with length scales, as is usually the case for non-local differential approaches in solid mechanics. As a consequence, in the CM non-locality attains to the physical phenomenon in its complex, and not necessarily to some type of material model.

Although it is not enriched with length scales, the material description of the CM is non-local in any case. This is due to the association of the configuration variables with the geometrical elements of one cell complex and the association of the source variables with the geometrical elements of a second cell complex, which is the dual of the first cell complex and

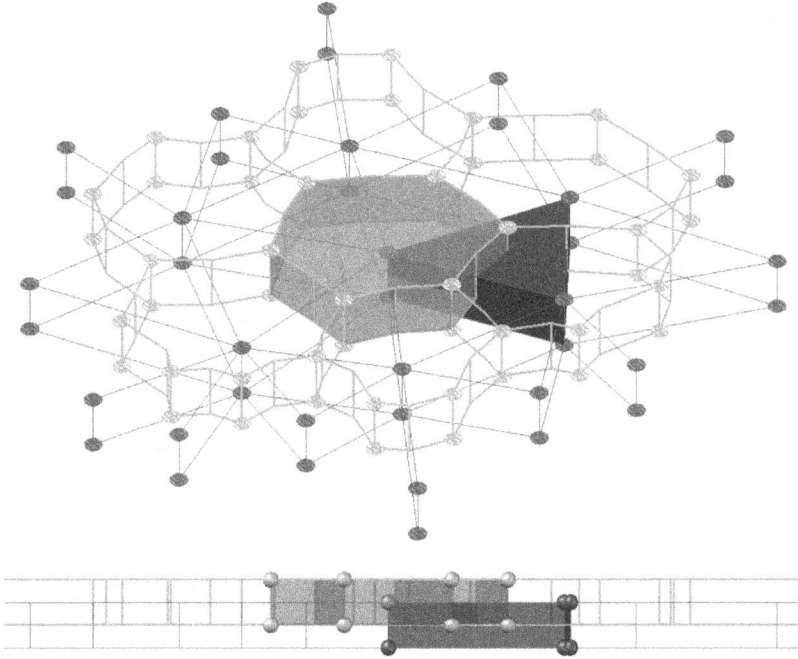

Figure 6.11. Staggering in 2D cell complexes with barycentric dual cells.

is staggered in the space. In particular, in order to preserve the geometrical content of cells, the primal mesh is provided with a thickness also in two-dimensional domains, which is a unit thickness (Fig. 6.11, Section 4.3). Moreover, since each dual volume must contain one node of the primal complex, in two-dimensional domains primal and dual cell complexes turn out to be shifted along the body thickness (Fig. 6.11), resembling the use of imbricate elements in FEM analysis (Section 6.3.1).

In Section 6.3.1, we have already pointed out that staggering is a computational technique used for achieving non-locality and avoiding spurious solutions in differential formulation. In the special case of the CM, staggering also takes on an energetic meaning, which allows us to provide an explanation of why it regularizes material instabilities. In fact, by recalling that the product between a configuration and a source variable is an energetic variable (Section 4.1), it follows immediately that having two staggered cell complexes, where the configuration variables are associated with the primal cell complex and the source variables are associated with the dual cell complex, implies that primal and dual elements always define a volume (in three-dimensional cell-complexes) or a surface (in two-dimensional cell-complexes), which has the meaning of an energy. Since these volumes and surfaces never degenerate until primal and dual elements are staggered, we can never incur in problems of localization with zero dissipated energy, as is instead the case of the differential formulation.

Staggering and the use of global variables are not the only reasons for the non-local nature of the CM computation. In fact, also the two-step coboundary process that characterizes any topological equation of the algebraic formulation plays a decisive role in this respect. In order to clarify this latter statement, let us consider, for example, the algebraic formulation of continuum mechanics (Fig. 6.12). The primal topological equations of the fundamental problem are kinematic equations, while the dual topological equations are equilibrium equations. In particular,

Displacements

Strains

Surface forces

Volume forces

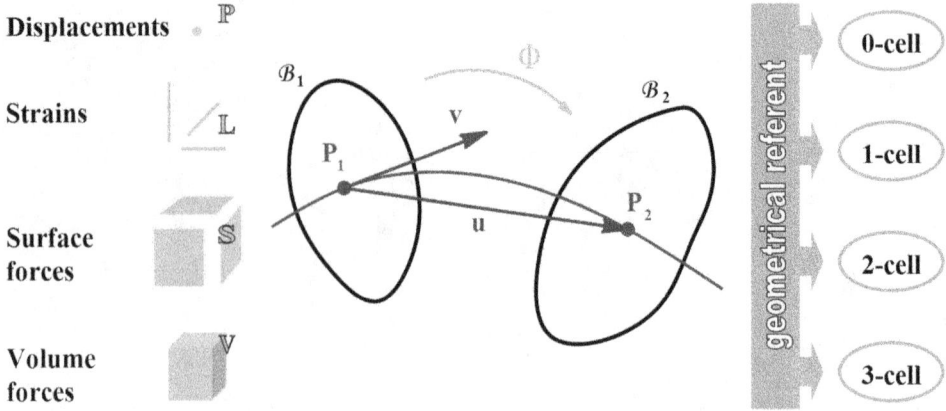

Figure 6.12. Association between space elements and variables in continuum mechanics.

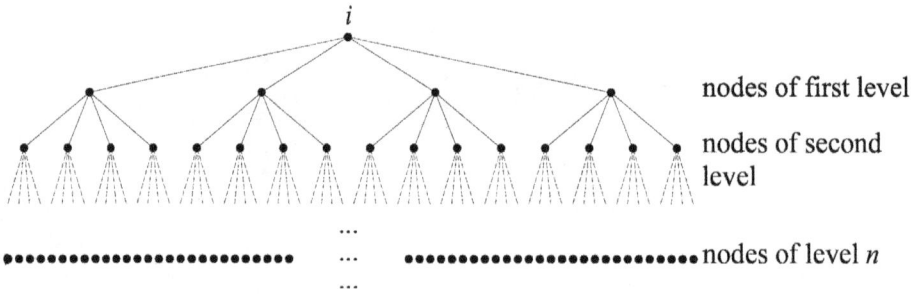

nodes of first level

nodes of second level

...

...

...

nodes of level n

Figure 6.13. Connections between several levels of the node-to-node interaction.

the equilibrium is a coboundary processes on a discrete p-form of degree two, the flux (Section 3.7), which can be described as follows:

- First step (on each 2-cell): spreading the normalized fluxes associated with all 2-cells to their cofaces, each multiplied by the relative incidence number.
- Second step (on each 3-cell): performing the algebraic sum of the normalized fluxes coming from the first step.

As a consequence, when we enforce balance in the CM, the global source variables involved in the balance of one 3-cell are also involved in the balance of all the surrounding 3-cells. This happens since each 3-cell is common to more than one 2-cell. Therefore, the balance at a given 3-cell does not depend on the current values, or previous history, of the global source variables at that 3-cell only, but on the current values, and previous history, taken by the global source variables in all the surrounding 3-cells. This establishes a sort of chain of interactions between nodes (Fig. 6.13), with the strain at each node influencing the stress at the given node by an amount which is proportional to its position into the chain, and interactions between nodes diminish with increasing distance. The existence of a chain of interactions gives non-local properties to the balance equations of the algebraic formulation, while the balance equations of the differential formulation are local.

Achieving non-locality in the CM balance equations is very important, as it enriches the description of physics given by the CM, compared to the descriptions given by any other numerical method, even those known as discrete methods. This enrichment follows on from the structure of the coboundary process given to the balance equations and from more than one 2-cell sharing the same 3-cell. That is, it is a consequence of the coboundary process, and there is no need to modify the balance equations in any way to provide them with non-local properties.

Being coboundary processes, even the kinematic equations are provided with non-local properties in the CM, while the kinematic equations of the differential formulation are local. We can therefore conclude that not only the global variables, but even all the governing equations of the algebraic formulation are non-local, since the CM enriches all the governing equations involved in a physical theory. In particular:

- The staggering between the elements of the primal cell complex, on which we compute the configuration variables, and the dual cell complex, on which we compute the source variables, provides non-locality to the algebraic relationships between configuration and source variables.
- The coboundary process provides non-locality to the algebraic balance and kinematic equations.

In conclusion, the CM enriches the computational description of all the physical theories of the macrocosm[208] by using notions of algebraic topology, as well as of mathematics. The consequence is that, by using the CM, there is no need to recover[209] non-locality a-posteriori, as for differential formulation. Non-locality is—we could say—intrinsic to algebraic formulation and is the result of using global instead of field variables, something that distinguishes the CM from any other numerical method, at present.

The new non-local formulation is desirable from a numerical point of view, since the numerical solution is reached sooner using discrete operators than with differential operators. Moreover, while a non-local differential formulation applicable to any inelastic constitutive model with strain-softening is not available, at present, and the non-local parameters need to be calibrated on the single physical phenomenon, the CM gives a unified formulation and does not need any parameter for providing a non-local description of physics. This achievement is of extreme importance just in the special case where the material under consideration is strain-softening. In fact, the intrinsic non-locality of the CM allows us to employ any local law[210]

---

[208] With the exception of the reversible thermodynamics, because it is the science of energy.

[209] Note that, if the incorporation of a length scale into the constitutive relationships is an attempt to recover the geometrical information that has been lost in performing the limit process, we have no guarantee of the two processes, performing the limit process and recovering the scale length, being perfectly dual as stands the geometrical content of the variables we are dealing with. This is a further reason for which preserving the length scales with an algebraic formulation is preferable to recovering one or more length scales with the differential formulation.

[210] See References Section for a list of papers by the author of this book, where a new local monotone non-decreasing material law, the **effective law**, has been successfully employed for modeling the size-effect in so-called strain-softening materials. The effective law is confirmed by experimental and analytical considerations. Its main contribution to the understanding of the failure mechanics of quasi-brittle materials,[205] in general, and concrete, in particular, is having reopened the question of strain-softening, whose existence and mathematical well-posedness[204] seemed to be no longer under discussion after the outcomes of the displacement controlled compression tests and the numerical successes of non-local differential approaches. The effective law is based on the idea that strain-softening is not a real material

(*continues next page*)

for describing the material behavior. In doing so, we can take into account both the local and the non-local effects and, according to what found by Bažant and Chang (1984) and Jirásek and Rolshoven (2002), this is sufficient for avoiding numerical instabilities in strain-softening modeling (Section 6.3.1). Therefore, the CM is a numerical method that provides a non-local description of physics without abandoning the principle of the local action altogether.

It is worth noting that, in the first differential theories of non-local elasticity, developed by Eringen and Edelen, non-locality was a property of the elastic problem in its complex, and not solely of its constitutive relations. In these non-locality theories, there was already the idea that non-locality is a property of the physical variables. In particular, the theories of non-local elasticity advanced by Eringen and Edelen in the early 1970s attributed a non-local character to body forces, mass, entropy, and internal energy. These are all global variables whose geometrical referent is a volume. It is thus clear that they, like all variables whose geometrical referent is more than zero-dimensional, cannot be properly described in a context, the differential formulation, in which all variables are related to points.

The idea of the non-local nature of the physical variables was not developed further, since the theories of non-local elasticity were too complicated to be calibrated and verified experimentally, let alone to be applied to any real problems. Treating only the stress–strain relationships as non-local, while the equilibrium and kinematic equations and their corresponding boundary conditions retain their standard form, was something needed later, to provide a practical formulation of these early theories. Consequently, incorporating the length scale into the constitutive relations only is the practical simplification of a more general theory and has no evident justification from a physical point of view. In this sense, we can state that the CM provides a physically more appealing non-local formulation, if compared to non-local differential approaches.

As far as electrodynamics is concerned, the close relationship between the CM space/time tesseract and the Minkowski space[61] (Section 3.8), with space and time that are intermingled in a four-dimensional space-time in both cases, allows us to use the space/time coboundary processes of the CM (Sections 5.1.1–5.1.4) for providing an algebraic description of electrodynamics that, just like the treatment of Minkowski, is based on a geometrical view of space/time. Therefore, this algebraic description can be framed in the set of the direct interaction theories[183] and, as a consequence, it is able to explain apparent action at a distance (quantum non-locality) by appealing exclusively to geometry, in contrast with Maxwell's field theory.

It is also worth noting that, in his lecture series on "The Meaning of Relativity", Einstein hopes in a purely algebraic unifying gravitational theory. He wrote:

"One can give good reasons why reality cannot at all be represented by a continuous field. From the quantum phenomena it appears to follow with certainty that a finite

---

(*continues from previous page*)

property, such as argued in several theoretical papers of last century, particularly of the 1980s (Hadamard, 1903; Hudson et al., 1971; Dresher and Vardoulakis, 1982; Bergan, 1983; Hegemier and Read, 1983; Sandler and Wright, 1983; Wu and Freud, 1983). The identification procedure of the effective law does not consist of a mere scale factor: the material is separated from the structure scale and the constitutive behavior is no more the mirror image of a structural problem at a lower scale. This results in a size-effect insensitive effective law between **effective strain** and **effective stress**.

system of finite energy can be completely described by a finite set of numbers (quantum numbers). This does not seem to be in accordance with a continuum theory and must lead to an attempt to find a purely algebraic theory for the representation of reality. But nobody knows how to find the basis for such a theory."

Well, if time, space, and energy are secondary features derived from a substrate below the Planck scale,[211] then Einstein's hypothetical algebraic system might resolve the EPR paradox (Section 6.2), although Bell's theorem would still be valid.

In conclusion, the classification diagram of the CM algebraic formulation puts into evidence the common mathematical structure that underlies classical and relativistic physical theories of the macrocosm and leads to a formulation that is non-local in itself. The classification has made it possible since the combinations between space and time elements are in finite number for the physical variables of the macrocosm (Fig. 4.1), exactly as the set of numbers that completely describes a finite system of finite energy is a finite set, in the words of Einstein. In the spirit of Einstein's hope in a purely algebraic description of reality, we can expect that the algebraic formulation of the CM can be extended to the microcosm, but this matter, although it sheds light on the potentialities of the CM, is outside the scope of this book.

---

[211] In particle physics and physical cosmology, the **Planck scale**, named after Max Karl Ernst Ludwig Planck (April 23, 1858–October 4, 1947), is an energy scale around 1.22×10^{19} GeV (which corresponds by the mass-energy equivalence to the Planck mass 2.17645×10^{-8} kg) at which quantum effects of gravity become strong. At this scale, present descriptions and theories of sub-atomic particle interactions in terms of quantum field theory break down and become inadequate, due to the impact of the apparent non-renormalizability of gravity within current theories.

REFERENCES

Alireza, H. M., S. Vijaya, and F. H. William. 1991. "Computation of Electromagnetic Scattering and Radiation Using a Time-Domain Finite-Volume Discretization Procedure." *Computer Physics Communications* 68 (1): 175–196.

Alotto, P., F. Freschi, and M. Repetto. 2010. "Multiphysics Problems via the Cell Method: The Role of Tonti Diagrams." *IEEE Transactions on Magnetics* 46 (8): 2959–2962. doi: http://dx.doi.org/10.1109/TMAG.2010.2044487

Alotto, P., F. Freschi, M. Repetto, and C. Rosso. 2013a. "Tonti diagrams." In *The Cell Method for Electrical Engineering and Multiphysics Problems, Lecture Notes in Electrical Engineering* 230. Springer-Verlag Berlin Heidelberg. doi: http://dx.doi.org/10.1007/978-3-642-36101-2

Alotto, P., F. Freschi, M. Repetto, and C. Rosso. 2013b. "Topological equations." In *The Cell Method for Electrical Engineering and Multiphysics Problems, Lecture Notes in Electrical Engineering* 230. Springer-Verlag Berlin Heidelberg. doi: http://dx.doi.org/10.1007/978-3-642-36101-2

Anderson, T. L. 2004. *Fracture Mechanics, Fundamentals and Applications* 3rd ed. Taylor & Francis Group.

Antman, S. S. 1995. "Nonlinear problems of elasticity." In *Applied Mathematical Sciences* 107. New York, Springer-Verlag.

Artin, M. 1991. *Algebra.* New Jersey, Prentice Hall.

Aspect, A., P. Grangier, and G. Roger. 1982. "Experimental Realization of Einstein-Podolsky-Rosen-Bohm Gedankenexperiment: A New Violation of Bell's Inequalities." *Physical Review Letters* 49 (2): 91–94. doi: http://dx.doi.org/10.1103/PhysRevLett.49.91

Ayres, F., and E. Mandelson. 2009. *Calculus* (*Schaum's Outlines Series*) 5th ed. Mc Graw Hill.

Bain, J. 2006. "Spacetime structuralism: §5 Manifolds *vs.* geometric algebra." In *The Ontology of Spacetime*, edited by Dennis Dieks. Elsevier.

Barut, A. O. 1964. *Electrodynamics and Classical Theory of Fields and Particles.* Courier Dover Publications.

Baylis, W. E. 1996. *Clifford (Geometric) Algebra with Applications to Physics, Mathematics, and Engineering.* Birkhäuser. doi: http://dx.doi.org/10.1007/978-1-4612-4104-1

Baylis, W. E. 2002. *Electrodynamics: A Modern Geometric Approach* 2nd ed. Birkhäuser.

Bažant, Z. P. 1991. "Why Continuum Damage is Nonlocal: Micromechanics Arguments." *Journal of Engineering Mechanics* 117 (5): 1070–1087. doi: http://dx.doi.org/10.1061/(ASCE)0733-9399(1991)117:5(1070)

Bažant, Z. P. 1994. "Nonlocal Damage Theory based on Micromechanics of Crack Interactions." *Journal of Engineering Mechanics* 120 (3): 593–617. doi: http://dx.doi.org/10.1061/(ASCE)0733-9399(1994)120:3(593)

Bažant, Z. P., and F-B. Lin. 1988a. "Nonlocal Smeared Cracking Model for Concrete Fracture." *Journal of Structural Engineering* 114 (11): 2493–2510. doi: http://dx.doi.org/10.1061/(ASCE)0733-9445(1988)114:11(2493)

Bažant, Z. P., and F-B. Lin. 1988b. "Nonlocal Yield-Limit Degradation." *International Journal for Numerical Methods in Engineering* 26: 1805–1823. doi: http://dx.doi.org/10.1002/nme.1620260809

Bažant, Z. P., and G. Pijaudier-Cabot. 1988. "Nonlocal Continuum Damage, Localization Instability and Convergence." *Journal of Applied Mechanics* 55: 287–293. doi: http://dx.doi.org/10.1115/1.3173674

Bažant, Z. P., and G. Pijaudier-Cabot. 1989. "Measurement of the Characteristic Length of Nonlocal Continuum." *Journal of Engineering Mechanics* 113: 2333–2347.

Bažant, Z. P., and J. Ožbolt. 1990. "Nonlocal Microplane Model for Fracture, Damage, and Size Effect in Structures." *Journal of Engineering Mechanics* 116 (11): 2485–2505. doi: http://dx.doi.org/10.1061/(ASCE)0733-9399(1990)116:11(2485)

Bažant, Z. P., and M. Jirásek. 2002. "Nonlocal Integral Formulations of Plasticity and Damage: Survey of Progress." *Journal of Engineering Mechanics* 128 (11): 1119–1149. doi: http://dx.doi.org/10.1061/(ASCE)0733-9399(2002)128:11(1119)

Bažant, Z. P., and T. P. Chang. 1984. "Is Strain-Softening Mathematically Admissible?" *Proceedings 5th Engineering Mechanics Division* 2: 1377–1380.

Bažant, Z. P., M. R. Tabbara, M. T. Kazemi, and G. Pijaudier-Cabot. 1990. "Random Particle Model for Fracture of Aggregate or Fiber Composites." *Journal of Engineering Mechanics* 116 (8): 1686–1705.

Bažant, Z. P., T. B. Belytschko, and T. P. Chang. 1984. "Continuum Model for Strain Softening." *Journal of Engineering Mechanics* 110 (12): 1666–1692. doi: http://dx.doi.org/10.1061/(ASCE)0733-9399(1984)110:12(1666)

Bellina, F., P. Bettini, E. Tonti, and F. Trevisan. 2002. "Finite Formulation for the Solution of a 2D Eddy-Current Problem." *IEEE Transaction on Magnetics* 38 (2): 561–564. doi: http://dx.doi.org/10.1109/20.996147

Belytschko, T., Y. Krongauz, D. Organ, M. Fleming, and P. Krysl. 1996. "Meshless Methods: An Overview and Recent Developments." *Computer Methods in Applied Mechanics Engineering* 139: 3–47. doi: http://dx.doi.org/10.1016/S0045-7825(96)01078-X

Belytschko, T., Z. P. Bažant, Y-W. Hyun, and T-P. Chang. 1986. "Strain Softening Materials and Finite-Element Solutions." *Computers & Structures* 23: 163–180. doi: http://dx.doi.org/10.1016/0045-7949(86)90210-5

Bergan, P. G. 1983. "Record of the Discussion on Numerical Modeling." *IUTAM W. Prager Symposium*, Northwestern University, Evanston, Ill.

Birdsall, C. K., and A. B. Langdon. 1985. *Plasma Physics via Computer Simulations*. McGraw-Hill Book Company.

Bishop, R., and S. I. Goldberg. 1980. *Tensor Analysis on Manifolds*. Courier Dover Publications.

Blázquez, A., and F. París. 2011. "Effect of numerical artificial corners appearing when using BEM on contact stresses." *Engineering Analysis with Boundary Elements* 35 (9): 1029–1037.

Boe, C., J. Rodriguez, C. Plazaola, I. Banfield, A. Fong, R. Caballero, and A. Vega. 2013. "A Hydrodynamic Assessment of a Remotely Operated Underwater Vehicle Based on Computational Fluid Dynamic – Part 1 – Numerical Simulation." *CMES: Computer Modeling in Engineering & Sciences* 90 (2): 165–177.

Bohm, D., B. J. Hiley, and A. Stuart. 1970. "On a New Mode of Description in Physics." *International Journal of Theoretical Physics* 3 (3): 171–183. doi: http://dx.doi.org/10.1007/BF00671000

Bolotin, V. V. 1963. *Nonconservative Problems of the Theory of Elastic Stability*. Oxford, Pergamon.

Borino, G., P. Fuschi, and C. Polizzotto. 1999. "A Thermodynamic Approach to Nonlocal Plasticity and Related Variational Approaches." *Journal of Applied Mechanics* 66: 952–963.

Bott, R., and L. W. Tu. 1982. *Differential Forms in Algebraic Topology*. Berlin, New York, Springer-Verlag.

Bourbaki, N. 1988. *Algebra*. Berlin, New York, Springer-Verlag.

Bourbaki, N. 1989. *Elements of Mathematics, Algebra I*. Springer-Verlag.

Brace, W. F., B. W. Paulding, and C. Scholz. 1966. "Dilatancy in the Fracture of Crystalline Rocks." *Journal of Geophysical Research* 71 (16): 3939–3953. doi: http://dx.doi.org/10.1029/JZ071i016p03939

Branin, F. H. Jr. 1966. "The Algebraic Topological Basis for Network Analogies and the Vector Calculus." *Proceedings Symposium on Generalized Networks*: 435–487, Brooklyn Polytechnic Institute.

Bryant, R. L., S. S. Chern, R. B. Gardner, H. L. Goldschmidt, and P. A. Griffiths. 1991. *Exterior Differential Systems*. Springer-Verlag. doi: http://dx.doi.org/10.1007/978-1-4613-9714-4

Burke, W. L. 1985. *Applied Differential Geometry*. Cambridge, Cambridge University Press. doi: http://dx.doi.org/10.1017/CBO9781139171786

Butcher, J. C. 2003. *Numerical Methods for Ordinary Differential Equations*. New York, John Wiley & Sons.

Cai, Y. C., J. K. Paik, and S. N. Atluri. 2010. "Locking-Free Thick-Thin Rod/Beam Element for Large Deformation Analyses of Space-Frame Structures, Based on the Reissner Variational Principle and a von Karman Type Nonlinear Theory." *CMES: Computer Modeling in Engineering & Sciences* 58 (1): 75–108.

Cai, Y. C., L. G. Tian, and S. N. Atluri. 2011. "A Simple Locking-Free Discrete Shear Triangular Plate Element." *CMES: Computer Modeling in Engineering & Sciences* 77 (3–4): 221–238.

Carslaw, H. S., and J. C. Jaeger. 1959. *Conduction of Heat in Solids* 2nd ed. Oxford University Press.

Catoni, F., D. Boccaletti, and R. Cannata. 2008. *Mathematics of Minkowski Space*. Birkhäuser Verlag, Basel.

Chandrasekhar, S. 1950. *Radiation Transfer*. Oxford University Press, London.

Chang, C–W. 2011. "A New Quasi-Boundary Scheme for Three-Dimensional Backward Heat Conduction Problems." *CMC: Computer Materials and Continua* 24 (3): 209–238.

Chen, C. Y., and C. Atkinson. 2009. "The Stress Analysis of Thin Contact Layers: A Viscoelastic Case." *CMES: Computer Modeling in Engineering & Sciences* 48 (3): 219–240.

Chen, J–S., C–T. Wu, and T. Belytschko. 2000. "Regularization of Material Instabilities by Meshfree Approximations with Intrinsic Length Scales." *International Journal for Numerical Methods in Engineering* 47: 1303–1322. doi: http://dx.doi.org/10.1002/(SICI)1097-0207(20000310)47:7<1303::AID-NME826>3.3.CO;2-X

Chen, Y., J. Cui, Y. Nie, and Y. Li. 2011. "A New Algorithm for the Thermo-Mechanical Coupled Frictional Contact Problem of Polycrystalline Aggregates Based on Plastic Slip Theory." *CMES: Computer Modeling in Engineering & Sciences* 76 (3): 189–206.

Clauser, J. F., M. A. Horne, A. Shimony, and R. A. Holt. 1969. "Proposed Experiment to Test Local Hidden-Variable Theories." *Physical Review Letters* 23 (15): 880–884. doi: http://dx.doi.org/10.1103/PhysRevLett.23.880

Conway, J. H., H. Burgiel, and C. Goodman-Strass. 2008. *The Symmetries of Things*. A. K. Peters Ltd.

Cosmi, F. 2000. "Applicazione del Metodo delle Celle con Approssimazione Quadratica." *Proceedings AIAS 2000*, Lucca, Italy: 131–140.

Cosmi, F. 2001. "Numerical Solution of Plane Elasticity Problems with the Cell Method." *CMES: Computer Modeling in Engineering & Sciences* 2 (3): 365–372.

Cosmi, F., and F. Di Marino. 2001. "Modelling of the Mechanical Behaviour of Porous Materials: A New Approach." *Acta of Bioengineering and Biomechanics* 3 (2): 55–66.

Courant, R., K. Friedrichs, and H. Lewy. 1956[1928]. *On the partial difference equations of mathematical physics*. AEC Research and Development Report, NYO-7689, New York, AEC Computing and Applied Mathematics Centre – Courant Institute of Mathematical Sciences.

Coxeter, H. S. M. 1973. *Regular Polytopes*. New York, Dover Publications, Inc.

Cromwell, P. R. 1999. *Polyhedra*. Cambridge University Press.

Curtis, C. W. 1968. *Linear Algebra*. Boston, Allyn & Bacon.

Davey, B. A., P. M. Idziak, W. A. Lampe, and G. F. McNulty. 2000. "Dualizability and Graph Algebras." *Discrete Mathematics* 214 (1): 145–172. doi: http://dx.doi.org/10.1016/S0012-365X(99)00225-3

Delić, D. 2001. "Finite Bases for Flat Graph Algebras." *Journal of Algebra* 246 (1): 453–469. doi: http://dx.doi.org/10.1006/jabr.2001.8947

Dong, L., and S. N. Atluri. 2011. "A Simple Procedure to Develop Efficient & Stable Hybrid/Mixed Elements, and Voronoi Cell Finite Elements for Macro- & Micromechanics." *CMC: Computer Material & Continua* 24 (1): 61–104.

Doran, C., and A. Lasenby. 2003. *Geometric Algebra for Physicists*. Cambridge University Press.

Dorst, L. 2002. *The Inner Products of Geometric Algebra*. Boston, MA, Birkhäuser Boston.

Dorst, L., D. Fontijne, and S. Mann. 2007. *Geometric Algebra for Computer Science: An Object-Oriented Approach to Geometry*. Amsterdam, Elsevier/Morgan Kaufmann.

Dresher, A., and I. Vardoulakis. 1982. "Geometric Softening in Triaxial Tests on Granular Material." *Geotechnique* 32 (4): 291–303. doi: http://dx.doi.org/10.1680/geot.1982.32.4.291

Drugan, W. J., and J. R. Willis. 1996. "A Micromechanics-Based Nonlocal Constitutive Equation and Estimates of Representative Volume Element Size for Elastic Composites." *Journal of the Mechanics and Physics of Solid* 44 (4): 497–524. doi: http://dx.doi.org/10.1016/0022-5096(96)00007-5

Duhem, P. 1893. "Le Potentiel Thermodynamique et la Pression Hydrostatique." *Annales Scientifiques De l'Ecole Normale Superieure* 10: 183–230.

Dummit, D. S., and R. M. Foote. 2003. *Abstract Algebra* 3rd ed. Wiley.

Edelen, D. G. B., A. E. Green, and N. Laws. 1971. "Nonlocal Continuum Mechanics." *Archive for Rational Mechanics and Analysis* 43 (1): 36–44. doi: http://dx.doi.org/10.1007/BF00251544

Ehle, B. L. 1969. *On Padé approximations to the exponential function and A-stable methods for the numerical solution of initial value problems.* Report 2010, University of Waterloo.

Einstein, A. 1948. "Quanten-Mechanik Und Wirklichkeit [Quantum Mechanics and Reality]." *Dialectica* 2 (3): 320–324. doi: http://dx.doi.org/10.1111/j.1746-8361.1948.tb00704.x

Einstein, A.; B. Podolsky, and N. Rosen. 1935. "Can Quantum-Mechanical Description of Physical Reality be Considered Complete?" *Physical Review* 47 (10): 777–780. doi: http://dx.doi.org/10.1103/PhysRev.47.777

Eringen, A. C. 1966. "A Unified Theory of Thermomechanical Materials." *International Journal of Engineering Science* 4 (2): 179–202. doi: http://dx.doi.org/10.1016/0020-7225(66)90022-X

Eringen, A. C. 1972. "Linear Theory of Nonlocal Elasticity and Dispersion of Plane Waves." *International Journal of Engineering Science* 10 (5): 425–435. doi: http://dx.doi.org/10.1016/0020-7225(72)90050-X

Eringen, A. C. 1983. "Theories of Nonlocal Plasticity." *International Journal of Engineering Science* 21 (7): 741–751. doi: http://dx.doi.org/10.1016/0020-7225(83)90058-7

Eringen, A. C., and B. S. Kim. 1974. "Stress Concentration at the Tip of a Crack." *Mechanics Research Communications* 1 (4): 233–237. doi: http://dx.doi.org/10.1016/0093-6413(74)90070-6

Eringen, A. C., and D. G. B. Edelen. 1972. "On Nonlocal Elasticity." *International Journal of Engineering Science* 10 (3): 233–248. doi: http://dx.doi.org/10.1016/0020-7225(72)90039-0

Eringen, A. C., C. G. Speziale, and B. S. Kim. 1977. "Crack-Tip Problem in Nonlocal Elasticity." *Journal of the Mechanics and Physics of Solids* 25 (5): 339–355. doi: http://dx.doi.org/10.1016/0022-5096(77)90002-3

Evans, L. C. 1998. *Partial Differential Equations.* American Mathematical Society.

Fenner, R. T. 1996. *Finite Element Methods for Engineers.* Imperial College Press, London. doi: http://dx.doi.org/10.1142/9781860943751

Fernandes, M. C. B., and J. D. M. Vianna. 1999. "On the generalized phase space approach to Duffin–Kemmer–Petiau particles." *Foundations of Physics* 29 (2): 201–219. doi: http://dx.doi.org/10.1023/A:1018869505031

Ferretti, E. 2001. *Modellazione del Comportamento del Cilindro Fasciato in Compressione.* Ph.D. Thesis (in Italian), University of Lecce, Italy.

Ferretti, E. 2003. "Crack Propagation Modeling by Remeshing using the Cell Method (CM)." *CMES: Computer Modeling in Engineering & Sciences* 4 (1): 51–72.

Ferretti, E. 2004a. "Crack-Path Analysis for Brittle and Non-Brittle Cracks: A Cell Method Approach." *CMES: Computer Modeling in Engineering & Sciences* 6 (3): 227–244.

Ferretti, E. 2004b. "A Cell Method (CM) Code for Modeling the Pullout Test Step-Wise." *CMES: Computer Modeling in Engineering & Sciences* 6 (5): 453–476.

Ferretti, E. 2004c. "A Discrete Nonlocal Formulation using Local Constitutive Laws." *International Journal of Fracture* (Letters section) 130 (3): L175–L182. doi: http://dx.doi.org/10.1007/s10704-004-2588-1

Ferretti, E. 2004d. "A Discussion of Strain-Softening in Concrete." *International Journal of Fracture* (Letters section) 126 (1): L3–L10. doi: http://dx.doi.org/10.1023/B:FRAC.0000025302.13043.4a

Ferretti, E. 2004e. "Experimental Procedure for Verifying Strain-Softening in Concrete." *International Journal of Fracture* (Letters section) 126 (2): L27–L34. doi: http://dx.doi.org/10.1023/B:FRAC.0000026384.55711.4a

Ferretti, E. 2004f. "On Poisson's Ratio and Volumetric Strain in Concrete." *International Journal of Fracture* (Letters section) 126 (3): L49–L55. doi: http://dx.doi.org/10.1023/B:FRAC.0000026587.43467.e6

Ferretti, E. 2004g. "On Strain-Softening in Dynamics." *International Journal of Fracture* (Letters section) 126 (4): L75–L82. doi: http://dx.doi.org/10.1023/B:FRAC.0000031188.52201.de

Ferretti, E. 2004h. "Modeling of the Pullout Test through the Cell Method." In G. C. Sih, L. Nobile. *RRRTEA - International Conference of Restoration, Recycling and Rejuvenation Technology for Engineering and Architecture Application*: 180–192. Aracne.

Ferretti, E. 2005a. "A Local Strictly Nondecreasing Material Law for Modeling Softening and Size-Effect: A Discrete Approach." *CMES: Computer Modeling in Engineering & Sciences* 9 (1): 19–48.

Ferretti, E. 2005b. "On nonlocality and locality: Differential and discrete formulations." *11th International Conference on Fracture - ICF11* 3: 1728–1733.

Ferretti, E. 2009. "Cell Method Analysis of Crack Propagation in Tensioned Concrete Plates." *CMES: Computer Modeling in Engineering & Sciences* 54 (3): 253–282.

Ferretti, E. 2012. "Shape-Effect in the Effective Law of Plain and Rubberized Concrete." *CMC: Computer Materials and Continua* 30 (3): 237–284.

Ferretti, E. 2013a. "The Cell Method: An Enriched Description of Physics Starting from the Algebraic Formulation." *CMC: Computer Materials and Continua* 36 (1): 49–72.

Ferretti, E. 2013b. "A Cell Method Stress Analysis in Thin Floor Tiles Subjected to Temperature Variation." *CMC: Computer Materials and Continua* 36 (3): 293–322.

Ferretti, E., A. Di Leo, and E. Viola. 2003a. A novel approach for the identification of material elastic constants. *CISM Courses and Lectures N. 471, Problems in Structural Identification and Diagnostic: General Aspects and Applications*: 117–131. Springer, Wien – New York.

Ferretti, E., A. Di Leo, and E. Viola. 2003b. Computational Aspects and Numerical Simulations in the Elastic Constants Identification. *CISM Courses and Lectures N. 471, Problems in Structural Identification and Diagnostic: General Aspects and Applications*: 133–147. Springer, Wien – New York.

Ferretti, E., and A. Di Leo. 2003. "Modelling of Compressive Tests on FRP Wrapped Concrete Cylinders through a Novel Triaxial Concrete Constitutive Law." *SITA: Scientific Israel – Technological Advantages* 5: 20–43.

Ferretti, E., and A. Di Leo. 2008. "Cracking and Creep Role in Displacement at Constant Load: Concrete Solids in Compression." *CMC: Computer Materials and Continua* 7 (2): 59–80.

Ferretti, E., E. Casadio, and A. Di Leo. 2008. "Masonry Walls under Shear Test: A CM Modeling." *CMES: Computer Modeling in Engineering & Sciences* 30 (3): 163–190.

Flanders, H. 1989. *Differential Forms with Applications to the Physical Sciences*. Dover Publications.

Fleming, W. 1987. *Functions of Several Variables* 3rd ed. New York, Springer-Verlag.

Freschi, F., L. Giaccone, and M. Repetto. 2008. "Educational value of the algebraic numeri-cal methods in electromagnetism." *COMPEL–The International Journal for Computation and Mathematics in Electrical and Electronic Engineering* 27 (6): 1343–1357. doi: http://dx.doi.org/10.1108/03321640810905828

Frescura, F. A. M., and B. J. Hiley. 1980a. "The Implicate Order, Algebras, and the Spinor." *Foundations of Physics* 10 (1–2): 7–31. doi: http://dx.doi.org/10.1007/BF00709014

Frescura, F. A. M., and B. J. Hiley. 1980b. "The Algebraization of Quantum Mechanics and the Implicate Order." *Foundations of Physics* 10 (9–10): 705–722. doi: http://dx.doi.org/10.1007/BF00708417

Frescura, F. A. M., and B. J. Hiley. 1981. "Geometric Interpretation of the Pauli Spinor." *American Journal of Physics* 49 (2): 152.

Frescura, F. A. M., and B. J. Hiley. 1984. "Algebras, Quantum Theory and Pre-Space." *Revista Brasileira de Fisica: 49–86. Volume Especial, Los 70 anos de Mario Schonberg*.

Fung, Y. C. 1977. *A First Course in Continuum Mechanics* 2nd ed. Prentice-Hall, Inc.

Gao, H., and Y. Huang. 2001. "Taylor-Based Nonlocal Theory of Plasticity." *International Journal of Solids and Structures* 38 (15): 2615–2637. doi: http://dx.doi.org/10.1016/S0020-7683(00)00173-6

Gardner, J. W., and R. Wiegandt. 2003. *Radical Theory of Rings*. Chapman & Hall/CRC Pure and Applied Mathematics.

George, P. L. 1995. "Automatic Mesh Generator using the Delaunay Voronoi Principle." *Surveys for Mathematics for Industry* 239–247.

Gilbarg, D., and N. Trudinger. 2001. *Elliptic Partial Differential Equations of Second Order*. Springer.

Gosset, T. 1900. *On the Regular and Semi-Regular Figures in Space of n Dimensions*. Messenger of Mathematics, Macmillan.

Griffiths, D. J. 1999. *Introduction to Electrodynamics* 3rd ed. New Jersey, Prentice Hall.

Grothendieck, A. 1973. *Topological Vector Spaces*. New York, Gordon and Breach Science Publishers.

Grünbaum, B. 2003. *Convex Polytopes*. Graduate Texts in Mathematics 221 2nd ed. Springer.

Gurson, A. L. 1977. "Continuum Theory of Ductile Rupture by Void Nucleation and Growth: Part I – Yield Criteria and Flow Rules for Porous Ductile Media." *Journal of Engineering Materials and Technology,* ASME 90: 2–15. doi: http://dx.doi.org/10.1115/1.3443401

Gurtin, M. E. 1981. *An Introduction to Continuum Mechanics.* New York, Academic Press.

Hadamard, J. 1903. *Leçons sur la Propagation des Ondes – Chapter VI.* Paris, France.

Hairer, E., C. Lubich, and G. Wanner. 2006. *Geometric Numerical Integration: Structure-Preserving Algorithms for Ordinary Differential Equations* 2nd ed. Springer Series in Computational Mathematics 31, Berlin, New York, Springer-Verlag.

Hallen, E. 1962. *Electromagnetic Theory*, Chapman & Hall.

Halmos, P. 1974. *Finite Dimensional Vector Spaces.* Springer.

Han, Z. D., H. T. Liu, A. M. Rajendran, and S. N. Atluri. 2006. "The Applications of Meshless Local Petrov-Galerkin (MLPG) Approaches in High-Speed Impact, Penetration and Perforation Problems." *CMES: Computer Modeling in Engineering & Sciences* 14 (2): 119–128.

Hartmann, S., R. Weyler, J. Oliver, J. C. Cante, and J. A. Hernández. 2010. "A 3D Frictionless Contact Domain Method for Large Deformation Problems." *CMES: Computer Modeling in Engineering & Sciences* 55 (3): 211–269.

Hazewinkel, M, ed. 2001a. "Contravariant Tensor." *Encyclopedia of Mathematics.* Springer.

Hazewinkel, M, ed. 2001b. "Covariant Tensor." *Encyclopedia of Mathematics.* Springer.

Hegemier, G. A., and H. E. Read. 1983. "Some Comments on Strain-Softening." *DARPA-NSF Workshop*, Northwestern University, Evanston, Ill.

Herstein, I. N. 1996. *Abstract Algebra* 3rd ed. Wiley.

Hesse, M. B. 1955. "Action at a Distance in Classical Physics." *Isis* 46 (4): 337–53. doi: http://dx.doi.org/10.1086/348429

Hesse, M. B. 1966. *Models and Analogies in Science* (revised ed.). Notre Dame, Indiana, Notre Dame University Press.

Hestenes, D. 1966. *Space-Time Algebra.* New York, Gordon and Breach.

Hestenes, D. 1999. *New Foundations for Classical Mechanics: Fundamental Theories of Physics* 2nd ed. Springer.

Hilbert, D. 1968. *Grundlagen der Geometrie* 10th ed. Stuttgart, Teubner. doi: http://dx.doi.org/10.1007/978-3-322-92726-2

Hiley, B. J. 2001. "A note on the role of idempotents in the extended Heisenberg algebra, Implications." *Scientific Aspects of ANPA* 22: 107–121, Cambridge.

Hiley, B. J. 2003. "Algebraic quantum mechanics, algebraic spinors and Hilbert space, Boundaries." In *Scientific Aspects of ANPA,* edited by K. G. Bowden. London: 149–186.

Hodgkin, A. L. 1964. *The Conduction of Nervous Impulses.* Thomas, Springfield, Ill.

Horodecki, R., P. Horodecki, M. Horodecki, and K. Horodecki. 2007. "Quantum entanglement." *Reviews of Modern Physics* 81 (2): 865–942. doi: http://dx.doi.org/10.1103/RevModPhys.81.865

Hu, X. Z., and F. H. Wittmann. 2000. "Size Effect on Toughness Induced by Cracks Close to Free Surface." *Engineering Fracture Mechanics* 65 (2–3): 209–211. doi: http://dx.doi.org/10.1016/S0013-7944(99)00123-X

Hudson, J. A., E. T. Brown, and C. Fairhurst. 1971. Shape of the Complete Stress-Strain Curve for Rock. *13th Symposium on Rock Mechanics*, University of Illinois, Urbana, Ill.

Huebner, K. H. 1975. *The Finite Element Method for Engineers.* Wiley.

Huerta, A., and G. Pijaudier-Cabot. 1994. "Discretization Influence on Regularization by two Localization Limiters." *Journal of Engineering Mechanics* 120 (6): 1198–1218. doi: http://dx.doi.org/10.1061/(ASCE)0733-9399(1994)120:6(1198)

Imai, R., and M. Nakagawa. 2012. "A Reduction Algorithm of Contact Problems for Core Seismic Analysis of Fast Breeder Reactors." *CMES: Computer Modeling in Engineering & Sciences* 84 (3): 253–281.

Jarak, T., and J. Sori\'c. 2011. "On Shear Locking in MLPG Solid-Shell Approach." *CMES: Computer Modeling in Engineering & Sciences* 81 (2): 157–194.

Jirásek, M. 1998a. "Embedded crack models for concrete fracture." In *Computational Modelling of Concrete Structures*, edited by R. de Borst, N. Bićanić, H. Mang, and G. Meschke: 291–300. Balkema, Rotterdam.

Jirásek, M. 1998b. "Nonlocal Models for Damage and Fracture: Comparison of Approaches." *International Journal of Solids Structures* 35 (31–32): 4133–4145. doi: http://dx.doi.org/10.1016/S0020-7683(97)00306-5

Jirásek, M. 1999. "Computational Aspects of Nonlocal Models." *Proceedings ECCM '99*: 1–10. München, Germany.

Jirásek, M., and B. Patzák. 2002. "Consistent Tangent Stiffness for Nonlocal Damage Models." *Computers & Structures* 80 (14–15): 1279–1293. doi: http://dx.doi.org/10.1016/S0045-7949(02)00078-0

Jirásek, M., and S. Rolshoven. 2003. "Comparison of Integral-Type Nonlocal Plasticity Models for Strain-Softening Materials." *International Journal of Engineering Science* 41 (13–14): 1553–1602. doi: http://dx.doi.org/10.1016/S0020-7225(03)00027-2

Jirásek, M., and T. Zimmermann. 1998. "Rotating Crack Model with Transition to Scalar Damage." *Journal of Engineering Mechanics* 124 (3): 277–284. doi: http://dx.doi.org/10.1061/(ASCE)0733-9399(1998)124:3(277)

Jirásek, M., and Z. P. Bažant. 1995. "Macroscopic Fracture Characteristics of Random Particle Systems." *International Journal of Fracture* 69: 201–228. doi: http://dx.doi.org/10.1007/BF00034763

Jirásek, M., and Z. P. Bažant. 2001. *Inelastic Analysis of Structures*. John Wiley and Sons.

Johnson, N. W. 1966. *The Theory of Uniform Polytopes and Honeycombs*. Ph.D. Dissertation, University of Toronto.

Jost, J. 2002. *Riemannian Geometry and Geometric Analysis*. Berlin, Springer-Verlag. doi: http://dx.doi.org/10.1007/978-3-662-04672-2

Kakuda, K., S. Obara, J. Toyotani, M. Meguro, and M. Furuichi. 2012. "Fluid Flow Simulation Using Particle Method and Its Physics-based Computer Graphics." *CMES: Computer Modeling in Engineering & Sciences* 83 (1): 57–72.

Kakuda, K., T. Nagashima, Y. Hayashi, S. Obara, J. Toyotani, N. Katsurada, S. Higuchi, and S. Matsuda. 2012. "Particle-based Fluid Flow Simulations on GPGPU Using CUDA." *CMES: Computer Modeling in Engineering & Sciences* 88 (1): 17–28.

Kelarev, A. V. 2003. *Graph Algebras and Automata*. New York, Marcel Dekker.

Kelarev, A. V., and O. V, Sokratova. 2003. "On congruences of automata defined by directed graphs." *Theoretical Computer Science* 301 (1–3): 31–43. doi: http://dx.doi.org/10.1016/S0304-3975(02)00544-3

Kelarev, A. V., and O. V. Sokratova. 2001. "Directed Graphs and Syntactic Algebras of Tree Languages." *Journal of Automata, Languages & Combinatorics* 6 (3): 305–311.

Kelarev, A. V., M. Miller, and O. V. Sokratova. 2005. "Languages Recognized by Two-Sided Automata of Graphs." *Proceedings of the Estonian Akademy of Science* 54 (1): 46–54.

Kiss, E. W., R. Pöschel, and P. Pröhle. 1990. "Subvarieties of varieties generated by graph algebras." *Acta Scientiarum Mathematicarium (Szeged)* 54 (1–2): 57–75.

Köthe, G. 1969. *Topological vector spaces*. Grundlehren der mathematischen Wissenschaften 159, New York, Springer-Verlag

Kröner, E. 1968. "Elasticity Theory of Materials with Long-Range Cohesive Forces." *International Journal of Solids Structures* 3: 731–742. doi: http://dx.doi.org/10.1016/0020-7683(67)90049-2

Krumhansl, J. A. 1965. "Generalized continuum field representation for lattice vibrations." In *Lattice dynamics*, edited by R. F. Wallis: 627–634. Pergamon, London. doi: http://dx.doi.org/10.1016/B978-1-4831-9838-5.50096-0

Kunin, I. A. 1966. "Theory of Elasticity with Spatial Dispersion." *Prikladnaya Matematika i Mekhanika* (in Russian) 30: 866.

Lam, T.-Y. 2005. *Introduction to Quadratic Forms over Fields*. Graduate Studies in Mathematics 67.

Lambert, J. D. 1977. "The initial value problem for ordinary differential equations." In *The State of the Art in Numerical Analysis*, edited by D. Jacobs: 451–501. New York, Academic Press.

Lambert, J. D. 1992. *Numerical Methods for Ordinary Differential Systems*. New York, Wiley.

Lang, S. 1972. *Differential manifolds*. Reading, Massachusetts–London–Don Mills, Ontario, Addison-Wesley Publishing Company, Inc.

Lang, S. 1987. *Linear algebra*. Berlin, New York, Springer-Verlag. doi: http://dx.doi.org/10.1007/978-1-4757-1949-9

Lang, S. 2002. *Algebra*. Graduate Texts in Mathematics 211 Revised 3rd ed. New York, Springer-Verlag.

Larson, R., R. P. Hosteler, and B. H. Edwards. 2005. *Calculus* 8th ed. Houghton Mifflin.

Lasenby, J., A. N. Lasenby, and C. J. L. Doran. 2000. "A Unified Mathematical Language for Physics and Engineering in the 21st Century." *Philosophical Transactions of the Royal Society of London* A358: 1–18. doi: http://dx.doi.org/10.1098/rsta.2000.0517

Latorre, D. R., J. W. Kenelly, I. B. Reed, and S. Biggers. 2007. *Calculus Concepts: An Applied Approach to the Mathematics of Change*. Cengage Learning.

Lax, P. 1996. *Linear Algebra*. Wiley-Interscience.

Leblond, J. B., G. Perrin, and J. Devaux. 1994. "Bifurcation Effects in Ductile Metals Incorporating Void Nucleation, Growth and Interaction." *Journal of Applied Mechanics*: 236–242. doi: http://dx.doi.org/10.1115/1.2901435

Lee, S.-M. 1988. "Graph algebras which admit only discrete topologies." *Congressus Numerantium* 64: 147–156.

Lee, S.-M. 1991. "Simple Graph Algebras and Simple Rings." *Southeast Asian Bulletin of Mathematics* 15 (2): 117–121.

Liu, C.-S. 2012. "Optimally Generalized Regularization Methods for Solving Linear Inverse Problems." *CMC: Computer Materials and Continua* 29 (2): 103–128.

Liu, C.-S., and S. N. Atluri. 2011. "An Iterative Method Using an Optimal Descent Vector, for Solving an Ill-Conditioned System Bx = b, Better and Faster than the Conjugate Gradient Method." *CMES: Computer Modeling in Engineering & Sciences* 80 (3–4): 275–298.

Liu, C.-S., H–H. Dai, and S. N. Atluri. 2011. "Iterative Solution of a System of Nonlinear Algebraic Equations F(x)=0, Using dot $\mathbf{x} = \lambda.[\alpha\mathbf{R} + \beta\mathbf{P}]$ or dot $\mathbf{x} = \lambda.[\alpha\mathbf{F} + \beta\mathbf{P}^*]$ R is a Normal to a Hyper-Surface Function of F, P Normal to R, and P* Normal to F." *CMES: Computer Modeling in Engineering & Sciences* 81 (3): 335–363.

Liu, C.-S., H–K. Hong, and S. N. Atluri. 2010. "Novel Algorithms Based on the Conjugate Gradient Method for Inverting Ill-Conditioned Matrices, and a New Regularization Method to Solve Ill-Posed Linear Systems." *CMES: Computer Modeling in Engineering & Sciences* 60 (3): 279–308.

Livesley, R. K. 1983. *Finite Elements, an Introduction for Engineers*. Cambridge University Press.

Lounesto, P. 2001. *Clifford Algebras and Spinors*. Cambridge, Cambridge University Press.

Lovelock, D., and R. Hanno. 1989 [1975]. *Tensors, Differential Forms, and Variational Principles*. Dover Publications. doi: http://dx.doi.org/10.1016/S0022-5096(00)00031-4

Luciano, R., and J. R. Willis. 2001. "Nonlocal Constitutive Response of a Random Laminate Subjected to Configuration-Dependent Body Force." *Journal of the Mechanics and Physics of Solids* 49 (2): 431–444.

Mac Lane, S., and G. Birkhoff. 1999. *Algebra*. AMS Chelsea.

Macdonald, A. 2011. *Linear and Geometric Algebra*. Charleston, CreateSpace.

Marrone, M. 2001a. "Convergence and Stability of the Cell Method with Non Symmetric Constitutive Matrices." *Proceedings 13th COMPUMAG*. Evian, France.

Marrone, M. 2001b. "Computational Aspects of Cell Method in Electrodynamics." *Progress in Electromagnetics Research*: 317–356. PIER 32 (Special Volume on *Geometrical Methods for Computational Electromagnetics*).

Marrone, M., V. F. Rodrìguez-Esquerre, and H. E. Hernàndez-Figueroa. 2002. "Novel Numerical Method for the Analysis of 2D Photonic Crystals: the Cell Method." *Optics Express* 10 (22): 1299–1304. doi: http://dx.doi.org/10.1364/OE.10.001299

Matoušek, J. 2002. *Lectures in Discrete Geometry*. Graduate Texts in Mathematics 212, Springer.

Mattiussi, C. 1997. "An Analysis of Finite Volume, Finite Element, and Finite Difference Methods using some Concepts from Algebraic Topology." *Journal of Computational Physics* 133: 289–309.

Mattiussi, C. 2000. "The finite volume, finite difference, and finite elements methods as numerical methods for physical field problems." In *Advances in Imaging and Electron Physics*, edited by P. Hawkes. 113: 1–146.

Mattiussi, C. 2001. "The Geometry of Time-Stepping." *Progress in Electromagnetics Research*: 123–149. PIER 32 (Special Volume on *Geometrical Methods for Computational Electromagnetics*).

Mavripilis, D. J. 1995. *Multigrid Techniques for Unstructured Meshes. Lecture, Series 1995–02, Computational Fluid Dynamics*, Von Karman Institute of Fluid Dynamics.

McNulty, G. F., and C. R. Shallon. 1983. Inherently nonfinitely based finite algebras. *Universal algebra and lattice theory* (*Puebla, 1982*): 206–231. Lecture Notes in Mathematics 1004, Berlin, New York, Springer-Verlag.

Micali, A., R. Boudet, and J. Helmstetter. 1989. *Clifford Algebras and their Applications in Mathematical Physics*, Workshop Proceedings: 2nd (Fundamental Theories of Physics), Kluwer.

Misner, C. W., K. S. Thorne, and J. A. Wheeler. 1973. *Gravitation*. W. H. Freeman.

Morton, K. W., and S. M. Stringer. 1995. Finite Volume Methods for Inviscid and Viscous Flows, Steady and Unsteady. *Lecture, Series 1995–02, Computational Fluid Dynamics*, Von Karman Institute of Fluid Dynamics.

Naber, G. L. 1992. *The Geometry of Minkowski Spacetime*. New York, Springer-Verlag. doi: http://dx.doi.org/10.1007/978-1-4757-4326-5

Nappi, A., and F. Tin-Loi. 1999. "A discrete formulation for the numerical analysis of masonry structures." In *Computational Mechanics for the Next Millennium*, edited by C. M. Wang, K. H. Lee, and K. K. Ang: 81–86. Elsevier, Singapore. doi: http://dx.doi.org/10.12989/sem.2001.11.2.171

Nappi, A., and F. Tin-Loi. 2001. "A Numerical Model for Masonry Implemented in the Framework of a Discrete Formulation." *Structural Engineering and Mechanics* 11 (2): 171–184.

Nappi, A., S. Rajgelj, and D. Zaccaria. 1997. "Application of the Cell Method to the Elastic-Plastic Analysis." *Proceedings Plasticity* 97: 14–18.

Nappi, A., S. Rajgelj, and D. Zaccaria. 2000. "A discrete formulation applied to crack growth problem." In *Mesomechanics 2000*, edited by G. G. Sih: 395–406. Tsinghua University Press, Beijing, P. R. China.

Needleman, A., and V. Tvergaard. 1998. "Dynamic Crack Growth in a Nonlocal Progressively Cavitating Solid." *European Journal of Mechanics A–Solids* 17 (3): 421–438. doi: http://dx.doi.org/10.1016/S0997-7538(98)80053-3

Newton, I. 1687. *Philosophiae Naturalis Principia Mathematica*.

Nilsson, C. 1994. "On Local Plasticity, Strain Softening, and Localization." *Report No. TVSM-1007*, Division of Structural Mechanics, Lund Institute of Technology, Lund, Sweden.

Nilsson, C. 1997. "Nonlocal Strain Softening Bar Revisited." *International Journal of Solids and Structures* 34: 4399–4419. doi: http://dx.doi.org/10.1016/S0020-7683(97)00019-X

Noll, W. 1972. A new Mathematical Theory of Simple Materials. *Archive for Rational Mechanics and Analysis* 48 (1): 1–50. doi: http://dx.doi.org/10.1007/BF00253367

Oates-Williams, S. 1984. "On the variety generated by Murskiĭ's algebra." *Algebra Universalis* 18 (2): 175–177. doi: http://dx.doi.org/10.1007/BF01198526

Okada, S., and R. Onodera. 1951. "Algebraification of Field Laws of Physics by Poincaré Process." *Bulletin of Yamagata University – Natural Sciences* 1 (4): 79–86.

Oseen, C. W. 1933. "The Theory of Liquid Crystals." *Transaction of Faraday Society* 29: 883–899.

Ožbolt, J., and Z. P. Bažant. 1996. "Numerical Smeared Fracture Analysis: Nonlocal Microcrack Interaction Approach." *International Journal for Numerical Methods in Engineering* 39: 635–661.

Pani, M., and F. Taddei. 2013. "The Cell Method: Quadratic Interpolation with Tetraedra for 3D Scalar Fields." To appear in *CMES: Computer Modeling in Engineering & Sciences*.

Penfield, P., and H. Haus. 1967. *Electrodynamics of Moving Media*. Cambridge MA, M.I.T. Press.

Penrose, R. 2007. *The Road to Reality*. Vintage books.

Penrose, R., and W. Rindler. 1986. *Spinors and Space-Time: Volume 1, Two-Spinor Calculus and Relativistic Field*. Cambridge University Press.

Pijaudier-Cabot, G., and Z. P. Bažant. 1987. "Nonlocal Damage Theory". *Journal of Engineering Mechanics* 113: 1512–1533. doi: http://dx.doi.org/10.1061/(ASCE)0733-9399(1987)113:10(1512)

Pimprikar, N., J. Teresa, D. Roy, R. M. Vasu, and K. Rajan. 2013. "An Approximately H1-optimal Petrov-Galerkin Meshfree Method: Application to Computation of Scattered Light for Optical Tomography." *CMES: Computer Modeling in Engineering & Sciences* 92 (1): 33–61.

Planas, J., G. V. Guinea, and M. Elices. 1996. "Basic Issues on Nonlocal Models: Uniaxial Modeling." *Technical Report No. 96-jp03*, Departamento de Ciencia de Materiales, ETS de Ingenieros de Caminos, Universidad Politécnica de Madrid, Ciudad Universitaria sn., 28040 Madrid, Spain.

Planas, J., M. Elices, and G. V. Guinea. 1993. "Cohesive Cracks versus Nonlocal Models: Closing the Gap." *International Journal of Fracture* 63: 173–187. doi: http://dx.doi.org/10.1007/BF00017284

Plenio, M. B., and S. Virmani. 2007. "An Introduction to Entanglement Measures." *Quantum Information & Computation* 1: 1–51.

Porteous, I. R. 1995. *Clifford Algebras and the Classical Groups*. Cambridge, Cambridge University Press.

Pöschel, R. 1989. "The Equational Logic for Graph Algebras." *Zeilschrift für mathematische Logik und Grundlagen der Mathematik* 35 (3): 273–282.

Qian, Z. Y., Z. D. Han, and S. N. Atluri. 2013. "A Fast Regularized Boundary Integral Method for Practical Acoustic Problems." *CMES: Computer Modeling in Engineering & Sciences* 91 (6): 463–484.

Rayleigh, O. M. 1918. "Notes on the Theory of Lubrication." *Philosophical Magazine* 35 (205): 1–12.

Reaz Ahmed, S., and S. K. Deb Nath. 2009. "A Simplified Analysis of the Tire-Tread Contact Problem using Displacement Potential Based Finite-Difference Technique." *CMES: Computer Modeling in Engineering & Sciences* 44 (1): 35–64.

Reinhardt, H. W., and H. A. W. Cornelissen. 1984. "Post-Peak Cyclic Behavior of Concrete in Uniaxial Tensile and Alternating Tensile and Compressive Loading." *Cement Concrete Research* 14: 263–270.

Robertson, A. P., and W. J. Robertson. 1964. *Topological Vector Spaces*. Cambridge Tracts in Mathematics 53, Cambridge University Press.

Rogula, D. 1965. "Influence of Spatial Acoustic Dispersion on Dynamical Properties of Dislocations. I." *Bulletin de l'Académie Polonaise des Sciences, Séries des Sciences Techniques* 13: 337–343.

Rogula, D. 1982. "Introduction to nonlocal theory of material media." In *Nonlocal theory of material media, CISM courses and lectures*, edited by D. Rogula ed. 268: 125–222. Springer, Wien. doi: http://dx.doi.org/10.1007/978-3-7091-2890-9

Roman, S. 2005. *Advanced Linear Algebra*. Graduate Texts in Mathematics 135 2nd ed. Berlin, New York, Springer-Verlag.

Rosenberg, S. 1997. *The Laplacian on a Riemannian manifold*. Cambridge University Press. doi: http://dx.doi.org/10.1017/CBO9780511623783

Rousselier, G. 1981. "Finite deformation constitutive relations including ductile fracture damage." In *Three-Dimensional Constitutive Relations and Ductile Fracture*, edited by S. Nemat-Nasser: 331–355.

Rudin, W. 1991. *Functional Analysis*. McGraw-Hill Science/Engineering/Math.

Sandler, I., and J. P. Wright. 1983. Summary of Strain-Softening. *DARPA-NSF Workshop*, Northwestern University, Evanston, Ill.

Saouridis, C. 1988. *Identification et Numérisation Objectives des Comportements Adoucissants: une Approche Multiéchelle de l'Endommagement du Béton*, Ph.D. Thesis, Univ. Paris VI, France.

Saouridis, C., and J. Mazars. 1992. "Prediction of the Failure and Size Effect in Concrete via a Biscale Damage Approach." *Engineering Computations* 9: 329–344. doi: http://dx.doi.org/10.1108/eb023870

Schaefer, H. H. 1971. *Topological vector spaces*. GTM 3, New York, Springer-Verlag.

Schey, H. M. 1996. *Div, Grad, Curl, and All That*. W. W. Norton & Company.

Schlangen, E., and J. G. M. van Mier. 1992. "Simple Lattice Model for Numerical Simulation of Fracture of Concrete Materials and Structures." *Materials and Structures* 25 (9): 534–542. doi: http://dx.doi.org/10.1007/BF02472449

Schnack, E., W. Weber, and Y. Zhu. 2011. "Discussion of Experimental Data for 3D Crack Propagation on the Basis of Three Dimensional Singularities." *CMES: Computer Modeling in Engineering & Sciences* 74 (1): 1–38.

Schönberg, M. 1954. "On the Hydrodynamical Model of the Quantum Mechanics." *Il Nuovo Cimento* 12 (1): 103–133. doi: http://dx.doi.org/10.1007/BF02820368

Schönberg, M. 1958. "Quantum Mechanics and Geometry." *Anais da Academia Brasileria de Ciencias* 30: 1–20.

Schouten, J. A. 1951. *Tensor Calculus for Physicists*. Clarendon Press, Oxford.

Schutz, B. 1980. *Geometrical Methods of Mathematical Physics*. Cambridge University Press.

Schutz, B. 1985. *A First Course in General Relativity*. Cambridge, UK, Cambridge University Press.

Selvadurai, A. P. S., and S. N. Atluri. 2010. *Contact Mechanics in Engineering Sciences*. Tech Science Press.

Sewell, M. J. 1987. *Maximum and Minimum Principles, a Unified Approach, with Applications*. Cambridge, Cambridge University Press. doi: http://dx.doi.org/10.1017/CBO9780511569234

Snygg, J. 2012. *A New Approach to Differential Geometry Using Clifford's Geometric Algebra*. Birkhäuser. doi: http://dx.doi.org/10.1007/978-0-8176-8283-5

Soares, D. Jr. 2010. "A Time-Domain Meshless Local Petrov-Galerkin Formulation for the Dynamic Analysis of Nonlinear Porous Media." *CMES: Computer Modeling in Engineering & Sciences* 66 (3): 227–248.

Solomentsev, E. D., and E. V. Shikin. 2001. Laplace–Beltrami equation. In *Encyclopedia of Mathematics*, edited by Hazewinkel, Michiel. Springer.

Sternberg, S. 1964. *Lectures on Differential Geometry*. New Jersey, Prentice Hall.

Stevens, D., and W. Power. 2010. "A Scalable Meshless Formulation Based on RBF Hermitian Interpolation for 3D Nonlinear Heat Conduction Problems." *CMES: Computer Modeling in Engineering & Sciences* 55 (2): 111–146.

Strömberg, L., and M. Ristinmaa. 1996. "FE Formulation of a Nonlocal Plasticity Theory." *Computer Methods in Applied Mechanics and Engineering* 136: 127–144. doi: http://dx.doi.org/10.1016/0045-7825(96)00997-8

Taddei, F., M. Pani, L. Zovatto, E. Tonti, and M. Viceconti. 2008. "A New Meshless Approach for Subject-Specific Strain Prediction in Long Bones: Evaluation of Accuracy." *Clinical Biomechanics* 23 (9): 1192–1199. doi: http://dx.doi.org/10.1016/j.clinbiomech.2008.06.009

Tegmark, M., and J. A. Wheeler. 2001. "100 Years of the Quantum." *Scientific American* 284: 68–75.

Theilig, H. 2010. "Efficient fracture analysis of 2D crack problems by the MVCCI method." *Structural Durability and Health Monitoring* 6 (3–4): 239–271.

Tonti, E. 1972. "On the Mathematical Structure of a Large Class of Physical Theories." *Rendiconti dell'Accademia Nazionale dei Lincei* 52: 48–56.

Tonti, E. 1974. "The Algebraic–Topological Structure of Physical Theories." *Conference on Symmetry, Similarity and Group Theoretic Methods in Mechanics*: 441–467. Calgary (Canada).

Tonti, E. 1975. *On the formal structure of physical theories*. Monograph of the Italian National Research Council.

Tonti, E. 1976. "The Reason for Analogies between Physical Theories." *Applied Mathematical Modelling* 1 (1): 37–50. doi: http://dx.doi.org/10.1016/j.clinbiomech.2008.06.009

Tonti, E. 1995. "On the geometrical structure of the electromagnetism." In *Gravitation, Electromagnetism and Geometrical Structures, for the 80th birthday of A. Lichnerowicz*, edited by G. Ferrarese: 281–308. Pitagora, Bologna.

Tonti, E. 1998. "Algebraic Topology and Computational Electromagnetism." *Fourth International Worksop on the Electric and Magnetic Field: from Numerical Models to industrial Applications*: 284–294. Marseille.

Tonti, E. 2001a. "A Direct Discrete Formulation of Field Laws: The Cell Method." *CMES: Computer Modeling in Engineering & Sciences* 2 (2): 237–258.

Tonti, E. 2001b. "A Direct Discrete Formulation for the Wave Equation." *Journal of Computational Acoustics* 9 (4): 1355–1382. doi: http://dx.doi.org/10.1142/S0218396X01001455

Tonti, E. 2001c. "Finite Formulation of the Electromagnetic Field." *Progress in Electromagnetics Research*: 1–44. *PIER 32* (Special Volume on *Geometrical Methods for Computational Electromagnetics*).

Tonti, E. 2001d. "Finite Formulation of the Electromagnetic Field." *International COMPUMAG Society Newsletter* 8 (1): 5–11.

Tonti, E. 2002. "Finite Formulation of the Electromagnetic Field." *IEE Transactions on Magnetics* 38 (2): 333–336. doi: http://dx.doi.org/10.1109/20.996090

Tonti, E. 2013. *The Mathematical Structure of Classical and Relativistic Physics*. Birkhäuser. doi: http://dx.doi.org/10.1007/978-1-4614-7422-7

Tonti, E., and F. Zarantonello. 2009. "Algebraic Formulation of Elastostatics: the Cell Method." *CMES: Computer Modeling in Engineering & Sciences* 39 (3): 201–236.

Tonti, E., and F. Zarantonello. 2010. "Algebraic Formulation of Elastodynamics: the Cell Method." *CMES: Computer Modeling in Engineering & Sciences* 64 (1): 37–70.

Trèves, F. 1967. *Topological Vector Spaces, Distributions, and Kernels*. Academic Press.

Truesdell, C., and W. Noll. 2004. *The non-linear field theories of mechanics: Third edition*. Springer.

Tvergaard, V., and A. Needleman. 1995. "Effects of Nonlocal Damage in Porous Plastic Solids." *International Journals of Solids and Structures* 32: 1063–1077.

Twiss, R. J., and E. M. Moores. 1992. "§2.1 The orientation of structures." In *Structural geology* 2nd ed. Macmillan.

van Dantzing, D. 1956. "On the Relation Between Geometry and Physics and the Concept of Space-Time." *Helvetica Physica Acta*, Supplement IV: 48–53.

van Mier, J. G. M. 1997. *Fracture Processes of Concrete*, CRC, Boca Raton, Fla.

Veblen, O., and J. H. C. Whitehead. 1932. *The Foundations of Differential Geometry*. Cambridge Tracts 29. London, Cambridge University Press.

Vermeer, P. A., and R. B. J. Brinkgreve. 1994. "A new effective non-local strain measure for softening plasticity." In *Localization and Bifurcation Theory for Solis and Rocks*, edited by R. Chambon, J. Desrues, and I. Vardoulakis: 89–100. Balkema, Rotterdam, The Netherlands.

Wheeler, J. A., C. Misner, and K. S. Thorne. 1973. *Gravitation*. W.H. Freeman & Co.

Wilder, R. L. 1965. *Introduction to Foundations of Mathematics*. John Wiley and Sons.

Willner, K. 2009. "Constitutive Contact Laws in Structural Dynamics." *CMES: Computer Modeling in Engineering & Sciences* 48 (3): 303–336.

Wu, F. H., and L. B. Freud. 1983. *Report MRL-E-145*. Brown University, Providence, RI, USA.

Wu, H.-C. 2005. *Continuum Mechanics and Plasticity*. CRC Press.

Wu, J.-Y., and C.-W. Chang. 2011. "A Differential Quadrature Method for Multi-Dimensional Inverse Heat Conduction Problem of Heat Source." *CMC: Computer Materials and Continua* 25 (3): 215–238.

Yee, K. 1966. "Numerical solution of initial boundary value problems involving Maxwell's equations in isotropic media." *IEEE Transactions on Antennas and Propagation* 14 (3): 302–307. doi: http://dx.doi.org/10.1109/TAP.1966.1138693

Yeih, W., C.-S. Liu, C.-L. Kuo, and S. N. Atluri. 2010. "On Solving the Direct/Inverse Cauchy Problems of Laplace Equation in a Multiply Connected Domain, using the Generalized Multiple-Source-Point Boundary-Collocation Trefftz Method & Characteristic Lengths." *CMC: Computer Materials and Continua* 17 (3): 275–302.

Yun, C., C. Junzhi, N. Yufeng, and L. Yiqiang. 2011. "A New Algorithm for the Thermo-Mechanical Coupled Frictional Contact Problem of Polycrystalline Aggregates Based on Plastic Slip Theory." *CMES: Computer Modeling in Engineering & Sciences* 76 (3): 189–206.

Zhang, Y., N.-A. Noda, and K. Takaishi. 2011. "Effects of Geometry on Intensity of Singular Stress Fields at the Corner of Single-Lap Joints." *World Academy of Science, Engineering and Technology* 79: 911–916.

Zheng, G., and Z.-W. Li. 2012. "Finite Element Analysis of Adjacent Building Response to Corner Effect of Excavation." *Tianjin Daxue Xuebao (Ziran Kexue yu Gongcheng Jishu Ban)/Journal of Tianjin University Science and Technology* 45 (8): 688–699.

Zhou, Y.-T., X. Li, D.-H. Yu, and K.-Y. Lee. 2010. "Coupled Crack/Contact Analysis for Composite Material Containing Periodic Cracks under Periodic Rigid Punches Action." *CMES: Computer Modeling in Engineering & Sciences* 63 (2): 163–189.

Zhu, H. H., Y. C. Cai, J. K. Paik, and S. N. Atluri. 2010. "Locking-Free Thick-Thin Rod/Beam Element Based on a von Karman Type Nonlinear Theory in Rotated Reference Frames for Large Deformation Analyses of Space-Frame Structures." *CMES: Computer Modeling in Engineering & Sciences* 57 (2): 175–204.

Zhu, T., and S. N. Atluri. 1998. "A Modified Collocation Method and a Penalty Formulation for Enforcing the Essential Boundary Conditions in the Element Free Galerkin Method." *Computer Mechanics* 21 (3): 211–222. doi: http://dx.doi.org/10.1007/s004660050296

Ziegler, G. M. 1995. *Lectures on Polytopes*. Graduate Texts in Mathematics 152, Springer.

Zill, D. G., S. Wright, and W. S. Wright. 2009. *Calculus: Early Transcendentals* 3rd ed. Jones & Bartlett Learning.

Zovatto, L. 2001. "Nuovi Orizzonti per il Metodo delle Celle: Proposta per un Approccio Meshless." *Proceedings AIMETA – GIMC* (in Italian): 1–10. Taormina, Italy.

INDEX

THIS TITLE IS FROM OUR MECHANICAL ENGINEERING GROUP COLLECTION.
OTHER TITLES OF INTEREST MIGHT BE...

Automotive Sensors
By John Turner, Joe Watson

Centrifugal and Axial Compressor Control
By Gregory K. McMillan

Virtual Engineering
By Joe Cecil

Reduce Your Engineering Drawing Errors:
Preventing the Most Common Mistakes
By Ronald Hanifan

Chemical Sensors: Fundamentals of Sensing
Materials Volume 2: Nanostructured Materials
By Ghenadii Korotcenkov

Biomedical Sensors
By Deric P. Jones, Joe Watson

Chemical Sensors: Comprehensive Sensor
Technologies Volume 4: Solid State Devices
By Ghenadii Korotcenkov

Acoustic High-Frequency Diffraction Theory
By Frederic Molinet

Chemical Sensors: Comprehensive Sensor
Technologies Volume 5: Electrochemical
and Optical Sensors
By Ghenadii Korotcenkov

Chemical Sensors: Comprehensive Sensor
Technologies Volume 6: Chemical Sensors
Applications
By Ghenadii Korotcenkov

Bio-Inspired Engineering
By Chris Jenkins

Chemical Sensors: Simulation and Modeling
Volume 1: Microstructural Characterization
and Modeling of Metal Oxides
By Ghenadii Korotcenkov

The Essentials of Finite Element Modeling and
Adaptive Refinement: For Beginning Analysts to
Advanced Researchers in Solid Mechanics
By John O. Dow

Chemical Sensors: Simulation and Modeling
Volume 2: Conductometric-Type Sensors
By Ghendaii Korotcenkov

Aerospace Sensors
By Alexander Nebylov

Chemical Sensors: Simulation and Modeling
Volume 3: Solid-State Devices
By Ghenadii Korotcenkov

Chemical Sensors: Simulation and Modeling
Volume 4: Optical Sensors
By Ghenadii Korotcenkov

Classical and Modern Engineering Methods in Fluid
Flow and Heat Transfer: An Introduction
for Engineers and Students
By Abram Dorfman

PEM Fuel Cells: Thermal and Water
Management Fundamentals
By Yun Wang, Ken S. Chen, Sun Chan Cho

Chemical Sensors: Simulation and Modeling
Volume 5: Electrochemical Sensors
By Ghenadii Korotcenkov

Announcing Digital Content Crafted by Librarians

Momentum Press offers digital content as authoritative treatments of advanced engineering topics, by leaders in their fields. Hosted on ebrary, MP provides practitioners, researchers, faculty and students in engineering, science and industry with innovative electronic content in sensors and controls engineering, advanced energy engineering, manufacturing, and materials science. **Momentum Press offers library-friendly terms:**

- perpetual access for a one-time fee
- no subscriptions or access fees required
- unlimited concurrent usage permitted
- downloadable PDFs provided
- free MARC records included
- free trials

The **Momentum Press** digital library is very affordable, with no obligation to buy in future years.

For more information, please visit **www.momentumpress.net/library** or to set up a trial in the US, please contact **mpsales@globalepress.com**.

www.ingramcontent.com/pod-product-compliance
Lightning Source LLC
Chambersburg PA
CBHW082006190326
41458CB00010B/3087